P9-DUK-493

The Commons in the New Millennium

Politics, Science, and the Environment
Peter M. Haas, Sheila Jasanoff, and Gene Rochlin, editors

Shadows in the Forest: Japan and the Politics of Timber in Southeast Asia
Peter Dauvergne

Views from the Alps: Regional Perspectives on Climate Change
Peter Cebon, Urs Dahinden, Huw Davies, Dieter M. Imboden, and Carlo C. Jaeger, eds.

People and Forests: Communities, Institutions, and Governance
Clark C. Gibson, Margaret A. McKean, and Elinor Ostrom, eds.

Learning to Manage Global Environmental Risks. Volume 1: *A Comparative History of Social Responses to Climate Change, Ozone Depletion, and Acid Rain.* Volume 2: *A Functional Analysis of Social Responses to Climate Change, Ozone Depletion, and Acid Rain*
The Social Learning Group

Changing the Atmosphere: Expert Knowledge and Environmental Governance
Clark Miller and Paul N. Edwards, eds.

Bureaucratic Landscapes: Interagency Cooperation and the Preservation of Biodiversity (forthcoming, Fall 2002)
Craig W. Thomas

The Commons in the New Millennium: Challenges and Adaptation
Nives Dolšak and Elinor Ostrom, eds.

The Commons in the New Millennium

Challenges and Adaptation

Edited by
Nives Dolšak and Elinor Ostrom

The MIT Press
Cambridge, Massachusetts
London, England

This book was set in Sabon by SNP Best-set Typesetter Ltd., Hong Kong

Printed and bound in the United States of America.

Library of Congress Cataloging-in-Publication Data

The commons in the new millennium : challenges and adaptation / edited by Nives Dolšak and Elinor Ostrom.
p. cm.—(Politics, science, and the environment)
Based on papers given at the 2000 IASCP conference held in Bloomington, Indiana.
Includes bibliographical references and index.
ISBN 0-262-04214-2 (hc. : alk. paper)—ISBN 0-262-54142-4 (pbk. : alk. paper)
1. Commons. 2. Natural resources, Communal. 3. Natural resources, Communal—Management. I. Dolšak, Nives. II. Ostrom, Elinor. III. Series.

HD1286 .C64 2003
333.7—dc21 2002029396

To all of our colleagues in the International Association for the Study of Common Property from whom we have learned many lessons

Contents

Figures

Tables

Series Foreword

As our understanding of environmental threats deepens and broadens, it is increasingly clear that many environmental issues cannot be simply understood, analyzed, or acted upon. The multifaceted relationships between human beings, social and political institutions, and the physical environment in which they are situated extend across disciplinary as well as geopolitical confines, and cannot be analyzed or resolved in isolation.

The purpose of this series is to address the increasingly complex questions of how societies come to understand, confront, and cope with both the sources and the manifestations of present and potential environmental threats. Works in the series may focus on matters political, scientific, technical, social, or economic. What they share is attention to the intertwined roles of politics, science, and technology in the recognition, framing, analysis, and management of environmentally related contemporary issues, and a manifest relevance to the increasingly difficult problems of identifying and forging environmentally sound public policy.

Peter M. Haas
Sheila Jasanoff
Gene Rochlin

Foreword

One day, back in the mid-1980s, I went to my university library and found that there were no subject entries for "commons" except those that referred to the English House of Commons and a scattered few about the open-field farming systems of England. Those few did interest me, because I was researching the history of conflicts over access to oystering grounds in colonial and early postrevolutionary America and wanted to find references to studies done of the English agrarian commons and their "enclosure." But there was nothing in the broader sense of "commons" or "common property" as a particular facet of how human beings and their social institutions relate to the natural world. How surprising, given that Hardin's famous article "The Tragedy of the Commons" had appeared in the journal *Science* in 1968 and that several works had either celebrated or criticized his analysis by the early 1980s. Apparently the metaphor of "commons" was too archaic to appear in standard cataloguing, at least in North America.

The Library of Congress, on which most university library cataloguing is based, began to use "commons" as well as "natural resources—communal" as subject headings for some books concerned with conservation of natural resources in the 1990s. I take this cataloguing change as a telling sign of the rise of scholarly and practitioner engagement with environmental and social problems within the framework of "common property," "common-pool resources," "community-based management," and, of course, "the commons." *The Commons in the New Millennium* is a significant advance in this movement.

The spate of books that came out in the late 1980s and early 1990s that seem to have triggered this change in library cataloguing (which is notoriously conservative) are among the key works that shaped the intellectual foundations of the International Association for the Study of Common Property (IASCP). IASCP is a nonprofit professional organization founded in the late 1980s to foster understanding of the challenges of managing common-pool resources and the potentials of community-based as well as government- and market-based approaches. The IASCP provides access to information about "the commons" through a library at Indiana University and through a virtual library on its homepage <http://www.indiana.edu/~iascp>, has a quarterly publication, and sponsors large international "common-property" conferences with the help of the Ford Foundation, the Rockefeller Brothers Fund, and other organizations interested in the intersection of human institutions and natural environments. *The Commons in the New Millennium* is based on papers given at the 2000 conference that was held in Bloomington, Indiana.

Cataloguing practices aside, the question of the commons is not and has never been archaic, as *The Commons in the New Millennium* clearly shows. Many things, resources, processes used or owned in common continue to be dealt with that way. There are practical, tactical, moral, and ideological reasons for resisting the isolation and privatization of everything. Similarly, there are good reasons for tempering dependence on high levels of government when the problems that must be resolved occur in places and on scales that may be served better by lower levels of governance. These are the major messages of the "commons" literature of the 1980s and 1990s, and they continue to be the messages relevant to the new millennium.

Among the challenges of the new millennium, reflected in this volume, is that of recognizing the diversity and complexity of socionatural systems. Natural-resource management has traditionally focused on one species, one that is a commodity (for either commercial or recreational use), or on one habitat and one user group. And much of the work on "the commons" follows this practice. More often than not, however, there are many and sometimes competing user groups deliberating over a particular place or species. In addition, environmental organizations and others, using the force of laws such as the Endangered Species Act, challenge the focus on just one species or population. What about predators? What about prey?

An "ecosystem" perspective challenges the traditional natural-resource management focus on single species and commodities and raises issues for the understanding and design of institutions for the commons. Who are the "commoners," and how are different uses and interests to be balanced? What are the appropriate institutions for complex, dynamic, and often poorly understood ecological systems? What are the appropriate loci for decision making, monitoring, enforcement, and evaluation, given differences in scale? How are public interests to be balanced with local interests and expertise?

Another set of issues coalesces around the "property rights" question and in particular the question of privatizing rights to erstwhile common resources. Recent privatizing efforts in both fisheries and air pollution are based on the same notion, that with stronger protections for private property, market mechanisms can help restore balance between private interests and public and environmental ones. Is this so? Under what conditions? And at what costs? These are the questions addressed in this book.

A third major topic of the new millennium is the development of social capital in managing common-pool resources. "Microcredit" has rightfully achieved renown, mostly in India, for the outstanding results achieved by various projects using the tools of microcredit. It certainly deserves greater attention as a "common-property resource" that can make a great difference to hundreds of thousands, millions, of people. There is also the question of how local communities gain political influence using social capital.

Earlier I noted that the question of the commons if labeled archaic, is so labeled mistakenly. The question of the commons is also *post*archaic. It is clearly with us, and it is with us more than ever as we increase our dependence on the World Wide Web for information and communication, as we accept the compelling evidence of the boundary-crossing nature of many cultural and environmental processes (such as climate change), as we recognize just how "common" the genetic heritage of all human beings is and how precious genetic resources might be, and as we try to find ways to balance public claims to the right of access with needs for stewardship and protection at all levels, from local to global.

Bonnie J. McCay, Rutgers University

Preface

We began working on the question of the challenges to traditional and modern commons in the new millennium in 1997 when we were asked to be cochairs of the eighth biennial meeting of the International Association for the Study of Common Property (IASCP). We started our initial plans by proposing a list of themes for the 2000 meetings and circulating these to colleagues for discussion in the period prior to the seventh biennial IASCP conference in 1998 in Vancouver. We are especially thankful to Bonnie McCay, the president of the IASCP at that time, for her input. In light of suggestions from many colleagues and the approval of the IASCP council, the theme that we chose for the conference was "Constituting the Commons: Crafting Sustainable Commons in the New Millennium."

In February and March of 1999, we started a comprehensive literature review to identify an international mix of the colleagues who would cover the subthemes that had been proposed. In April 1999, we sent out invitations to our colleagues to serve on the program committee. We extend our gratitude to our colleagues who accepted our invitations. They come from a wide range of disciplines and world regions: Minoti Chakravarty-Kaul (Lady Shri Ram College for Women, New Delhi, India), Clark Gibson (Department of Political Science, University of California, San Diego), Susan Hanna (Agricultural and Resource Economics, Oregon State University, Corvallis, Oregon), Charlotte Hess (Workshop in Political Theory and Policy Analysis, Indiana University, Bloomington), Ruth Meinzen-Dick (International Food Policy Research Institute, Washington, D.C.), Calvin Nhira (International Development Research Center, Regional Office South Africa, Johannesburg, South Africa), Susan Stonich (Department of Anthropology, University of

California, Santa Barbara), Cristian Vallejos (Forest Stewardship Council, A.C., Oaxaça, Mexico), James Walker (Department of Economics, Indiana University, Bloomington), and Lini Wollenberg (Center for International Forestry Research, Bogor Barat, Indonesia). With their help, we built an exciting conference program.

The response to the call for papers for the 2000 meeting was overwhelming. We were simply swamped with paper proposals. We had 441 individual paper submissions and 15 panel submissions. The Conference Program Committee gave us invaluable help in selecting the papers for presentation in over seventy panels, covering topics such as fisheries, forests, global commons, land use and land tenure, historical developments, micro-macro linkages, natural resources and their interlinkages, new commons, private property and the commons, resilience, spatial scales, and theoretical approaches.

Without the generous support of multiple cosponsors, we would not have been able to organize this large meeting. We would like to thank the Ford Foundation, the Rockefeller Brothers Fund, the National Science Foundation, and Indiana University for their support of the 2000 IASCP meeting. Thanks to immense effort and long hours of the staff at the Workshop in Political Theory and Policy Analysis, the Center for the Study of Institutions, Population, and Environmental Change (CIPEC), and many volunteers, the conference was well organized and ran efficiently. We would especially like to thank our colleagues on the Local Arrangements Committee: Susan Baer, Charla Britt, Michelle Curtain, Paula Cotner, Ray Eliason, Gayle Higgins, Bob Lezotte, Linda Smith, Nicole Todd, Laura Wisen, and Patty Zielinski. We thank as well the many volunteers who helped in planning and organizing all important logistical aspects of this meeting and handling all unplanned, but crucial, details.

The stimulus to edit this volume came from Ken Wilson of the Ford Foundation. Ken, an ardent supporter of the IASCP efforts, suggested that we should work on a publication that examines challenges of governance of natural commons that resource users will face in the near future. The chapters for this volume were selected from over 300 papers presented at the 2000 IASCP meeting. We extend our special thanks to Ken for his moral support. The Ford Foundation, the National Science Foundation, the Workshop in Political Theory and Policy Analysis, and

CIPEC have provided additional funding for the compilation and editing of this volume.

This book is only one of multiple intellectual outcomes of the 2000 meeting. With colleagues at the National Research Council (NRC) on Human Dimensions of Global Change, we engaged in a broader theoretical endeavor reviewing theoretical advancements we have made in understanding commons. This multidisciplinary volume, titled *The Drama of the Commons*, reviews the lessons acquired since 1985 when the IASCP was established, again with strong support from the NRC, the National Science Foundation, the Ford Foundation, and the Rockefeller Brothers Fund.

Editing this volume has been an intellectually productive and enriching experience. It has helped us to better understand the new challenges resource managers and researchers will face in the new millennium. We thank Clay Morgan, the MIT acquisitions editor, for his unstinting support and valuable input and for identifying three excellent reviewers. Their reviews made this volume much stronger. We appreciate the care and thoughtfulness of the volume editor, Michael Harrup. The contributors to this volume undertook multiple revisions. We would like to thank them for their hard work and dedication to this common endeavor. We wish to extend special gratitude to Patty Zielinski, the technical editor of this volume, whose patience and skills helped greatly strengthen the quality of the volume. We also thank Shaun McMahon and Johanna Hanley for their assistance at various stages in the manuscript process, and Evelyn Lwanga for preparing the final index for the book. We dedicate this book to our IASCP colleagues—without this network, we would not have had the rich pool of papers from which to draw and the many helpful suggestions on how to make the volume better.

Contributors

James M. Acheson Professor of Anthropology and Marine Sciences, University of Maine, Orono

C. Leigh Anderson Associate Professor, Daniel J. Evans School of Public Affairs, University of Washington, Seattle

Regina Birner Assistant Professor, Institute of Rural Development, University of Göttingen, Göttingen, Germany

Jennifer F. Brewer Ph.D. candidate, Graduate School of Geography, Clark University, Worcester, Massachusetts

Eduardo S. Brondizio Assistant Professor, Department of Anthropology; Assistant Director, Anthropological Center for Training and Research on Global Environmental Change (ACT); Faculty Associate, Center for the Study of Institutions, Population, and Environmental Change (CIPEC), Indiana University, Bloomington

Lars Carlsson Associate Professor, Division of Political Science, Department of Business Administration and Social Sciences, Luleå University of Technology, Luleå, Sweden

David W. Cash Belfer Center for Science and International Affairs, John F. Kennedy School of Government, Harvard University, Cambridge, Massachusetts

Christopher M. Dewees Marine Fisheries Specialist, Department of Wildlife and Fish Conservation Biology, University of California at Davis

Nives Dolšak Assistant Professor, Masters of Arts in Policy Studies, Interdisciplinary Arts and Sciences, University of Washington, Bothell

Einar Eythórsson Researcher, Norwegian Institute for Urban and Regional Research, Alta, Norway

Alexander E. Farrell Research Engineer, Department of Engineering and Public Policy, Carnegie Mellon University, Pittsburgh, Pennsylvania

Martha E. Geores Associate Professor, Department of Geography, University of Maryland, College Park

Clark C. Gibson Associate Professor of Political Science, Department of Political Science, University of California at San Diego

Susan Hanna Professor of Marine Economics, Department of Agricultural and Resource Economics, Oregon State University, Corvallis

Matthew J. Hoffmann Department of Political Science and International Relations, University of Delaware, Newark

Anna Knox Senior Research Fellow, International Center for Tropical Agriculture, Cali, Colombia

Rita Lindayati Natural Resource Management Consultant, Quebec, Canada

Laura A. Locker Department of Political Science, The Johns Hopkins University, Baltimore, Maryland

Ruth S. Meinzen-Dick Senior Research Fellow, International Food Policy Research Institute, Washington, D.C.

M. Granger Morgan Professor and Head, Department of Engineering and Public Policy, Carnegie Mellon University, Pittsburgh, Pennsylvania

Rachel A. Nugent Fogarty International Center DITR, National Institutes of Health, Bethesda, Maryland

Elinor Ostrom Arthur F. Bentley Professor of Political Science and Codirector, Workshop in Political Theory and Policy Analysis, and Center for the Study of Institutions, Population, and Environmental Change, Indiana University, Bloomington

Heidi Wittmer Senior Researcher, Department of Economics, Sociology and Law, Center for Environmental Research, Leipzig, Germany

Tracy Yandle Assistant Professor, Department of Environmental Studies, Emory University, Atlanta, Georgia

I

Introduction

1
The Challenges of the Commons

Nives Dolšak and Elinor Ostrom

New and Old Challenges to Governing Common-Pool Resources

When either of us has given a presentation related to the analysis of common-pool resources or common-property institutions, someone in the audience has frequently remarked: "Why are you wasting your time with the study of small, unimportant, local resources and outdated institutions? Don't you know that local resources are boring? Common-property institutions are a thing of the past. Common-property institutions will wither away within the next few decades." (The estimated time for the demise of common-property institutions varies from five to twenty-five years.) The basic message of the challenger is that common-pool resources are insignificant in modern times and that common-property institutions are not worth studying because they will not survive.

On the other hand, it is certainly the case that common-pool resources will continue to be a core type of resource of major theoretical and policy significance for as long as humans continue to rely on water, air, and the atmosphere. It is a myth that common-pool resources are all small. In addition to the prototypic local resources, the oceans, the gene pool, and the atmosphere are all common-pool resources. No one-to-one relationship exists, of course, between any kind of good or resource and the type of institutions that are used to govern and manage that resource. Thus, common-pool resources might continue for a very long time and be governed entirely by national governments or private property—an unlikely, but possible, future world. Will common-property institutions survive into the next century?

Our answer is yes! Further, in addition to existing, robust common-property institutions, new forms of common property are frequently created. The modern corporation, for example, is viewed by some as the epitome of private property. A publicly held corporation, however, is more properly thought of as *common property* than as strictly *private property*. A large number of shareholders, managers, employees, and customers hold identifiable rights in the corporation, but no one person or family holds all of the relevant property rights. To make decisions related to the allocation of resources from a corporation treasury, multiple institutional arrangements must be evoked for each and every kind of decision. No single decision maker can make all of these decisions based entirely on his or her own preferences or values. The impact of decisions made by one set of actors affects all of the others who have an economic interest in the profitability of the corporation or an interest in the environmental impacts of the production processes adopted by the corporation.

In addition to the modern corporation, many urban families now live in condominiums—a combination of private and common property—as well as owning rights to use a vacation residence with ten or twenty other families. The number of common-property institutions appears to be growing rather than diminishing over time.

On the resource side, we are also creating new common-pool resources. The Internet and the multiplicity of servers providing accessing to Web sites exhibit characteristics of a common-pool resource. It is difficult to prevent users from accessing a Web site, and too large a number of users at a given time can result in collapse of the server. Difficulties that the Federal Aviation Administration is facing in designing a system for allocating landing slots for airplanes at U.S. airports illustrate how challenging it is to allocate a common-pool resource when the demand for the resource grossly exceeds supply. These are only a few of the modern common-pool resources with many difficult and unresolved governance structures. Studying institutions regulating long-standing common-pool resources at various scales can provide important lessons for governing these new common-pool resources.

This volume can be thought of as a long answer to those who think common-pool resources involve only small-scale, insignificant, local resources or that common-property institutions are a thing of the past. It had its origin in a large international conference for the International Association for the Study of Common Property (IASCP) held in

Bloomington, Indiana, on May 31–June 4, 2000. The conference focused on new commons and on new challenges to traditional commons.

The chapters in this volume contribute in particular to our understanding of the issues involved in answering the following questions: (1) What contemporary developments challenge traditional common-property institutions, and how are these institutions adapting? (2) How is the ever-increasing scale of human interactions affecting the governance of larger-scale common-pool resources? and (3) What progress is being made in the design of institutions that "privatize" some rights that individuals have to the use of a common-pool resource? Though all the chapters provide a theoretical background for the analyses they present, they all investigate a common-pool resource governance problem from the real world and examine challenges that resource users face in governing their common-pool resources and how they adapt to them.

An important challenge in governing common-pool resources lies in the fact that stocks and flows of these resources are often difficult to define with great precision. These resources can be fugitive (e.g., fisheries and wildlife) and often cannot be stored. They are often used at different geographic scales, and these uses are often in conflict: local forest users accrue benefits when the forest is used for timber production, for example, whereas global users of forests benefit from standing trees as they sequester a major global pollutant (Young 1999). Further, the use of common-pool resources often presents negative externalities to those who do not necessarily benefit from such use. For example, harvesting timber may lead to the deterioration of water quality downstream from the location where the timber is harvested (Bruce 1999). The chapters in part II of this volume address the challenges of governing *complex*, larger-scale common-pool resources, which requires a shift from strategies for single-species resource governance to those more suited to multiple-species, habitat, and multiple-use common-pool resource governance.

Most traditional common-pool resources have already been governed by one regime or another. These regimes have performed with various levels of success. The challenge in managing such resources is to devise more effective institutions when the remnants of the previous regimes are still present. For example, many policymakers have acted in accordance with the assertion of policy analysts that two alternative "solutions" exist to the presumed ever-present "tragedy of the commons": common-pool resources, according to these analysts, were either to be privatized or to be appropriated by the national government and managed by a governmental agency. After decades of poor management and deterioration of common-pool resources, regimes developed in accordance with this dichotomous thinking are being changed. The challenge now is to devise institutions that reallocate the common-pool resource in the presence of political action by those who would lose in the process of reallocation. Chapters in part III of this volume analyze various difficulties of devising private rights to access and withdrawal from fisheries and to the depositing of pollutants in the air.

Protecting a common-pool resource from overuse requires that users or external authorities create rules that regulate its use. Devising such rules requires the joint effort of a large proportion of the resource users. Given that it is costly and that all resource users will benefit from new rules protecting the resource, creating such rules itself requires that the users overcome collective-action dilemmas. Groups with longer traditions of mutual trust and close-knit communities that enable resource users to reciprocate in behavior are more likely than other groups to succeed in devising and sustaining successful institutions (Lam 1998).

Common-pool resources and their users are not only found in isolated communities lacking connections to the external world. On the contrary, today most users of common-pool resources interact with other people in an institutional environment that is external to the one regulating the common-pool resource and imposes constraints on the regime governing it. They are forced to seek external legal authorities to protect the institutions governing the common-pool resource. An external political process determines how much support such users will receive from their national government in enforcing a self-organized regime. This is particularly important when "outsiders" begin to use a common-pool resource illegally or would like to gain access to it. The challenges of developing this *social capital* and mobilizing it to affect the external regulatory environment are addressed by the chapters in part IV.

The questions raised in the preceding paragraphs are addressed in this volume from the perspectives of multiple disciplines. One task of this first chapter is therefore to provide a common vocabulary for the enterprise. Consequently, in the next section of the chapter, we will be addressing the following issues: What are the core meanings of terms such as "common-pool resources" and "common-property institutions"? Further, individual authors will be examining some relationships and not others. In doing so, they gain analytical strength by focusing on a limited set of questions. But given the multiplicity of factors that affect the patterns of resource use, their analyses focus only on a part of these more complex patterns. To orient the reader to the larger picture, we therefore develop, in the chapter's third section, an analytical framework that enables us to see a broader view of the complex set of relationships that affects the use of common-pool resources. In the closing section, we provide a brief introduction to the individual chapters in the volume.

Toward a Common Vocabulary

Considerable confusion exists in the literature about terms such as "common-pool resources," "common-property institutions," "the commons," and "commoners," as well as many terms related to these. If scholars from multiple disciplines are going to collaborate with one another, it is essential that they develop an analytical approach to their topic and a technical vocabulary. In this volume, when we refer to a resource as a common-pool resource, we mean by this that the resource shares two characteristics with other resources. The first shared characteristic, subtractability or rivalness, has to do with the idea that what one person harvests from or deposits in a resource subtracts from the ability of others to do the same. The tons of fish or acre-feet of water withdrawn from a particular water resource by one user are no longer available to others using the same resource. The absorptive capacity of an airshed or watershed is reduced each time a user emits pollutants into the air or water. It is this characteristic that can lead to the overuse, congestion, or even destruction of a common-pool resource (Ostrom, Gardner, and Walker 1994).

The second shared characteristic relates to the cost of excluding potential beneficiaries from access to the resource. Common-pool resources and public goods share a similar problem in that potential beneficiaries of each face the temptation to be a free rider. They may be able to gain benefits without contributing to the costs of providing, maintaining, and regulating the resource involved. The institutions that individuals craft to govern any common-pool resource must therefore deal with the threats of overuse and of free riding. With well-designed institutions, it is possible to implement ways of excluding potential users from any common-pool or public resource. (It is much easier, of course, to design such institutions with respect to wheat and rice than to the oceans or the atmosphere—or even a local watercourse.) Mechanisms of exclusion do not come without a cost, however, and regimes for governance of common-pool resources must all deal with these costs in some fashion.

In addition to having these two shared attributes, common-pool resources differ from one another in regard to a wide variety of other attributes, such as their size, whether the valued units produced by the resource are mobile like water or wildlife or stable like a forest stand, whether these units are stored, and how variable their production over time and space is (Schlager, Blomquist, and Tang 1994). Thus, any analysis of common-pool problems needs to focus both on the commonalities and on the differences among common-pool resources.

Institutions that are crafted in efforts to increase the efficiency and sustainability of resource use over time can be thought of as the "dos and don'ts" commonly understood in regard to the entry, harvesting, and management of a resource and how individuals acquire or transfer rights to use a resource (Bromley et al. 1992; Schlager and Ostrom 1992; McCay and Acheson 1987). Three broad forms of ownership can govern a common-pool resource: government, private, or common-property ownership. Many variants of these three broad types of ownership regimes exist. Further, there is no consistent evidence that any one of these regimes is best suited for all types of common-pool resources (Ostrom, Dietz et al. 2002), even though considerable debate about the relative advantages of one regime or the other exists in the academic literature.

One frequently recommended solution to the tragedy of the commons is to privatize the resource. Even though a common-property regime also is a form of private property, "privatization" usually refers to the case in which the rights to a resource are allocated to individuals rather than groups and individuals are free to sell these rights to others. Privatization is, however, hardly ever complete; individual users or firms are not assigned all rights, but only the right to access and use the resource and to transfer their right to another party (Cole 1999). Many common-pool resources have been privatized, some more successfully than others. The challenge in privatization of common-pool resources is to devise an institutional design that ensures sustainability and efficiency in managing resources with specific characteristics in a given external legal and regulatory environment. We cannot simply transfer an institutional design that worked well for managing one type of common-pool resource in one region of the world to another type of resource in another region and expect to repeat the success. For example, private individual transferable rights may work efficiently to reduce air pollution in the United States, but this solution may not work efficiently for other resources in other countries with political and institutional environments that differ from that in the United States.

Many common-pool resources exhibit high uncertainty in the flow of their stocks and in the effects that resource withdrawals have on resource stocks. For example, estimating the current stock of fish in a particular fishery is always an uncertain task (Wilson 2002). Further, estimating the effect of the withdrawal of a given quantity of fish on fish stocks is an even more uncertain task given that many other factors that are exogenous to policy design affect the stocks (e.g., weather and presence/absence of predators). If policymakers specify a total allowable catch—allocated as individual fishing quotas among fishermen—in an effort to ensure the sustainable use of the fishery, they may succeed in accomplishing this objective or they may not. Successful implementation of such a scheme requires that the total allowable catch, and thus the individual fishing quotas, be reestimated over time. Such policy flexibility, however, reduces the security of rights allocated to individual users. Some institutions are more capable of handling these types of uncertainty than others (Rose 2002). The pertinent question, then, is how to devise institutions that allow for the necessary flexibility without reducing the individual's incentives for resource stewardship. Let us now embed that question in a broader framework.

Analytical Framework

Resource use depends on multiple groups of factors. Linkages among these factors and common-pool resource use are presented in figure 1.1

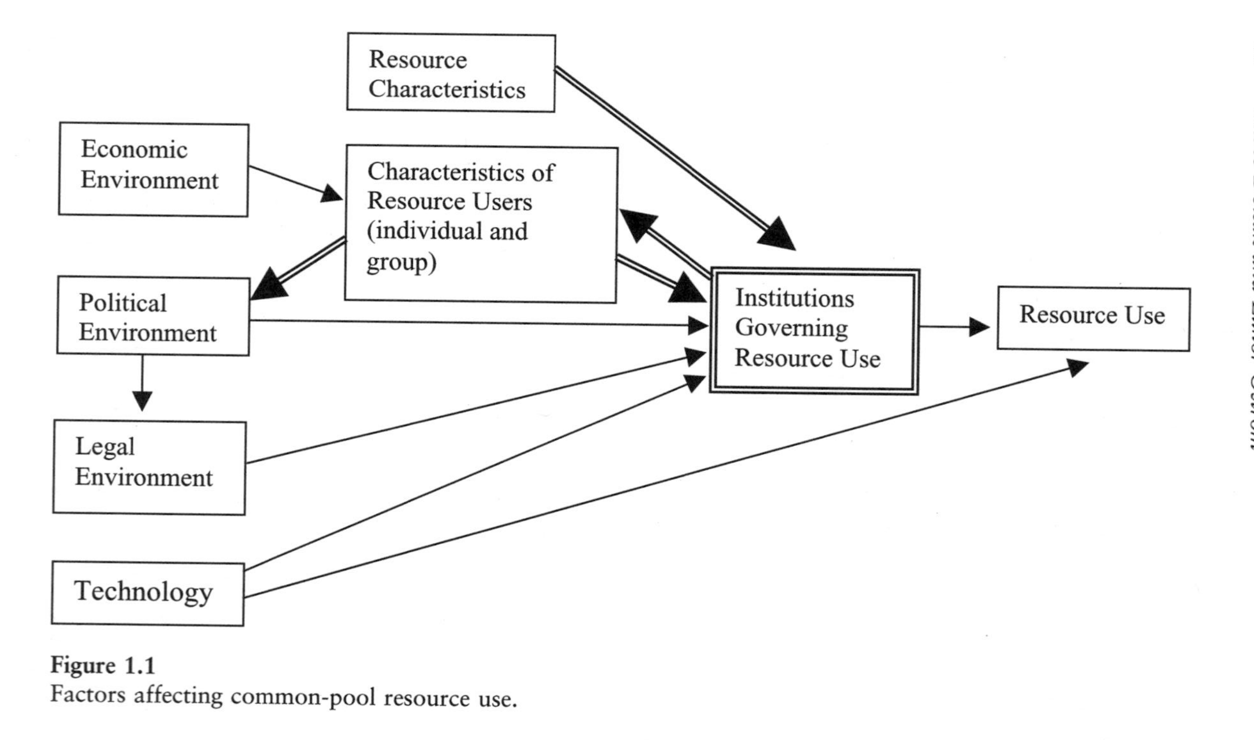

Figure 1.1
Factors affecting common-pool resource use.

(see also Thomson 1992; Oakerson 1992). Contributions of this volume to this framework are depicted with double lines. The external economic environment affects users' preferences and assets. Two external markets are especially important in this regard: markets for goods and services originating from the common-pool resource and markets providing alternative sources of income to resource users (Jodha 1985). Political and legal environments affect the institutions that govern common-pool resource use. Actors in the external legal environment can devise institutions governing common-pool resources, assign legitimacy to common-pool resource users that enables them to devise their own institutions and implement them successfully, and define how various institutional levels interact. The external political environment, with its processes of selecting the political party in power, affects the policies selected and devised in the legal environment. Technology affects the institutions governing common-pool resource use indirectly, by providing methods for monitoring such use, as well as directly, by providing the means employed in common-pool resource use and extraction.

Characteristics specific to a particular common-pool resource and its users affect institutions governing the use of that resource. The more uniform, simple, and small-scale the resource is, the easier it is to design institutions governing it and to prevent its overuse and deterioration. By the same token, complex resources with interactive use and negative externalities are especially difficult to manage. Chapters in part II of this volume examine how the latter characteristics affect institutions governing the resource use. Characteristics of individual common-pool resource users, such as their preferences and assets, and characteristics of the group (group cohesion, trust, homogeneity, size) affect institutions governing the resource. Common-pool resource use is then affected by the institutions governing it and the available technology. The previous linkages have been analyzed in the common-pool resource literature (see, e.g., Bromley et al. 1992; McCay and Acheson 1987; Gibson, McKean, and Ostrom 2000).

In addition to these linkages, there are others that are not widely examined in the existing literature on common-pool resources and are the focus of parts III and IV of this volume. Institutional design is usually conceptualized as a function of multiple factors including characteristics of resource users. Chapters in part III of this volume examine, however, how the characteristics of resource users can, in turn, be affected by the chosen institutional design. In particular, they examine how a selection of a particular institutional design, such as private transferable rights to use a resource, motivates some resource users to cooperate in efforts to prevent overuse of the resource, whereas other users opt not to participate in these efforts. Further, these chapters examine how adopting of private transferable rights affects the initial resource users as individuals and as groups. Chapters in part IV of this volume challenge an analytical approach in the common-pool resource literature that treats external political environment as exogenous to local common-pool resource governance. They indicate how common-pool resource users can bring their social capital to bear to affect the external political and legal environments in addition to affecting resource use. We now turn to a detailed discussion of the framework presented in figure 1.1.

Characteristics of a Common-Pool Resource and Their Effect on Institutional Design

Research in the last two decades has explored which characteristics of the biophysical world (i.e., of the common-pool resource) are conducive to establishing and sustaining institutions that prevent overuse and deterioration of common-pool resources. Key variables have been identified that, holding all other factors constant, promote successful common-pool resource management (McCay and Acheson 1987; E. Ostrom 1990, 2001). The relationships between these variables and management outcomes, however, are not deterministic. Some characteristics have a stronger effect on some rules and a weaker effect on others (Ostrom, Gardner, and Walker 1994). Multiple institutions have been designed that overcome some difficulties conditioned by the physical world (Dolšak 2000). Further, in the complex world of natural common-pool resources, it is a challenge to apply an analytical approach of varying one variable (i.e., the resource characteristics) while holding others constant. However, the following characteristics of common-pool resources have been identified as conducive to successful governance of such resources: small size, stable and well-delineated resource boundaries, relatively small negative externalities resulting from resource use, ability of resource users to monitor resource stocks and flows, moderate level of resource use (i.e., the extent of the resource use is significant so that the resource stocks do not seem abundant, yet the resource is not overused beyond the point at which it is still possible to prevent its deterioration), and well-understood (by the users) dynamics of the resource.

Common-pool resources of *smaller size* are considered by many scholars to be more conducive to the establishment and maintenance of institutions for managing the resource successfully (see discussion in Baland and Platteau 1996). At this point, however, we do not have a good definition of the size of a common-pool resource. Researchers usually group common-pool resources into local, regional, and global resources. Also, it is not clear how the size itself affects institutional design. Rather, size may interact with other variables. Smaller common-pool resources are easier to monitor (Tucker 2000). Stocks of small resources can usually be monitored by relatively simple methods with high reliability. Small common-pool resources usually have fewer users, in which case the monitoring of resource flows is simpler and the effects each unit of resource flow has on resource stocks and levels of compliance with the resource use limits are easier to measure than for large common-pool resources. On the other hand, stocks and flows from large-scale common-pool resources, such as those on regional or global scale, require more sophisticated measurement techniques.

Even though it is more difficult to design and enforce institutions to manage regional and global resources, several have been successfully managed and protected (Young 1994). Devising and implementing these institutions requires the effort of a larger group of resource users and more sophisticated technologies than those required for resources on a smaller scale. This calls for devising institutions at various levels of spatial aggregation and linking them (McGinnis and Ostrom 1996). The question then arises as to what we give up by devising institutions at any given level. Alternatively, how do we link various levels so that we combine advantages of institutions at various levels (Young 2002; Berkes 2002)?

Common-pool resources *with well-delineated and stable boundaries* are found to be more conducive to the emergence and sustenance of institutions for their management than those that do not have such boundaries. If boundaries of a particular resource are well understood and do not change over time, then it is easier to determine the users of the resource and the extent of their resource withdrawal. On the other hand, if the resource generates units that move substantial distances—such as migrating fish—the number of users of the resource drastically increases, making it more difficult to create and maintain institutions to prevent overuse of the resource.

Relatively simple common-pool resources with a *limited extent of negative externalities* are easier to manage than those that are part of a complex, interactive system of resources. The more complex the system of resources, the more difficult it is for resource users to agree on rules addressing these externalities. In any analysis of externalities, however, we need to define carefully the type of externality we are examining, as different types of externalities require different rules. Use of a common-pool resource by one user creates externalities for other users of that resource. (For example, in fisheries, withdrawals by one user create negative externalities for other users.) Withdrawing one unit of a resource reduces the number of resource units available to other resource users, thereby increasing the costs of withdrawal. Further, if the resource exhibits heterogeneity in spatial resource distribution (more-productive versus less-productive fishing spots, for example), resource users must devise rules that assign access to particularly productive spots. If the technology of resource withdrawal exhibits heterogeneity (for example, trawlers versus smaller boats, use of dynamite versus nets, etc.), then the use of one technology imposes negative externalities on users of other technologies. Again, rules must be devised that address these different types of externalities (Ostrom, Gardner, and Walker 1994).

When a common-pool resource is part of a *complex system*, and the information on the interactions among resources in the system becomes available only over time, with a continuous use of these resources, then more complex institutions that manage the interconnected system resources must be devised. For example, the harvest level of one fish species may be affected by levels of harvest of other species (reducing the number of predator species may increase the stock of a given species; reducing the amount of species that constitute an important link in the food chain of a given species may reduce the stocks of that species). Further, the level of fish stocks may be affected by the quality of the water in which they live, which is a function of the use of water as a pollution sink (Olsen and Shortle 1996). In this case, we need to devise institutions regulating multiple species or even ecosystems. This may significantly increase the number of resource users that have to be regulated, as well as their heterogeneity. In chapter 3, Hanna examines how these challenges are addressed in the new regulations for fisheries in the United States.

A further difficulty in measuring resource stocks and flows is exhibited in resources that have nonuniform pattern of effects of flows on the stocks. For example, a unit of deposited air pollutant in one area may have more detrimental effects on stocks of a particular resource than a unit deposited in other areas. Research on using the atmosphere as a sink for air pollutants suggests this is a major problem in devising rules for maintaining air quality, especially for pollutants that remain in close proximity to the emission point, for example, nitrous oxides or lead (Tietenberg 1974, 1980, 1990; Hahn and Hester 1989). Empirical analyses comparing management of various common-pool resources suggest, however, that the problems of nonuniformity of resource use can be significantly reduced by choosing an appropriate institutional design. Examples include introducing exchange ratios for the use of air as a pollution sink at various localities or choosing to regulate the product causing the air pollution (e.g., gasoline) rather than air-polluting activities (e.g., driving) (Dolšak 2000).

Common-pool resources with a *moderate to heavy level of resource use* are more likely to be regulated in an attempt to prevent overuse than resources that are either highly abundant or so near destruction that no institutional design regulating their use is likely to bring about a significant improvement (Ostrom 2001). There are some resources, however, that have been regulated only as they approached serious deterioration or extinction. This predominantly holds true for resources with a very large number of users with very heterogeneous interests in resource preservation. Such resources include the ozone layer and endangered species.

Common-pool resources whose stocks are time dependent are even more difficult to manage than those that exhibit no time dependence. Experimental research (Herr, Gardner, and Walker 1997) has examined emergence and performance of institutions to manage time-independent and time-dependent common-pool resources in a laboratory. Results in the laboratory indicate a much more pessimistic picture for management of time-dependent than of time-independent resources; the former were destroyed in as few as four repetitions of an experiment. Empirical research suggests that users of renewable resources pay close attention to the withdrawal rate and replacement rate. They are less likely to devise institutions to manage common-pool resources if they estimate that the replacement rate grossly exceeds the withdrawal rate, or if they estimate that the withdrawal rate exceeds the replacement rate by so much that the common-pool resource is close to destruction already.

The relationship between the replacement rate and withdrawal rate for a particular resource may change over time because of many factors. Replacement rates may change as a result of factors that are beyond the effects of policy design, and it is possible that these changes may go unobserved for some time before the common-pool resource users become alarmed. Further, technological development can increase withdrawal rates and alter the status of some resources from renewable to nonrenewable. Examples of such developments are depletion of the ozone layer and global warming (Young 1993).

Several new, man-made common-pool resources fall into the category of instantly renewable. These include the electromagnetic spectrum, airplane landing slots, and the Internet. An important characteristic of these systems is that overuse of them has little impact over time once the overuse ends. The problem is one of crowding, rather than degrading the characteristics of the system itself. Just as it appears that nonrenewable resources generate different indicators of their condition to users than do those that are renewable over a moderate timeframe, it is likely that instantly renewable resources also provide different indicators. For example, systems of instantly renewable resources are by definition forgiving, in that they instantly reward changes in user behavior. Because there is little danger of overshooting and collapse, there may be more willingness on the part of those responsible for governing such systems to experiment with novel approaches to their management, because the costs of institutional mistakes are small. On the other hand, since the consequences of poorly designed and enforced institutions are reversible, there is less incentive to take any serious action with respect to governance of such resources.

Empirical research also suggests that common-pool resources with stocks whose dynamics are well understood are more conducive to the creation and maintenance of institutions to manage them (Ostrom

2001). Theoretical analyses yield similar findings (Olsen and Shortle 1996; Pindyck 1984, 1991). To be able to devise successful rules for regulating stocks and flows of a common-pool resource, resource managers have to understand the variables that affect the stocks of a common-pool resource over time, especially variables that can be manipulated by institutional design.

If benefits of using a common-pool resource accrue to users at multiple levels, users of the resource have to devise institutions at all these multiple levels and link them. Berkes (2002) and Young (2002) suggest that, rather than those that select one given level at which a common-pool resource with users at multiple levels will be governed, institutional designs developed at interconnected multiple levels work better. The difficulty arises when users of a common-pool resource at various levels are in conflict. This issue is addressed by Geores (chapter 4) and Singleton (1998).

Characteristics of Common-Pool Resource Appropriators

Research in the past twenty years has substantially increased our understanding of the effect of the number of resource appropriators and their heterogeneity on the performance of institutions for managing common-pool resources. Users who trust each other are more likely to restrain their use of the common-pool resource and comply with agreed-upon limits of resource use. Further, users who are connected by multiple issues and over a longer period of time can use issue linkages and reciprocity to induce cooperation. Building social capital, however, requires time and resources. In chapter 9, Anderson, Locker, and Nugent study how financial capital can be used to build social capital (as defined by Putnam [2000] and conceptualized as a public good) and prevent resource deterioration. A different concept of social capital is employed in chapter 10 by Birner and Wittmer. They develop their notion of social capital based on the work of Bourdieu (1992), who defines social capital as a private resource resulting from the individual's position in a network.

External Economic Environment

It would be outdated to view common-pool resources and the communities using them as isolated and autarchic. On the contrary, many common-pool resource users rely on external markets both for their alternative sources of income and for a market in which they can sell products originating from use of the resource as well. These linkages affect the extent of the common-pool resource use.

The common-pool resource literature illustrates that access to a market and commercialization of the commons does not have a uniform effect across resources studied. On the one hand, commercialization is seen as destroying the social fabric of communities, replacing traditional principles of cooperation with those of competition and causing resource deterioration (Sengupta 1995; Long et al. 1999; McCay and Jentoft 1998). Commercialization and access to markets shifts cultivation from traditional species to cash crops. Commercialization also increases income differentiation in communities. Households without sufficient labor to produce cash crops are left behind. Such a shift does not occur in communities without access to a market (Long et al. 1999).

On the other hand, commercialization is seen as protecting the commons by generating sufficient financial resources for investments in resource regeneration, technology advancements, and, thereby, institutional sustenance (Morrow and Watts Hull 1996; Ascher 1995). Access to markets requires, however, that community members (or a trusted council) be trained to understand the dynamics of market economies so that they are not cheated.

Economic globalization provides opportunities for users of local common-pool resources to access larger markets in which they can sell goods or services originating from the use of those resources. In an environment that supports cash crop cultivation, taking advantage of these opportunities may result in common-pool resource deterioration and even destruction. The leveling of forests to make way for coffee plantations offers an example of such deterioration and destruction (Southworth and Tucker 2001). Increased economic globalization, however, when linked to "global environmental values" (Wapner 1995), may actually protect local common-pool resources. Globalization of environmental concerns may result in increased demand for local resource preservation, financed by individuals or businesses in foreign countries. Consumers in developed countries may be willing to pay a premium for products that are harvested and/or produced in a sustainable manner. Local agencies may then be established in developing countries to issue certificates for sustainable forest products (Viana et al. 1996; O'Hara et al. 1994). This increases the benefits of sustainable common-pool resource management that accrue for local common-pool resource users. Further, debt-for-nature swaps may provide financial resources at a country level that may be linked to resource preservation.

Globalization and access to markets with a high demand for protection may actually protect common-pool resources in countries with a lower ability to postpone current resource use with the goal of protection (Shultz and Holbrook 1999). It is obvious, then, that we cannot discuss one single way in which commercialization affects resource use; rather, we need to study a number of factors affecting this outcome and, potentially, to suggest how commercialization can be used to the advantage of maintenance of common-pool resources.

Globalization also means that international donor agencies become involved in developing countries' resource use. Therefore, we can no longer discuss self-sufficient, isolated communities with respect to common-pool resources. Common-property regimes in developing countries can benefit from funds available from national and international donor agencies. In some cases, common-property regimes are even initiated by these donors. This brings into the equation a set of important new actors and dynamics that all pose challenges for the governance of common-pool resources. In particular, external funders follow different time frames and usually operate on a much shorter cycle than is required for the adaptive development of successful institutions (Morrow and Watts Hull 1996). When common-property regimes are initiated with external donors' funding, a danger exists that the devised rules will not correspond to the social customs, norms, and value orientations of those on whom they are imposed. Further, the user community may not be given authority to change the rules governing the resource; rather, this authority may be vested in the donor or national government of the country hosting the project. International donors prefer engaging in projects with national governments rather than with the recipient community (Ostrom, Gibson et al. 2002). The branch of the national government overseeing the project may not view it as legitimate and deserving of funds and may impede the project and even appropriate the funds made available for one project to other projects that it finds to be of higher importance (Martens et al. 2002). On the other hand, the involvement of powerful international donors may bring legitimacy to communities that would otherwise, because of the local power structure, not be given the authority to govern the resource (Ascher 1995). Further, the financial capital made available by external donors may affect the social fabric of the communities (Anderson, Locker, and Nugent, chapter 9).

External Legal Environment

The external legal environment devises institutions for governing common-pool resources and/or gives legitimacy to users to enable them to devise their own institutions and implement them successfully. When both external regulatory agencies and resource users create and enforce rules regulating a resource, conflict may erupt among competing rule systems that can potentially bring ruin to the resource, as presented in chapter 2 by Acheson and Brewer.

Regardless of who governs a particular common-pool resource, it is essential to regulate access to the resources and to enforce the rules formulated to govern its use. Many national agencies that govern common-pool resources lack sufficient resources to enforce entry rules; a de jure state-owned resource turns into a de facto open-access resource. Such an inability to manage resources successfully, along with pressures of international donor agencies, has motivated many national governments to transfer governance of common-pool resources to communities in or nearby these resources (Agrawal and Ostrom 2001).

Processes of devolution of governance pose a special challenge to users and to resource preservation. If devolution of authority over a common-pool resource results in withdrawal of governmental resources that were previously available for financing works to maintain the resources, the community may fail to manage the resource successfully. Community members, who are used to receiving these services free from the government, may lack access to credit from other sources and may not be able to pay for maintenance works or access to new technologies (Long et al. 1999). It is therefore important that communities given governance authority over common-pool resources gain access to external funding (Mahamane et al. 1995). Chapter 9 by Anderson, Locker, and Nugent addresses how external funding can be used for building not only physical capital but also social capital. Further, devolution of authority for governing common-pool resources to communities will not necessarily result in resource protection if there are markets for cash crops whose harvest would result in deterioration (cutting timber) or even in destruction (converting forest into coffee plantations) of the resources entrusted to the communities and the cash obtained through the raising and marketing of those crops proves too tempting.

Institutional Design

The ability of common-pool resource appropriators to communicate, devise rules for appropriating the resource, and penalize rule-breaching behavior is considered to be an essential element of successful institutions for common-pool resource management. Designing institutions for managing a common-pool resource (i.e., designing rules regulating resource allocation, monitoring, and enforcement) is a costly effort. Resource users will devise new institutions for managing that resource or change existing rules governing its use when the perceived benefits of the change in the rules exceed the costs associated with creating the rules and with the change of the resource use pattern. Previous sections of this chapter have reviewed which characteristics of the resource and its users are more conducive to the creation of new rules that prevent overuse and deterioration of the resource. These characteristics enter the calculus of resource users when they are deciding whether or not to change resource use rules. Creation, monitoring, and enforcement of rules, however, involves an aspect of providing a public good. Even those who might be interested in designing new rules may decide to free ride on the efforts of others (Ostrom 2001).

This economic motivation for changing the rules governing use of a resource is, however, not always sufficient to bring about such a change. Some users of a common-pool resource who are highly motivated to change the rules governing its use may not have the authority to do so, as defined by the external environmental regulatory environment and the constitutional-choice rules of the institutions regulating the resource. Even if they have the authority to change the rules, they may not be able to place the discussion on an agenda of a legislative body with the authority to modify rules. And, even if the discussion of the change of rules is placed on the agenda, the voting rules used in that body may prevent the proposed change of the rules from being accepted at the group level.

Given the large variation in common-pool resources, their patterns of use, and their users, researchers agree that no single institutional design can be devised that will work in all of the many different common-pool resource situations. Researchers also agree, however, that we can discuss a set of general principles that increase performance of an institutional design (E. Ostrom 1990; Tucker 1999; Bardhan 1999):

1. Rules are devised and managed by resource users.
2. Compliance with rules is easy to monitor.
3. Rules are enforceable.
4. Sanctions are graduated.
5. Adjudication is available at low cost.
6. Monitors and other officials are accountable to users.
7. Institutions to regulate a given common-pool resource may need to be devised at multiple levels.
8. Procedures exist for revising rules.

In the following paragraphs, we analyze the importance of some of these design principles that are addressed by chapters in this volume.

When *rules are devised and managed by resource users*, they will better reflect the characteristics of the resource, and the users will be more familiar with the rules and thereby less likely to fail to comply with them (Acheson and Brewer, chapter 2). Resource users may be authorized to devise rules at several levels. First, rules at the so-called constitutional-choice level (V. Ostrom 1990, 1997) are those that pertain to formulation, governance, and modification of the regime. If users of the resource have the authority to define these rules, these users control whose resource use is limited and define the broad characteristics of the institutional design. At the second level, users of the resource can be authorized to devise only the so-called collective-choice level rules, those pertaining to policymaking processes and resource management. Resource users may, however, have authority only at a third level, to devise the so-called operational-choice level rules. These are the rules that pertain to appropriation, monitoring, and enforcement.

Compliance with rules that are easy to monitor and enforce increases the environmental effectiveness of institutional designs. Monitoring of rules and sanctioning of those who break them, however, are similar to providing public goods. One theory would predict that rational individuals would free ride and not contribute resources to ensure monitoring and sanctioning. Experimental results suggest, however, that resource users are often willing to invest resources to ensure monitoring and sanctioning (Ostrom, Gardner, and Walker 1994; Fehr and Gächter 2000).

Resources of large size, resources that exhibit high complexity, and resources whose use results in extensive negative externalities may require that *institutions be devised at multiple levels that are connected* (McGinnis and Ostrom 1996). Linking institutions vertically can result in tensions between benefits and costs of institutional arrangements at various levels. These tensions depend on the characteristics of the resource (the more heterogeneous the resource, that is, providing a variety of benefits to a number of users at different costs, the higher the tensions) and on characteristics of the resource users (higher levels of institutions governing the resource tend to satisfy economically and politically powerful actors).

When benefits of common-pool resource use accrue to users at different levels (local, national, regional, global), institutions are likely to be developed at these various levels to govern the resource (Young 2002). The question, then, becomes how these institutions are to be linked. Such linkages can be loose or strong, hierarchical or decentralized. The choice of the type of linkage will depend on (1) the nature of the common-pool resource, (2) the flow of information across levels, and (3) the power of individual levels.

The nature of a good can be conceptualized in terms of two characteristics: homogeneity and the extent of negative externalities caused by the use of the good. If a common-pool resource has very similar characteristics across all levels of resource use, then institutions can be developed at a high level that effectively govern the use of the resource. If the use of a common-pool resource results in negative externalities, there are two solutions. The first is that those who cause negative externalities and those who bear them can negotiate the acceptable level of such externalities (Coase 1960). This is the only possible solution when it comes to international common-pool resources, as no international hierarchical institution exists that would be able to impose and enforce an institutional design. This solution, however, is accompanied by very high transaction costs.

A second solution is to develop hierarchical institutions at a higher level of resource use that regulate such common-pool resources (as in the case of regional air pollution in the United States). If the flow of information across levels of resource use is efficient, hierarchical linkages among levels may be more efficient than strictly decentralized linkages (Hausken 1995). Whether a hierarchical linkage will actually be established, however, will depend on the relationship of powers of actors at different levels of a hierarchy. If the actors at the lower levels are more powerful than actors at the higher levels, a hierarchical system may not be designed, even though it would be more efficient than a large number of local institutions.

With increased connectedness of resource users in many spheres of their lives, such as natural environment and economic activity, institutions at various levels of resource use need to be nested. The question then arises as to whether increased connectedness requires unified institutional designs for governing common-pool resources across entire regions. Effective institutions for managing such resources are designed to take natural environments into account. If we then have environmental differences among subregions, we would expect to see differences in institutional designs in these subregions (Low et al. 2002). It is important, however, to design linkages across subregions that allow for effective communication among decision makers and coordination of adopted institutions.

When preferences as to how a common-pool resource is to be used differ across levels, institutions have to be designed to resolve such conflicts. In the case of hierarchical linkages among levels, resource users at the higher levels may be able to impose rules regulating resource users at lower levels. In the case of weak and decentralized linkages, on the other hand, resource users at higher levels may not be able to dictate a particular resource use schedule. Rather, they may offer to cover a part of the costs of changing the institutions governing the use of the resource. The question then arises as to what proportion of these costs the higher-level organizations should cover. International environmental organizations, such as Global Environmental Facility, have developed a principle of additionality to help in determining how these costs should be apportioned among the levels. Under this principle, an international organization covers any costs of changing institutions governing common-pool resource use at a lower level beyond what brings benefits to that lower level. In theory, this principle makes sense, but it is extremely difficult to operationalize it in reality.

Procedures for revising rules allow for a trial-and-error, incremental process to devising institutions for governing common-pool resources. This is especially important in the case of resources whose stocks and flows we cannot measure with high reliability, whose dynamics we do not yet understand, whose dynamics depend on factors that are beyond what institutional design can affect, or whose resource use patterns quickly adapt to changes in demand for the resource or in technologies used in resource appropriation. As Farrell and Morgan (chapter 7) illustrate for the global atmosphere, these procedures are not linear; changes in the incentives of various actors over time can result in major changes in governance institutions.

Privatization as a solution for preventing the overuse of common-pool resources has been recommended since the mid-1950s. Some mid-twentieth-century analysts suggested it as a solution to the limited effectiveness of governmental command-and-control instruments in managing common-pool resources (Gordon 1954; Dales 1968), others as a solution for preventing overuse of resources that were in open access (Hardin 1968). Tradable permits have also been suggested as a solution to reduce overexploitation of common-pool resources (Dales 1968; Crocker 1966; Montgomery 1972). When property rights are well-defined and easily enforced, markets efficiently determine what and how much is produced (based on the market prices), how it is most efficiently produced (based on the relationships between marginal productivity of inputs and their prices), how it is distributed (depending on individuals' income and preferences), and how consumption is allocated over time (based on differences in individuals' discount rates). In addition to static economic cost-efficiency, tradable permits also have dynamic advantages over command-and-control instruments (Jaffe and Stavins 1995).

Theoretical advantages of private tradable permits to use a common-pool resource may not often be fully realized because of characteristics of the common-pool resource and limitations of markets in these permits. Not all common-pool resources, for example, are easily amenable to the design of private transferable rights. It is difficult to design tradable permits for very complex systems with high variability in time and interconnectedness in space (such as ground-level ozone pollution). Assuring sustainable use of a common-pool resource in the presence of high uncertainty over time requires that permit rights be redefined as new information becomes available. This decreases the security of such rights and reduces economic motivation for investing in them, thereby reducing the effectiveness of any institutional design that depends on such rights. Common-pool resources with spatial negative externalities require that spatial extent of the rights be limited. This reduces the size of the market for these rights and again decreases the economic effectiveness of tradable permits based on these rights. This is an important challenge in designing a tradable permit system. Chapter 7 by Farrell and Morgan examines these issues in greater detail.

Allocating tradable permits to use a common-pool resource also has important social consequences. First, allocating these rights to existing users of a particular common-pool resource at no charge and making new users purchase the rights creates barriers to new users' benefiting from the resource. It is possible to design a system that treats both existing and new resource users equally. The political feasibility of such a system, however, would critically depend on the relative political power of the two groups of users. Second, trading these permits enables those who place the highest value on the resource to purchase the rights to use the resource (provided, of course, that they can afford to make the purchase). This way, the market theoretically ensures that the common-pool resource is used by those who value it most. One unwanted outcome of this process, however, is that users who place lower valuation on the resource lose their rights to use the resource. If these users have no alternative sources of income, losing the rights to use the resource can result in loss of means to provide for survival. For some common-pool resources, tradable permit systems have been shown to lead to a concentration of permits in the hands of large corporations, with small users losing access to common-pool resources. These issues are examined in greater detail in chapter 5 by Yandle and Dewees (on fisheries in New Zealand) and chapter 6 by Eythórsson (on fisheries in Iceland).

Technological Development

Technological development may or may not degrade a common-pool resource (Meyer and Turner 1990). This dichotomy can be explained by examining the level of technical development pertinent to each particular case. Initially, in the agrarian-based economies, population pressures and technology development resulted in land change that caused a loss of forest cover, depletion of species, soil erosion, degradation, and carbon emissions from land change. Advanced industrial economies, however, may have decreased the pressures on the commons by providing more efficient ways of using them because of the technological development.

Further, technological development enables us to monitor common-pool resource use more effectively and at lower costs. Use of citizens band communication equipment in fisheries has enabled fishermen to share information about cost-breaching behavior observed in the field. New emissions-monitoring equipment installed in smokestacks of electric utilities and improved computer communication and processing ability has enabled the U.S. Environmental Protection Agency to engage in continuous monitoring not only of the quantity of air emissions from these smokestacks, but also of the number of emissions permits that each power plant was issued, has purchased, or has sold.

The Volume and Its Organization

This volume analyzes new challenges that owners, managers, policymakers, and analysts face in managing natural commons, such as forests, water resources, and fisheries. It also examines challenges in managing commons caused by new findings about physical characteristics, about complexity and interconnectedness, about new institutional arrangements at both the micro (privatization) and the macro levels (economic and political changes in countries and regions), and about the role of culture and social capital in sustainable management of the commons. Practical applications of the issues raised are discussed in light of empirical analyses of various commons, and suggestions for sustainable management are presented.

Part II, "Managing Species, Habitats, and Landscapes at Multiple Scales," examines specific issues that arise when a common-pool resource has a high level of interconnectedness with other resources, its boundaries are difficult to define and protect, and it provides services at various levels and on various scales. Chapter 2 by Acheson and Brewer reviews the traditional and well-functioning territoriality system in the Maine lobster fishery and examines institutional changes that have resulted from an increased demand for the resource. Chapter 3 by Hanna reviews the key uncertainties in the transition of U.S. marine fisheries regulation from a single-species approach to one aimed at protection of the ecosystem. It addresses scientific uncertainties, the changing expectations of the public, the distribution of authority and rights, the role of institutional and ecological scale, and the role of technology as both a promoter and inhibitor of conservative use. Chapter 4 by Geores analyzes the issues related to selection of resource scale, as this is not merely an ecological or biological question, but also a political one and one pertaining to resource management. Depending on the scale considered, an analyst and a manager may or may not observe all processes affecting a common-pool resource, representation of spatial patterns of such processes will differ, and the methods required to observe the causal processes among the factors affecting the resource may differ.

Part III, "Privatization," examines how common-pool resources can be successfully privatized. Chapter 5 by Yandle and Dewees reviews the implementation of individual transferable quotas (ITQs) in fisheries in New Zealand over the span of a decade. The implementation of these quotas resulted in economic conflicts among the major stakeholders, such as fishermen, vessel owners, processors and marketing units, workers in those facilities, and fishing communities. There is a broad consensus, however, that the country's fishing industry is "better off" since the system has been implemented. Chapter 6 by Eythórsson addresses the survival of the ITQ system in Iceland, where most of the public supports the ITQ system, but there is individual disagreement about the principles for allocating these quotas. Chapter 7 by Farrell and Morgan examines privatization of the atmosphere as a sink for certain air pollutants (sulfur dioxide and nitrous oxides). Use of the atmosphere as a pollution sink can result in levels of atmospheric quality that differ significantly across large spatial areas. This upwind versus downwind situation may create important barriers to successful markets for emissions rights. Chapter 7 therefore focuses on the ways the spatially differentiated effects of emissions have been addressed in the U.S. air pollution emission markets.

The chapters in part IV, "Financial, Social, and Political Capital: Managing Common-Pool Resources and Shaping the Macropolitical and Macroeconomic Environment," examine the challenges of building and mobilizing social capital. Chapter 8 by Lindayati reviews changes in the external political environment in Indonesia and links them to the extent of political autonomy local forest users can regain. Increased democratization of the state creates options for resource users to mobilize their capital and use the pressure of international donor organizations to force changes in governance of local forests. These demands for changes, however, do not easily transfer into devolution of power. Their success is fairly limited: communities are given only a limited power to manage the forests that they used to govern entirely.

Chapter 9 by Anderson, Locker, and Nugent examines the role of social capital as a set of social networks and its importance for common-pool resource management. The authors develop a framework to analyze conditions under which microcredit may be available to individuals to build physical, human, and social capital and to prevent resource over-exploitation. They illustrate the use of the theoretical framework in their analysis of microcredit schemes in Vietnam. Chapter 10 by Birner and Wittmer examines the question of when social capital, conceptualized as a set of both horizontal and vertical networks, can be translated into the ability to affect political outcomes (political capital), such as common-pool resource preservation. Structural parameters of social and political systems determine which type of social capital actors can accumulate and to what extent they can transform this into political capital. A case study of vibrant forestry communities in northern Thailand relates how these communities converted their social capital into political capital to oppose a proposed national forestry bill supported by powerful economic actors and the forestry ministry.

As Dolšak, Brondizio, Carlsson, Cash, Gibson, Hoffmann, Knox, Meinzen-Dick, and Ostrom discuss in chapter 11 in the concluding part V, social and financial capital do not necessarily lead to better resource preservation. Further research needs to focus on operationalizing social capital and its links to resource management in comparative empirical studies.

References

Agrawal, Arun, and Elinor Ostrom. 2001. "Collective Action, Property Rights, and Decentralization in Resource Use in India and Nepal." *Politics & Society* 29(4) (December):485–514.

Ascher, William. 1995. *Communities and Sustainable Forestry in Developing Countries*. San Francisco: ICS Press.

Bardhan, Pranab. 1999. "Water Community: An Empirical Analysis of Cooperation on Irrigation in South India." Working paper. Berkeley: University of California, Department of Economics.

Baland, Jean-Marie, and Jean-Philippe Platteau. 1996. *Halting Degradation of Natural Resources: Is There a Role for Rural Communities?* Oxford: Oxford University Press.

Berkes, Fikret. 2002. "Cross-Scale Institutional Linkages: Perspectives from the Bottom Up." In *The Drama of the Commons,* ed. Elinor Ostrom, Thomas Dietz, Nives Dolšak, Paul C. Stern, Susan Stonich, and Elke Weber, 293–321. Washington, DC: National Research Council, National Academy Press.

Bourdieu, Pierre. 1992. "Ökonomisches Kapital–Kulturelles Kapital–Soziales Kapital" (Economic Capital—Cultural Capital—Social Capital). In *Die verborgenen Mechanismen der Macht* (*The Hidden Mechanisms of Power*). Schriften zu Politik und Kultur 1, 49–79. Hamburg: VSA-Verlag.

Bromley, Daniel W., David Feeny, Margaret McKean, Pauline Peters, Jere Gilles, Ronald Oakerson, C. Ford Runge, and James Thomson, eds. 1992. *Making the Commons Work: Theory, Practice, and Policy.* San Francisco: ICS Press.

Bruce, John W. 1999. *Legal Bases for the Management of Forest Resources as Common Property*. Vol. 14. Rome: Food and Agriculture Organization (FAO) of the United Nations.

Coase, Ronald H. 1960. "The Problem of Social Cost." *Journal of Law and Economics* 3:1–44.

Cole, Daniel H. 1999. "Clearing the Air: Four Propositions about Property Rights and Environmental Protection." *Duke Environmental Law & Policy Forum* 10(1):103–130.

Crocker, T. D. 1966. "The Structuring of Atmospheric Pollution Control Systems." In *The Economics of Air Pollution*, ed. Harold Wolozin, 61–86. New York: Norton.

Dales, J. 1968. *Pollution, Property, and Prices: An Essay in Policy-Making and Economics*. Toronto: University of Toronto Press.

Dolšak, Nives. 2000. "Marketable Permits: Managing Local, Regional, and Global Commons." Ph.D. diss. Bloomington: Indiana University.

Fehr, Ernst, and Simon Gächter. 2000. "Fairness and Retaliation: The Economics of Reciprocity." *Journal of Economic Perspectives* 14(3):159–181.

Gibson, Clark, Margaret McKcan, and Elinor Ostrom, eds. 2000. *People and Forests: Communities, Institutions, and Governance*. Cambridge: MIT Press.

Gordon, H. Scott. 1954. "The Economic Theory of a Common Property Resource: The Fishery." *Journal of Political Economy* 62 (April):124–142.

Hahn, R. W., and G. L. Hester. 1989. "Where Did All the Markets Go? An Analysis of EPA's Emissions Trading Program." *Yale Journal on Regulation* 6:109–153.

Hardin, Garrett. 1968. "The Tragedy of the Commons." *Science* 162 (December):1243–1248.

Hausken, Kjell. 1995. "Intra-Level & Inter-Level Interaction." *Rationality and Society* 7(4):465–488.

Herr, Andrew, Roy Gardner, and James Walker. 1997. "An Experimental Study of Time-Independent and Time-Dependent Externalities in the Commons." *Games and Economic Behavior* 19:77–96.

Jaffe, A. B., and R. N. Stavins. 1995. "Dynamic Incentives of Environmental Regulations: The Effects of Alternative Policy Instruments on Technology Diffusion." *Journal of Environmental Economics and Management* 29(3, suppl. part 2):S43–S63.

Jodha, N. S. 1985. "Market Forces and Erosion of Common Property Resources." In *Agricultural Markets in the Semi-Arid Tropics: Proceedings of the International Workshop, 24–28 October, 1983, ICRISAT Center, India*, ed. M. von Oppen, 263–277. Patancheru, India: International Crops Research Institute for the Semi-Arid Tropics.

Lam, Wai Fung. 1998. *Governing Irrigation Systems in Nepal: Institutions, Infrastructure, and Collective Action.* Oakland, CA: ICS Press.

Long, Chun-Lin, Jefferson Fox, Lu Xing, Gao Lihong, Cai Kui, and Wang Jieru. 1999. "State Policies, Markets, Land-Use Practices, and Common Property: Fifty Years of Change in a Yunnan Village, China." *Mountain Research and Development* 19(2):133–139.

Low, Bobbi, Elinor Ostrom, Carl Simon, and James Wilson. 2002. "Redundancy and Diversity: Do They Influence Optimal Management?" In *Navigating Social-Ecological Systems: Building Resilience for Complexity and Change*, ed. Fikret Berkes, Johan Colding, and Carl Folke. Cambridge: Cambridge University Press, forthcoming.

Mahamane, El Hadj Laouali, Pierre Montagne, Alain Bertrand, and Didier Babin. 1995. "Creation of New Commons as Local Rural Development Tools: Rural Markets of Wood Energy in Niger." Paper presented at the Fifth Conference of the International Association for the Study of Common Property, Bodoe, Norway, May 24–28.

Martens, Bertin, Uwe Mummert, Peter Murrell, and Paul Seabright. 2002. *The Institutional Economics of Foreign Aid.* Cambridge: Cambridge University Press.

McCay, Bonnie J., and James M. Acheson. 1987. *The Question of the Commons: The Culture and Ecology of Communal Resources.* Tucson: University of Arizona Press.

McCay, Bonnie J., and Svens Jentoft. 1998. "Market or Community Failure? Critical Perspectives on Common Property Research." *Human Organization* 57(1):21–29.

McGinnis, Michael, and Elinor Ostrom. 1996. "Design Principles for Local and Global Commons." In *The International Political Economy and International Institutions,* vol. 2, ed. Oran Young, 465–493. Cheltenham, UK: Edward Elgar.

Meyer, William B., and B. L. Turner II. 1990. "Editorial Introduction." In *The Earth as Transformed by Human Action: Global and Regional Changes in the Biosphere over the Past 300 Years,* ed. B. L. Turner II, 469–471. Cambridge: Cambridge University Press.

Montgomery, David W. 1972. "Markets in Licenses and Efficient Pollution Control Programs." *Journal of Economic Theory* 5:395–418.

Morrow, Christopher E., and Rebecca Watts Hull. 1996. "Donor-Initiated Common Pool Resource Institutions: The Case of Yanesha Forestry Cooperative." *World Development* 24(10):1641–1657.

Oakerson, Ronald J. 1992. "Analyzing the Commons: A Framework." In *Making the Commons Work: Theory, Practice, and Policy,* ed. Daniel W. Bromley et al., 41–59. San Francisco: ICS Press.

O'Hara, Jennifer, Mirei Endara, Ted Wong, Chris Hopkins, and Paul Maykish, eds. 1994. *Timber Certification: Implications for Tropical Forest Management.* New Haven: Yale University, Yale School of Forestry and Environmental Studies.

Olsen, J. R., and J. S. Shortle. 1996. "The Optimal Control of Emission and Renewable Resource Harvesting under Uncertainty." *Environmental & Resource Economics* 7(2):97–115.

Ostrom, Elinor. 1990. *Governing the Commons: The Evolution of Institutions for Collective Action.* New York: Cambridge University Press.

Ostrom, Elinor. 2001. "Reformulating the Commons." In *Protecting the Commons: A Framework for Resource Management in the Americas,* ed. Joanna Burger, Elinor Ostrom, Richard B. Norgaard, David Policansky, and Bernard D. Goldstein, 17–41. Washington, DC: Island Press.

Ostrom, Elinor, Thomas Dietz, Nives Dolšak, Paul C. Stern, Susan Stonich, and Elke Weber, eds. 2002. *The Drama of the Commons.* Washington, DC: National Research Council, National Academy Press.

Ostrom, Elinor, Roy Gardner, and James Walker. 1994. *Rules, Games, and Common-Pool Resources.* Ann Arbor: University of Michigan Press.

Ostrom, Elinor, Clark Gibson, Sujai Shivakumar, and Krister Andersson. 2002. *Aid, Incentives, and Sustainability. An Institutional Analysis of Development Cooperation: Main Report.* Sida Studies in Evaluation #02/01. Stockholm, Sweden: Swedish International Development Cooperation Agency (Sida).

Ostrom, Vincent. 1990. "An Inquiry Concerning Liberty and Equality in the American Constitutional System." *Publius* 20(2):35–52.

Ostrom, Vincent. 1997. *The Meaning of Democracy and the Vulnerability of Democracies: A Response to Tocqueville's Challenge.* Ann Arbor: University of Michigan Press.

Pindyck, Robert S. 1984. "Uncertainty in the Theory of Renewable Resource Markets." *Review of Economic Studies* 51(2):289–303.

Pindyck, Robert S. 1991. "Irreversibility, Uncertainty, and Investment." *Journal of Economic Literature* 29:1110–1148.

Putnam, Robert. 2000. *Bowling Alone: The Collapse and Revival of American Community*. New York: Simon and Schuster.

Rose, Carol M. 2002. "Common Property, Regulatory Property, and Environmental Protection: Comparing Community-Based Management to Tradable Environmental Allowances." In *The Drama of the Commons,* ed. Elinor Ostrom, Thomas Dietz, Nives Dolsak, Paul C. Stern, Susan Stonich, and Elke Weber, 233–257. Washington, DC: National Research Council, National Academy Press.

Schlager, Edella, William Blomquist, and Shui Yan Tang. 1994. "Mobile Flows, Storage, and Self-Organized Institutions for Governing Common-Pool Resources." *Land Economics* 70(3) (August): 294–317.

Schlager, Edella, and Elinor Ostrom. 1992. "Property-Rights Regimes and Natural Resources: A Conceptual Analysis." *Land Economics* 68(3) (August): 249–262.

Sengupta, Nirmal. 1995. "Common Property Institutions and Markets." *Indian Economic Review* 30(2):187–201.

Shultz, Clifford J. II, and Morris B. Holbrook. 1999. "Marketing and the Tragedy of the Commons: A Synthesis, Commentary, and Analysis for Action." *Journal of Public Policy and Marketing* 18(2):218–229.

Singleton, Sara. 1998. *Constructing Cooperation: The Evolution of Institutions of Comanagement.* Ann Arbor: University of Michigan Press.

Southworth, Jane, and Catherine M Tucker. 2001. "The Influence of Accessibility, Local Institutions, and Socioeconomic Factors on Forest Cover Change in the Mountains of Western Honduras." *Mountain Research and Development* 21(3):276–283.

Thomson, James. 1992. *A Framework for Analyzing Institutional Incentives in Community Forestry*. Vol. 10 of Community Forestry Notes. Rome: Food and Agriculture Organization of the United Nations.

Tietenberg, T. H. 1974. "The Design of Property Rights for Air Pollution Control." *Public Policy* 27(3):275–292.

Tietenberg, T. H. 1980. "Transferable Discharge Permits and the Control of Stationary Source Air Pollution: A Survey and Synthesis." *Land Economics* 56(4):392–416.

Tietenberg, T. H. 1990. "Economic Instruments for Environmental Regulation." *Oxford Review of Economic Policy* 6(1):17–33.

Tucker, Catherine M. 1999. "Common Property Design Principles and Development in a Honduran Community." *Praxis: The Fletcher Journal of Development Studies* 15:47–76.

Tucker, Catherine M. 2000. "Striving for Sustainable Forest Management in Mexico and Honduras: The Experience of Two Communities." *Mountain Research and Development* 20(2):116–117.

Viana, Virgilio M., J. Ervin, R. Donovan, C. Elliott, and H. Gholz, eds. 1996. *Certification of Forest Products: Issues and Perspectives*. Washington, DC: Island Press.

Wapner, P. 1995. "Politics without Borders: Environmental Activism and World Civic Politics." *World Politics* 47:311–340.

Wilson, James. 2002. "Scientific Uncertainty, Complex Systems, and the Design of Common-Pool Institutions." In *The Drama of the Commons,* ed. Elinor Ostrom, Thomas Dietz, Nives Dolšak, Paul C. Stern, Susan Stonich, and Elke Weber, 327–359. Washington, DC: National Research Council, National Academy Press.

Young, Oran. 1993. "Negotiating an International Climate Regime: Institutional Bargaining for Environmental Governance Systems." In *Global Accord: Environmental Challenges and International Responses,* ed. Nazli Choucri, 431–452. Cambridge: MIT Press.

Young, Oran. 1994. *International Governance: Protecting the Environment in a Stateless Society.* Ithaca: Cornell University Press.

Young, Oran. 1999. *Governance in World Affairs*. Ithaca: Cornell University Press.

Young, Oran. 2002. "Institutional Interplay: The Environmental Consequences of Cross-Scale Linkages." In *The Drama of the Commons,* ed. Elinor Ostrom, Thomas Dietz, Nives Dolšak, Paul C. Stern, Susan Stonich, and Elke Weber, 263–291. Washington, DC: National Research Council, National Academy Press.

II

Managing Species, Habitats, and Landscapes at Multiple Scales

2

Changes in the Territorial System of the Maine Lobster Industry

James M. Acheson and Jennifer F. Brewer

The state of Maine is currently home to several different ocean tenure systems that operate at different scales and involve different principles. These are, as Delaney and Leitner (1997, 93) say, a "nested hierarchy of bounded spaces of differing size." One exists at the local scale and is defined by local practice. Another exists at the level of the state and is codified into state law. Still a third has been imposed by the federal government since 1977 and the passage of the Fisheries Conservation and Management Act, which gave federal authorities power to manage fisheries out to 200 miles. In this chapter, we will be primarily concerned with changes in the local-level system. Of secondary concern is the state system, which has had some effect on the local-scale system. We will not discuss the territorial system operating at the federal scale, primarily because that system operates in offshore federal waters where comparatively little lobster fishing takes place. We recognize, however, that this scale has become increasingly important as the federal government gains more power, for better or for worse, over resource management (see chapter 3).

According to the law of Maine, all of the state's oceans, lakes, and rivers are public property. Ocean waters are held in trust by the state for all citizens. All ocean beaches to the high-tide mark are owned by the state, and all citizens have legal access to them.

In the lobster fishery, a different tradition prevails. Here, local fishing territories are the rule. To go lobstering one needs a state license, which ostensibly allows a person to fish anywhere in state waters. In reality, more is required. One also needs to gain admission to a group of people fishing from the same harbor—a group that Acheson calls a "harbor gang" that maintains a fishing territory for the use of its members.[1] This means that two different groups with different kinds of authority have laid claim to the same inshore waters. As Geores (chapter 4) points out, this situation often leads to conflict.

In the Maine lobster fishery, potential conflict between these two systems is kept to a minimum through the avoidance of actions among those in one system that would directly challenge those in the other system. Among state fishery officials, there has been a tacit acceptance of the traditional territorial system. Everyone knows it exists, but it has generally been accepted as long as violence and destruction of property are kept to a minimum and do not come to public attention. When they do, those cutting traps are prosecuted in court, long-standing tradition aside. In this sense the lobster industry has operated as an encapsulated political system—a system within a system—in which both public officials and lobster fishermen have made accommodations for each other. Both sides have historically treated the territorial system like a skeleton in the closet; everyone knew it was there, but no one wanted to admit to its existence.

While this system may seem exotic and unusual in a modern country, riparian rights and ownership of ocean areas is quite common in worldwide perspective (Acheson 1981, 280–81). In a large number of maritime societies, rights to exploit ocean areas are variously held by communities, kinship groups, or individuals under a wide variety of property rights regimes.

In recent years, the local-scale territorial system of the Maine lobster industry has undergone a number of important changes. Some have stemmed from technological and economic factors. Others have resulted from new management laws and enforcement activities of the government. In this chapter, we describe some of these changes in the local-scale system and the reasons for them. We will use the term "territoriality" to refer to the local-scale system. In studying the local-scale system, however, it should not be forgotten that territoriality exists at larger scales and that what goes on at the local scale is influenced by activities of officials operating at higher scales, especially in the area of law enforcement.

The Maine Lobster Industry: General Information

The Maine lobster industry throughout its history has been an inshore trap fishery. In 2002, the typical full-time lobster fisherman has a boat about thirty-five feet long equipped with a diesel or gas engine, which he operates with a helper or alone. He uses over 550 traps made of wood or wire, which are baited with fish remnants (Acheson and Acheson 1998). Each trap is equipped with funnel-shaped nylon "heads" that make it easy for lobsters to climb into the trap but difficult to find their way out. The traps are connected to a buoy made of wood or styrofoam by a warp line. The buoys are painted with a distinctive combination of colors registered with the state (Acheson 1988, 84–90). Each day fishermen sell their catches to one of the eighty private dealerships or seventeen cooperatives that are located along the coast (Acheson 1988, 115–132).

Maine's lobster fishery is one of the world's most successful. Lobster catches are at all time historic highs at present. From 1947, when the modern enumeration system was started, to 1988, Maine lobster landings were very stable, ranging from 15.1 million to 22.7 million pounds per year. Since 1988, lobster catches have been over 30 million pounds every year. In 1998 the catch was 46 million pounds, and in 2000, it was over 52 million pounds. The causes of these huge catches are uncertain (Acheson and Steneck 1997), but one important factor is the strong conservation ethic of the industry, which has resulted in a number of laws that have aided in conserving the resource. (Other causes of the large catches are almost certainly environmental factors.) These laws include both minimum and maximum lobster size measurements to protect juveniles and large reproductive-sized lobsters, a prohibition on taking egged lobsters, the "V-notch" law (to protect proven breeding females), the escape vent law allowing small lobsters to escape from traps, and a law requiring that lobsters be taken only with traps, a very selective type of gear. There is no quota on the number of lobsters that can be taken. These laws, which have been on the books for decades, are compatible with the existing territorial system. Recent legislation, especially a new zone management law,[2] is causing (along with other factors) profound changes in the territorial system. These laws are broadly supported by the Maine lobster industry and are largely self-enforcing.

Aspects of the Traditional Territorial System

In Maine, lobster fishing rights are held jointly by a group of people fishing from a particular harbor, or "harbor gang." Once one gains admission to a harbor gang, one is usually allowed to go fishing only in the territory of that gang. Interlopers are usually warned verbally or by minor molestation of their gear. If they persist, they are subject to retaliation in the form of having some of their lobstering gear destroyed.

Lobster fishing territories are typically quite small and are fished by small groups. Most territories are under 100 square miles, with the vast majority far smaller. These areas might be fished by as few as six or eight boats, and harbors containing over fifty boats are rare indeed.

The territorial boundaries near shore are delineated by features on shore: a cove, a ledge. Further offshore, boundaries are marked by reference to landmarks on shore or on islands. In recent decades Loran C lines (an electronic navigational device) have sometimes been used to define fishing locations and territories.

The delineation of territorial boundaries varies considerably with distance from shore. Thus, boundaries close to shore are known precisely and are well defended; further offshore, boundaries are less well defined. In the middle of large bays, men from four or five harbors might fish together, which is sometimes called "mixed fishing." If one goes ten miles from shore, there is no territoriality at all, and people have always been free to fish where they want in areas this far from shore, provided they stay away from the areas of island fishermen, who defend territories of their own.

Lobster fishing territorial boundaries are the result of competition among fishermen for fishing space. Boundaries exist because people from a particular harbor are able to occupy an area over time and prevent people from other harbors from placing traps there. Occasionally agreements are reached between harbors about the explicit location of boundaries, but these are not enforceable by any third party. People come to know where they are allowed to fish, and the vast majority usually do not deliberately fish outside that area. Local lobstering territories are always vulnerable, however, to the incursions of others. When push comes to shove, fishing areas are maintained by the willingness and ability of those who claim them to defend them.

One of the keys to maintaining fishing space is to occupy it. There is less to be gained by placing traps in an area saturated with the traps of people from another harbor than by placing them in unoccupied areas (assuming lobsters are to be had in both places). If people from a gang have enough traps in an area, they may not have to do anything more to defend it. An unoccupied area provides an opportunity for others to move in. For this reason, some fishermen continually make a practice of placing traps on the border of their fishing area and perhaps a little beyond, a process called "pushing the lines."

Only in very rare instances do local territorial conflicts result in physical fights, gunshots, or boats being sunk. When these serious offenses occur, and police or the Marine Patrol are involved, those found guilty in court are punished.

The vast majority of conflicts over fishing space involve destruction of fishing gear or threats to destroy gear. Destroying another person's lobster traps, after all, not only removes the symbol of someone else's incursion; it also removes his capacity to take lobsters from the area: the primary consideration. Although cutting traps is illegal and can result in the loss of license and a heavy fine, small-scale trap cutting can be done in comparative safety. It is notoriously difficult to successfully prosecute trap-cutting incidents. People who destroy traps of others rarely advertise the fact. It is very difficult to observe people in the commission of such crimes. It is hard to see what a fisherman is doing on his boat a few hundred yards away even in broad daylight. A lobster boat two miles away is almost invisible. Adding confusion to the situation is the fact that all fishermen lose a certain number of traps to storms, and traps are sometimes accidentally cut off by the propellers of passing boats. A fisherman who cannot find some traps may suspect they have been cut off and may have a good idea who was responsible, but usually the kind of evidence that will stand up in court is lacking. As a result, people who are missing traps usually do little but complain and perhaps move some of the remaining traps to a safer location.

Sometimes victims of trap-cutting incidents will defend themselves in kind. Such conflicts can escalate, with the guilty and innocent alike blindly retaliating against each other in a series of trap-cutting incidents.

Fishermen generally are cautious about touching others' gear for fear of retaliation. When traps owned by people from another harbor appear in "their" territory, they may do little but complain or warn the intruder. In some cases, two half hitches are placed on the buoy of the offending traps; in other instances the "heads" (nylon mesh holding in the captured lobsters) are cut out; occasionally, a note in a bottle is placed in the trap. If fishermen decide to destroy the intruding traps, they will usually take care not to destroy so many that their victim is provoked into an all-out cut war. Their own traps are vulnerable, after all. As one fisherman said, "the secret of driving someone from your area is to just destroy enough of his traps to make fishing unprofitable."

What is surprising is the lack of violence and conflict concerning territoriality at the local level. The rules concerning territoriality are well known, and most people are reluctant to violate them and deliberately cause trouble. Small-scale trap cutting is a problem all along the coast, but incidents in which large numbers of traps are cut are rare and becoming rarer.

Over the course of the past several decades, all local lobster fishing boundaries in Maine have moved somewhat, but some have moved surprisingly slowly. Whether boundaries remain stable or move is the result of a political process involving competition and conflict between groups of fishermen. Boundaries move when a group of fishermen—usually a small group from one harbor—successfully place traps in the area occupied by another harbor gang and are able to keep them there. Not all attempts to expand the amount of fishing space are successful, however, and some gangs have been very successful in defending their fishing territories for decades.

Movement of local-level boundaries is rarely the result of actions by a single individual. A single person who attempts to move into an area occupied by a group or to defend a boundary against a group of invaders may lose so much gear that he is forced to retreat in defeat. A successful defense against invaders or a successful invasion against opposition depends on the ability to organize an effective and coordinated team. Usually such teams are composed of a small group (three to eight people) whose forays are coordinated by one or two leaders. Their activities are usually kept quiet. In many cases, but not all, the fishermen involved are quite young.

Although teams are composed of more than one person, entire gangs are not involved. One older fisherman from an island that has been very successful in defending its boundaries explained the situation in these words: "You never attempt to defend lines by yourself. If you do, you will become a target. We just get the number of people it takes to do the job. If two or three people can hold the lines, fine. But if it takes more, we get them."

The major impediment to organizing such teams is overcoming a strong inclination among those affected to be a free rider. It is very tempting to let others do the dirty work of invading or defending boundaries while getting the benefits of their activities. After all, everyone in the harbor benefits when boundaries are defended, and it is only those involved in the defense who are generally in serious danger of prosecution or retaliation.[3] Some harbors have been much more successful in organizing such teams than others.

A decision to defend one's own lines against invasion or invade the area of others depends on the costs and benefits involved. Maintaining a local territory reduces the numbers of fishermen in that territory. This can result in two kinds of benefits: fewer snarls and increased catches per unit of effort. The primary costs of defending a territory are the threat of prosecution, the potential to lose some of one's own gear, and the psychic costs of being involved in conflict (not to be discounted). The fact that in mainland harbors a sense of territorial ownership is very strong near shore and nonexistent in offshore fishing areas can be explained in terms of the competition for productive fishing bottom and the way this affects the costs and benefits involved. Lobsters are concentrated in inshore areas in the summer when large numbers of them shed into larger sizes. The number of traps per unit of area is high, since there is relatively little of this kind of bottom and the numbers of fishermen is at its annual high, since large numbers of part-timers with small skiffs are in the fishery in these months. Under these conditions, excluding others will increase the proportion of traps one has on the bottom, augmenting catches per trap. It will also reduce snarls. In such areas, the benefits of holding territory outweigh the costs. The benefits of maintaining territorial rights to offshore areas, where lobsters are concentrated in the winter, are far less. At this time of year the number of traps per unit of area is less. There are more square miles of offshore fishing grounds. There are fewer boats fishing at this time of year, since only those with large boats are capable of exploiting these offshore grounds, and even many with large boats choose not to fish since the weather is often very bad, trap losses are high, and the fishing is far less productive. Since traps are not crowded in offshore zones, snarling is not generally a problem, and removing other traps will not increase one's own catches much, if at all. In these offshore waters the costs of defending an area generally outweigh the benefits to be had by defending territorial claims.

There are two different types of territories at the local scale. Mainland harbors have what Acheson (1988) calls nucleated territories. That is, there is a strong sense of territoriality near the harbor where the boats are anchored, and a stranger putting traps in this area is almost certain to be retaliated against. This sense of ownership grows progressively weaker the further from the harbor one goes, and the willingness to retaliate against interlopers is less. On the periphery (i.e., three to five miles away from the harbor) there is no strong sense of territoriality. (This is not to suggest that these more distant "mixed fishing" areas are open to anyone.) If one goes far enough away from one's own harbor, one inevitably enters an area that is defended by another gang. How far one can go without incurring retaliation depends greatly on one's personal characteristics. Older, experienced fishermen from large, well-established families who have a history of fishing in the area and a lot of "friends" are accorded more leeway. They have more allies, after all.

In the middle of the twentieth century, mainland harbors maintained some control over who was permitted to join the harbor gang. Usually only people from the town were permitted to go fishing, with members of long-established fishing families given preference. These restrictions have broken down in many places. People wanting to join the gang may be harassed for a time, but most will succeed in joining if they are persistent enough and obey the conservation laws. As a result, the number of people in most of these mainland gangs has increased dramatically, which has exacerbated the problem of trap congestion.

The island areas, particularly those in Penobscot Bay, exhibit what Acheson (1988) calls perimeter-defended territoriality. Here, boundaries are known to the yard and are strongly defended. No mixed fishing is allowed within the boundaries of such islands, although some mixed fishing is permitted outside.

Gangs utilizing perimeter-defended areas control entry into their harbor gangs. People who do not meet the criteria for membership in a particular gang are not permitted to fish in the territory of that gang. Depending on the island, various combinations of island land ownership, kinship, or residence are used to restrict access to the island's fishing area. In at least one instance, the owners of an island rent fishing rights to other fishermen on a kind of subcontract basis.

Fishermen exploiting perimeter-defended territories have been successful in establishing boundaries around their fishing areas and defending them. Their communities have most of the characteristics that make it possible to generate rules and enforce them successfully (Ostrom 1990, 188ff; 2000, 149ff). They are small, stable groups with a strong sense of community and a lot of social capital. Their inhabitants are very dependent on the lobster fishery, and they have an ideology stressing maintaining that fishery for future generations. In such groups, in which it is easy to monitor the behavior of others and one is dependent on the goodwill of other people, it is unwise to be seen as one who shirks one's responsibility to defend the most valuable resource the island has: its communal fishing grounds.

Recent Changes in the Maine Lobster Industry: Trap Escalation and Trap Limits

Over the course of the past several decades, there has been a steady increase in the number of lobster traps in Maine's waters, leading to trap congestion. This escalation in the number of traps was made possible by the adoption of a number of technological innovations, including the hydraulic trap hauler[4] beginning in the 1960s, the adoption of larger boats able to haul more traps, and nylon netting, which lasts far longer than the older hemp twine.

A number of nontechnological factors have also motivated people to greatly increase the numbers of traps fished. The first is competition among fishermen. In every harbor, some people would put more traps in the water to increase the proportion of traps they had on the bottom and their incomes. Others would follow suit in an attempt to keep even. This would stimulate people to fish even more traps to stay ahead of the competition. This cycle has gone on for the past fifty years. In some places, such as Casco and Penobscot Bays before 1995, it was common for people to fish 1,800 traps, and people with 3,000 traps were not unheard of.

The second nontechnological factor increasing the number of traps has been an increase in the number of full-time fishermen as people moved out of the beleaguered groundfisheries, scallop fisheries, etc., into the booming lobster fishery. Although the number of fishermen has remained approximately the same, a very large number of license holders who had been part-time fishermen earning their living in other industries have become full-time lobster fishermen using far more traps than they did as part-timers. In 1978, one study showed that approximately 20 percent of the lobster license holders were full-timers; another study done in 1998 revealed that 51 percent were full-time fishermen (Acheson and Acheson 1998). It comes as no surprise that full-time fishermen use far more traps than people who depend on other jobs or other fisheries for most of their income.

A third nontechnological factor in the increase in the number of traps has been the uncertainty brought about by recent laws and the threat of federal intervention in the fishery, which has motivated people to fish more traps in an attempt to grandfather themselves into the fishery. Since 1995, there has been a strong feeling that the federal government may force low trap limits on the fishery. Those who begin with a lot of traps, they believe, will end up with more gear than those who start at a lower base figure (Acheson 2001).

All three of these factors have worked in tandem to motivate large numbers of fishermen to purchase more traps. As a result, since 1930 there has been an increase in the number of traps used. In 1960, there were approximately 750,000 traps in Maine waters; in 1999, there were 3,045,000 trap tags issued (although the number of tags sold considerably exceeds the number of traps in use) (Maine Department of Marine Resources 2000).

The increases in the amount of gear finally led to a trap limit law. Many fishermen have wanted trap limits (a maximum number of traps that could be fished by a single license holder) for some thirty years. There was no coast-wide consensus on what the limit on the number of traps per license holder should be, however. What was considered an adequate limit in some areas was considered far too restrictive in others.

In 1995 the Maine legislature solved this problem by passing what has become known as the zone management law, which went into effect that year. This law stipulated that the entire coast of the state was divided into zones and that each zone was to be governed by a council composed of lobster fishermen elected by the license holders of that zone. These zone councils have the power to recommend rules for their zone on the number of traps to be used (a trap limit), the times of day when fishing will be permitted, and the number of traps that can be fished on a single warp line. If these rules are passed by referendum by two thirds of the voting license holders in the zone, they are referred to the Commissioner of Marine Resources, who can make them regulations enforceable by the wardens. In fact, seven zones were established, and by 1998, all of them had passed trap limits. Six of the seven zones passed an 800-trap limit for 2000; one passed a 600-trap limit.

In 1999, the legislature passed another law giving the zones the power to limit entry into their zone by imposing an in/out ratio.[5] In 2000, five of the seven zones passed limited-entry rules for their zones. Despite the trap limits and limited-entry rules imposed, the number of traps in use has continued to climb (Acheson 2001).

These changes have affected the local territorial system in two important ways. First, the increased number of traps has greatly increased trap congestion, motivating fishermen to exploit areas where they did not fish previously in a search for productive fishing bottom not saturated with traps. The general result is that full-time fishermen are exploiting more square miles of area, further from their home harbors, than they did previously. Second, although the zone management law has generally been very successful, its implementation has had some problems. Some of the most serious have resulted from conflicts over zone boundaries. As we shall see, both of these factors are having an effect on the territorial system.

Administrative Changes

Over the course of the past several decades, the Marine Patrol force has become increasingly professional and effective. Enforcement of laws has become much stricter. In part this is due to leadership in the state's Department of Marine Resources, increased enforcement budgets, and better training. But it is also due to changes in attitudes in the industry. People in the industry have become better educated and more committed to conservation. They are much more likely to report infractions of the law to Marine Patrol officers.

Increased law enforcement has increased the risks for those who cut traps. In the past, trap-cutting incidents were not likely to result in offenders' being apprehended by the law. It is still difficult to get evidence of trap cutting that will stand up in court, and some victims prefer to handle the situation through private retaliation rather than cooperation with the police and the Marine Patrol. Increasingly, however, victims of trap-cutting incidents are willing to report the incidents to the Marine Patrol, and the Marine Patrol officers are having far more success in enforcing such violations of the law.

One man from an island whose territory was being invaded by mainlanders was heard to remark, "In the good old days we would have taken care of the problem with the knife, but this isn't the good old days." Within the past three years, three men from the Penobscot Bay region were convicted of cutting off large numbers of traps. All three were fined and lost their licenses. One was sent to jail. These arrests would have been unlikely thirty years ago because of the code of silence that prevailed in the industry at that time.

Changes in the Territorial System

There have been four important changes in the territorial system in the past decade, all of which stem from one or a combination of the factors discussed above.

Offshore Fishing Areas

First, a high percentage of full-time fishermen are now fishing in offshore areas—especially in the winter—in areas that were never fished much in the past and were never incorporated into the territorial system. Moreover, there has been little attempt to extend the territorial system to incorporate these offshore areas. This means that a lot of lobster fishing is now taking place beyond the confines of the traditional territorial system.

In the past, offshore waters were not heavily exploited. To be sure, some lobsters could be caught in deep offshore waters in the winter months, but the catches were not big enough to tempt many people. Moreover, boats were generally not large enough to be used safely and comfortably in these offshore areas, where seas are often very rough in the winter months. It was common for people in the lobster industry to haul up all of their gear in the early winter, put their boat on shore, and devote the winter months to building traps or boats. Many fishermen still follow this pattern.

The increased amount of fishing in these offshore waters was motivated, in great part, by the increasing trap congestion and the search for productive fishing locations where one could place traps without their getting snarled with those of other fishermen. The adoption of larger, faster boats equipped with better electronic navigational devices reduced the amount of time necessary to travel to offshore grounds and made it possible to fish there more safely in winter. Then the widespread replacement of homemade wooden traps with factory-made wire traps freed up a lot of time that fishermen had profitably devoted to trap building in the winter.

In addition, there may be an ecological change making it possible to catch lobsters further from shore and on types of bottom that had previously never been productive. In the 1970s, it was widely believed that lobsters could not be caught on mud bottom, and fishermen always had some traps close to shore at all times. Lobsters can now be caught in numbers on mud bottom in deep water in the winter.

As a result of all these changes, many fishermen are placing large numbers of traps in offshore areas and fishing throughout the winter months. Fishermen from Friendship and Bremen are currently placing traps ten miles south of Monhegan Island where only a few Monhegan fishermen used to place traps twenty years ago. People from Spruce Head are fishing large numbers of traps in the winter south of Matinicus Rock, some thirty miles from their home harbor. Boats from Portland and other towns in Casco Bay have been going to offshore areas southeast of Cape Elizabeth. Interestingly enough, these fishermen have asserted their right to fish in such offshore areas, and, as we shall see, have been quick to assert their rights to fish there when the establishment of limited-entry rules threatened their ability to put large numbers of traps in offshore waters over the line in another zone. They have made no attempt, however, to incorporate these areas into their traditional fishing areas.

Why haven't such offshore areas been incorporated into existing territories? The basic reason is that the benefits of claiming territory in these offshore waters would not outweigh the costs. There would be few if any benefits to claiming territory in offshore waters, because traps in most of this region are generally not competitive with each other. The offshore fishing area is so vast relative to the numbers of fishermen who exploit these waters and traps are far enough apart that removing traps would not likely result in increased catches for those remaining. Furthermore, the costs of trying to establish a territory would likely be quite high. There is, first of all, the possibility of a trap-cutting foray, resulting in loss of license or being fined. Then, it is always difficult to coordinate people to defend an area. These areas are exploited, typically, by people from several harbors. There is no instance (of which we are aware) in which people from a number of different harbors have been able to organize territorial defense, since they are not likely to know each other well; they have little social capital built up; and most likely they have a history of antagonism, competition, and even conflict to overcome. If people from one harbor attempted to oust the men from other harbors fishing in a particular area, they would likely be overwhelmed by greater numbers of men from these other harbors. The fray that would follow would almost certainly attract the attention of law enforcement officers.

Even though the activities of fishermen are not being curtailed by traditional territorial rules in these offshore areas, their activities are being limited by the formal rules, which are likely to become more stringent in the future.

Inshore Territories

Second, important changes have occurred in areas that have long been incorporated in the traditional territorial system. Particularly in western parts of the coast, the amount of mixed fishing in inshore areas is increasing, and the amount of bottom that is the exclusive fishing area of particular harbors has decreased. This is due to the cost-benefit ratios faced by defenders and aggressors.

From the 1960s to the present, many traditional boundaries have been under extreme pressure from communities at the heads of bays and in coastal rivers. Until the 1950s or later, fishermen from these communities had small boats and gangs of traps and fished only during the summer and early fall months. In recent decades an increasing number of them have become full-time fishermen with larger boats and more traps. In pursuit of year-round fishing, they have entered territories formerly exclusive to peninsular harbors, closer to the mouths of bays and open ocean. These expansionists have been willing to sacrifice considerably to gain additional fishing space, since the alternative is to be locked into the upper reaches of the bays where winter and spring fishing is sparse or nonexistent. For fishermen in the open-water harbors that already had year-round grounds, repelling these "river rats" was often not worth the prospective cost.[6] As a result, fishermen from communities such as Bremen and Wiscasset now are able to fish up river in the summer and in the middle of bays or ocean areas in the fall and winter. Mixed fishing is hence more prevalent now in many areas.

The recent increases in trap congestion in combination with more effective law enforcement have changed the cost-benefit ratios of territorial defense in ways that have also exacerbated the trend to larger mixed-fishing areas. People who invade other areas have more to gain in terms of access to increased bottom and face less chance of effective retaliation then they did in the past. The people whose boundaries have been violated have little to gain from a defense attempt, in comparison with the possible losses.

Island Areas

Third, some of the island areas have been very successful in defending their territorial interests. Some, such as Criehaven, Green Island, Metinic, and other islands in Penobscot Bay have mounted an effective traditional defense. The traditional areas of these islands have been maintained largely intact. Nonetheless, these island areas are pressured by men from crowded mainland harbors who are willing to sacrifice a good deal to gain additional space in which to fish. Some of the lines of these island areas have been "pushed back."

Two islands have succeeded in defending fishing areas by going to the government. In 1984, fishermen from Swan's Island, under increasing pressure from mainland fishermen and experiencing increasing trap escalation, were able to get a consensus among the people on the island to request that the state form a "conservation zone" that encompassed the island and its surrounding waters.[7] They were successful in persuading people from adjacent harbors and the commissioner to support their efforts. In 1984, the Swan's Island conservation zone was formed, specifying that only 350 traps would be used in waters adjacent to the island. People from other harbors would be allowed to fish in Swan's Island waters, provided they obeyed the Swan's Island rules.

Monhegan Island was also successful in establishing a conservation zone after several years of conflict with fishermen from Friendship. Monhegan has long had a two-mile area around the island where fishing is allowed only from January to June. It has also had an informal trap limit and has limited entry into its own harbor gang as well. In 1995 fishermen from the mainland harbor of Friendship were setting traps south of the island in waters claimed by Monhegan. A large number of Friendship traps disappeared in a series of "killer fogs." Subsequently, Monhegan traps were destroyed, and a Monhegan boat was sunk deliberately at nearby Port Clyde. In 1996, the Commissioner of Marine Resources, Robin Alden, got both sides to come to an agreement setting up a conservation zone around Monhegan, providing that people from other harbors could fish there if they obeyed the special Monhegan rules. This agreement broke down when several Friendship fishermen signed up to fish in the Monhegan zone and then permission was retracted. In 1998, after months of trouble and endless debate in the newspapers and on docks, the Monhegan fishermen successfully lobbied the legislature to establish a law that only people who lived on the island could fish in the Monhegan conservation zone. The law also established a stringent trap limit, a limited-entry program, and a special apprenticeship program, and it continued the six-month fishing season. Monhegan now has the most conservative management regime in Maine.

A number of other islands, emboldened by Monhegan's success, were seriously discussing lobbying to have such conservation zones established around their own islands. The legislature responded by establishing a special commission to study the establishment of more "subzones." The final report of this commission, written in 1998, concluded that "subzones should be discouraged at this time" (Maine Department of Marine Resources 1998, 1). This conclusion was clearly one that was desired by officers of the Department of Marine Resources and members of the legislature, who had come under considerable constituent pressure concerning this issue. To date, no other island or harbor has managed to get the legislature to establish another special fishing zone. The areas around Swan's Island and Monhegan Island stand alone. It is important to note that these two islands have defended their traditional fishing areas by unusual means. They have maneuvered the legislature into formalizing traditional territories and using officers of the state Marine Patrol to defend those boundaries. They have also agreed to the most stringent conservation rules in existence in the industry.

Government-Imposed Boundaries

Fourth, activities of government units at higher scales than local are also affecting the use of fishing space. When Maine's zone management law was passed in 1995, few people thought the imposition of zone boundaries would cause a major problem. After all, the law stated that people could fish on both sides of a zone boundary line. If there was a difference in rules between the zones, they could fish on both sides, provided that they followed the rules of the most restrictive zone. The zone boundaries were set in 1996 by interim zone councils and were placed to coincide with some of the traditional boundaries.

By 2000, however, five of the seven zones established under the law were embroiled in zone boundary disputes, all essentially distributional in nature. Two of the three disputes did not reach crisis points until rules were imposed that restricted someone from fishing in a mixed-fishing area where he had fished traditionally. The first serious dispute took place off Pemaquid Point, the boundary between zones D and E. (The zones are designated A through G.) The dispute came to a crisis when the fishermen of zone E voted in a more restrictive trap limit (600 in 2000) than that in effect in zone D (with an 800-trap limit). This meant that fishermen with more than 600 traps (including many full-time fishermen) could not go fishing in the zone E territory (the most-restrictive-zone clause). This made it impossible for fishermen from harbors on Muscongus Bay (i.e., New Harbor, Round Pond, and Bremen) to access "shedder" bottom right off Pemaquid Point and productive deep-water bottom southwest of Pemaquid Point. These fishermen were very unhappy at being denied access to bottom they had used for many years. Some in zone E were pleased by this turn of events, since the law would allow them to go anywhere in zone D territory, whereas the zone D fishermen were prohibited from crossing the line into zone E. This dispute was settled late in 1999 and early in 2000 by fishermen from both sides agreeing to a "buffer zone" off Pemaquid Point where people from both zones could fish. This solution was accepted by the commissioner.

The next dispute occurred between zones C and D concerning an area close to Vinalhaven where fishermen from Wheeler's Bay (zone D) had successfully invaded an area previously fished by Matinicus and Vinalhaven. The Wheeler's Bay fishermen wanted to continue fishing in this contested area. Matinicus and Vinalhaven fishermen wanted them out, and they were pleased when the zone boundary was drawn in such a way as to place much of the contested area in zone C. This dispute simmered from 1995 to 1999, but there were few problems since zones C and D both had voted in an 800-trap limit. People from both zones could fish on both sides of the line with no restrictions. However, in 1999, a limited-entry-by-zone law was passed, which resulted in limitations on the numbers of traps that people from one zone could place in another. The logic behind the law is that if limited entry is to be enforceable, a zone limiting entry into its ranks must have assurance that fishermen from adjacent zones with no limited-entry rules will not be permitted to place large numbers of traps in its waters. As a result, the Commissioner of Marine Resources used his regulatory authority to enact what has become known as the 49/51 percent rule, which makes it illegal for license holders to place more than 49 percent of their traps in the waters of another zone. This rule suddenly made it impossible for fishermen to continue to place the numbers of traps they used to put in what they considered their traditional fishing grounds. For example, it prohibited big fishermen from the Spruce Head area from placing a large number of traps south of Matinicus Rock where they had put traps for the previous several years. At that point, the dispute heated up. It, too, was settled by the imposition of a buffer zone in the spring of 2000.

The third dispute is between zones F and G over waters in the Cape Elizabeth area south of Portland. Fishermen from zone F have long fished very large amounts of gear in the winter in offshore areas that are now part of zone G. The zone G people want to restrict the access of these zone F fishermen to the area. In 1997, the zone G fishermen passed a very restrictive 600-trap limit, which made it illegal for those from zone F to fish large gangs of gear across the line. The zone F fishermen were incensed and insisted on their right to fish in places in zone G where they had long placed traps, regardless of zone G trap limit rules. The dispute was put on hold when the zone G trap limit was nullified by a court ruling in 1998, but it heated up again with the passage of the limited-entry-by-zone law, which again will make it illegal for zone F fishermen to place large amounts of gear in zone G waters. It still has not been settled.

Since they were established in 1997, zone boundaries have increasingly influenced where lobster license holders can fish. They are changing traditional use patterns inshore by preventing some fishermen from using parts of their traditional area, and, in effect, ceding those areas to fishermen from other harbors. They are also limiting where fishermen can go offshore. The activities and fishing practices of fishermen exploiting deep-water offshore areas are not being limited by traditional boundaries, but since 1999 and the imposition of the 49/51 percent rule, they are limited by the percentage of traps that can be placed outside the boundaries of one's own zone. Again, this impedes some people from exploiting grounds they have used for years and benefits those from other harbors who are faced with less competition. All of a sudden, access to fishing ground depends not only on one's being able to exclude others from an area by coercion or by saturating it with traps, but also on one's ability to influence the position of zone boundaries in the political arena at the state scale.

In the near future, two others kinds of actions by Maine's government will very likely impose other boundaries at higher scales than the local ones that will further limit lobster fishing practices. In 2000, state officials were sued twice under the Endangered Species and Marine Mammal Acts by environmental groups for permitting lobster fishermen to use traps that ostensibly caused the deaths of at least two right whales and entanglement of several others. One suit was dismissed, but the other succeeded, with the result that the National Marine Fisheries Service introduced a plan with more stringent rules to protect the right whales. The plan that is evolving in 2002 will force the industry to remove traps from the water on short notice when concentrations of whales are in the area and maintain trap-free corridors when and where whales are migrating. It promises to be very expensive for the industry.

Moreover, in 2002, the management rules imposed by the Atlantic States Marine Fisheries Commission (ASMFC), which took over lobster management in federal waters in 1995, took effect and began to be enforced in earnest. The boundary between area 1 (Maine, New Hampshire, and Massachusetts north of Cape Cod) and area 3 (the outer Gulf of Maine) lies forty miles off the Maine coast. Since these two areas have different management rules, this new boundary will affect fishing practices at least in the offshore areas.

Summary and Conclusion

Traditional lobstering territories in Maine are changing. One source of change is trap escalation and the exploitation of new areas with bigger, better equipped boats. Another is the actions of the Maine state government, which has passed a zone management law and stiffened its enforcement of laws prohibiting gear molestation.

These factors are having four different effects on the territorial system and spatial strategies of lobster fishermen. First, many full-time fishermen with large boats are fishing offshore, especially in the winter months, in areas that have always been outside the traditional territorial system. It is unlikely that territorial rules will be imposed in these deep-water areas.

Second, these factors have altered the cost-benefit ratio of defending traditional boundaries, with the result that the amount of territory held exclusively by harbor gangs has decreased, and the number of areas where mixed fishing takes place have increased. Third, Monhegan Island and Swan's Island have employed an unusual strategy to defend their traditional fishing area from mainland boats: they have successfully lobbied the legislature and Department of Marine Resources to establish exclusive conservation zones around these islands.

Fourth, rules imposed by government units at higher scales than the local are affecting the lobster fishery. The new state zone management law has imposed boundaries and rules that are changing where people fish. In some cases, these boundaries reinforce local-scale zone boundaries; in others they help fishermen from some harbors access areas they did not hold under the traditional system and impede others from accessing traditional bottom. Lobster fishermen are aware of this and have begun to lobby, through the zone council process, to affect the placement of boundaries with their own interests in mind.

In the near future, the management rules imposed by the ASMFC at still a higher scale will impose a boundary forty miles offshore. The right whale plan will result in other boundary lines in offshore waters. This means that lobster fishermen, who are increasingly fishing in offshore waters unencumbered by local-scale territorial lines, will almost certainly be limited by lines imposed by government at a higher scale.

There are many in Maine who believe that we are witnessing a fundamental change in the local-scale territorial system. Some think someday there will be no traditional boundaries at all. This point of view may be overstated; traditional boundaries will likely continue to influence fishing practices. It is clear, however, that the rules governing the use of fishing space by the lobster industry at all scales are in a state of rapid flux.

We hope the case study presented in this chapter will make a contribution to the study of common-pool resources in general. In the literature, there is a growing consensus on a number of issues. One is that if common-pool resources are to be managed, boundaries must be maintained (Ostrom 2000, 146). Another is that privatization of resources will result in more efficient use and conservation. Neither is possible in the absence of territoriality. Unfortunately, there are few case studies in the literature about how territories come into being and why they change. This chapter provides such a case study.

Acknowledgments

The data on which this chapter is based were gathered during the course of a project entitled "Case Studies in Co-Management," supported by the University of Maine Sea Grant Program. The fieldwork was done in 1998 and 1999 primarily by the authors. Teresa Johnson, a graduate student in marine policy at the University of Maine, gathered some of the data on territories in the Penobscot Bay area. Ann Acheson and Patty Zielinski proofread the various drafts.

Notes

1. There is no commonly accepted word in the industry for these harbor groups. People talk about the "crowd" from Stonington, the "boys" from Five Islands, the Friendship "gang," etc. Acheson has been calling these groups "gangs" since the 1970s but has learned to do so with some trepidation. A gang in coastal Maine parlance neutrally denotes a group or bunch. One might speak of a gang of men or a gang of traps. The word connotes violence and illegality in most of the United States, however.

2. The zone management law was passed in 1995 in response to industry pressure. The industry has long wanted to have limits on the number of traps per fisherman to reduce trap congestion and reduce costs. There was no agreement along the coast, however, concerning the number of traps to be allowed. Dividing the coast into zones managed by elected zone councils allowed the members of each zone to devise the trap limits they wanted. Trap limits will not, however, reduce mortality on the lobster.

3. Organizing these political teams presents a classic communal-action dilemma. These are situations in which there is a divergence between what is rational for the individual and what is optimal for the society. In communal-action dilemmas, rational action by individuals leads to less than optimal results or disaster for the larger group.

4. Hydraulic haulers consist of a hydraulic pump operated by a belt attached to the boat's engine, which in turn is connected by hoses to a wheel mounted outside the cabin near the boat's steering post. When a fisherman wishes to retrieve a trap, he places the warp (i.e., rope attached to the trap) in a groove in the wheel. The hauler is then turned on and the spinning wheel rapidly pulls the trap to the surface.

5. An in/out ratio is a ratio of those who are allowed to enter to those who have given up their licenses. A one-to-three in/out ratio indicates that one license will be issued to a new fisherman for every three that are retired.

6. To be sure, excluding the "river rats" would result in slightly larger catches per trap for the defenders, but it would also result in a conflict that would be very costly in terms of traps lost and prosecution. The benefits would not outweigh the costs.

7. It is a matter of dispute as to whether the primary motivation for these zones is resource conservation or economic profit (Brewer 2001).

References

Acheson, James M. 1981. "Anthropology of Fishing." *Annual Review of Anthropology* 10:275–316.

Acheson, James M. 1988. *The Lobster Gangs of Maine*. Hanover, NH: New England University Press.

Acheson, James M. 2001. "Confounding the Goals of Management: The Response of the Maine Lobster Industry to a Trap Limit." *North American Journal of Fisheries Management* 21(2):404–416.

Acheson, James M., and Ann W. Acheson. 1998. "Lobster Zone Questionnaire Project: Selected Results as Requested by the Department of Marine Resources." Manuscript. Orono and Hallowell: University of Maine Sea Grant and Maine Department of Marine Resources.

Acheson, James M., and Robert S. Steneck. 1997. "Bust and Then Boom in the Maine Lobster Industry: Perspectives of Fishers and Biologists." *North American Journal of Fisheries Management* 17:826–847.

Brewer, Jennifer F. 2001. "Lines in the Water." Paper presented at the Association of American Geographers annual meeting, New York, February 27–March 3.

Delaney, D., and H. Leitner. 1997. "The Political Construction of Scale." *Political Geography* 16(2):93–97.

Maine Department of Marine Resources. 1998. "Final Report of the Sub-zone Task Force." Manuscript. Augusta: Maine Department of Marine Resources.

Maine Department of Marine Resources. 2000. "Summary of the Maine Lobster Fishery: 1880–1999." Manuscript. Augusta: Maine Department of Marine Resources.

Ostrom, Elinor. 1990. *Governing the Commons: The Evolution of Institutions for Collective Action*. New York: Cambridge University Press.

Ostrom, Elinor. 2000. "Collective Action and the Evolution of Social Norms." *Journal of Economic Perspectives* 14(3):137–158.

3

Transition in the American Fishing Commons: Management Problems and Institutional Design Challenges

Susan Hanna

Introduction

U.S. fishery management is in the midst of a transition that will significantly change the marine fishing commons. It is a transition that requires changes in traditional institutional approaches and in the scale, scope, and process of management. These changes are taking place in the face of management contention and prohibitions on the use of certain management tools, such as private property rights. The changes are layered over a management structure and historical context that carry their own momentum. All challenge the transition. This mixture of new requirements, current practice, and historical expectations characterizes the complexities of the transition and the challenges it presents.

Management Structure

The 1976 Fishery Conservation and Management Act (FCMA) was the first legislation providing comprehensive federal authority over fisheries within the U.S. exclusive economic zone (EEZ): the ocean area extending from the seaward boundary of each coastal state out to 200 miles offshore. It established a new fishery governance structure that distributed management authority among the federal government and the regions through a system of eight regional fishery management councils.

The regional fishery management councils are democratic decision-making bodies in which representatives of recreational and commercial user groups, states, tribes, and the federal government collaborate to develop fishery management plans and implement fishery regulations. Scientists, user groups, environmental organizations, and the public participate as advisors and provide testimony at public meetings. The idea of the council system is to develop regional approaches to fishery management within a framework of legally acceptable practice.

Two premises underlie the regional management approach: first, that people who have working knowledge of regional fisheries can make the most informed decisions about those fisheries; and second, that management of fisheries within state waters (in most cases out to three miles from shore) should be coordinated with management of fisheries in federal waters. The regional councils reflect recognition that fisheries are regional rather than national in scale and character.

Despite its recognition of regional fishery interests, the law also protects the national interest in fisheries. Regional fishery management councils recommend management plans and regulations to the Secretary of Commerce, who holds the ultimate authority for their approval and responsibility for their consistency with federal law. In most cases, the secretary delegates this approval authority to the National Marine Fisheries Service (Hanna et al. 2000). The eight councils take different approaches to decision making and management, as was anticipated and intended when the FCMA was first developed. Each council involves many actors who represent different interests and incentives. The fisheries they manage are diverse and complex.

Council composition is dictated by law: each council has a voting membership that includes the regional administrator of the National Marine Fisheries Service, directors of state fishery agencies, and public members who are usually, but not always, representatives of the commercial and recreational fishing industries. A tribal representative sits on one council. Nonvoting membership includes the area directors of the U.S. Fish and Wildlife Service, the U.S. Coast Guard, the Interstate Marine Fisheries Commissions, and the U.S. Department of State (Hanna et al. 2000).

The FCMA has undergone many amendments in response to changing fishery conditions. The most recent amendments are contained in the 1996 Sustainable Fisheries Act (SFA), which amended and renamed the FCMA as the Magnuson-Stevens Fishery Conservation and Management Act (MSFCMA). The SFA charged the regional management councils with new, stricter responsibilities for stewardship of the nation's marine fisheries.

Management Context

The changes embedded in the fishery management transition are not unique to fisheries but mirror changes in the management of other publicly owned resources in which new public values and ideas about sustainable management are expressed. The current stage in the evolution of U.S. fishery management results largely from a failure to effectively meet public expectations for sustainable management of marine ecosystems. This failure also exists worldwide, observable in the many marine fishery problems that signal a poor integration of people and ecosystems by management. These problems include diminished biological and economic productivity, increased conflict, costly management, and institutional fragmentation (NMFS 1999; Hanna et al. 2000). They are widely acknowledged as the basis for the need to reconstruct fishery management (Costanza et al. 1998; Iudicello, Weber, and Wieland 1999; National Research Council 1999a; Hanna et al. 2000).

Fishery management, through its failure to find a sustainable balance in fishery exploitation, has also created additional costs beyond those mentioned in the previous paragraph. Overexploited fishery resources mean reduced opportunities for those industries and communities dependent on them for economic well-being. This reduction in opportunities, combined with the need to rebuild stocks, expands the information requirements of management, adds complexity to regulations, and creates conflicts among user groups. These in turn increase management costs, undermine management legitimacy, and decrease management effectiveness.

The transition in fishery management has its origin in international agreements and national law. Its boundaries are well defined in legal terms. Its path, however, and the process of finding that path are less well defined. We know the properties of the institutional structure within which we now manage fisheries, but there is great uncertainty regarding the structure and process of institutional change. The institutional structure of fishery management within the United States and internationally is a legacy of the past, designed to promote growth within relatively simple conservation limits. It is not designed to encompass biodiversity and human complexity, control competing uses, or integrate ecological and human dynamics (Ostrom 1995; Hanna 1999). The redesign of this institutional structure is the greatest challenge of fishery management transition. The set of rights, responsibilities, and rules that constitute the institutional environment of U.S. fisheries is weak and outdated, but it is entrenched in practice and expectation.

Meanwhile, pressures on marine ecosystems intensify. International demand for seafood is increasing, driven by population growth, rising incomes, changing tastes for fish, and the limited availability of substitutes. Fish capture technology, in becoming less expensive and more sophisticated, has allowed intensive exploitation of wild fish populations. Innovations in fishing methods and fishing strategies have provided access to the last "frontiers" of unused resources (Hanna 1997). These pressures create a strong need for management institutions that simultaneously promote the health of ecosystems and economies.

This chapter addresses the transition U.S. fishery management is undergoing in the context of management problems and challenges of institutional design. It addresses the reconstruction required to promote a long-term balance between humans and marine ecosystems.

The Transition

The status of U.S. fisheries is, as with that of all natural resources, a product of its history, which has been one of development and growth, stress and decline. The sequence has been repeated in most fisheries, with only the timing of the sequence differing among them. The historical paths of American fisheries have included foreign dominance, domestic territorial claims, and eventual domestic control. They have included expansionary investment and contractionary disinvestment. The paths have varied widely in timing across regions under different approaches to management (Miles et al. 1982; Hennemuth and Rockwell 1987; Wise 1991; Hanna and Hall-Arber 2000; Hanna et al. 2000).

Regional influences have been and continue to be important drivers of American fisheries, but fisheries also function in contexts larger than regions. Policy choices made at national and international levels have affected how, why, and when fisheries have developed, the development path they have followed, and how well they have been sustained. The path has resulted in many successes, but it has also created problems.

Management Scope

A fishery combines economic, social, cultural, political, ecological, and physical components. The scope of fishery management may encompass some or all of these components and in doing so will be forced to accommodate the trade-offs among them. Management scope may be defined in terms of either regulatory control or planning focus. In terms of regulatory control (e.g., over geographic areas or gear types), managers in the post–World War II period have responded to increasing exploitation pressure by expanding the scope or management from localized input controls (e.g., minimum fish sizes) to more complex combinations of input-output controls (gear restrictions and quotas) and across previously independent authorities, such as states.

In terms of a planning focus, management scope presents a more variable picture. The number of fishery components included in long-term objectives—the vision for the fishery—can be confusingly vague. Are fisheries managed primarily for biological ends? For economic productivity? For social and cultural goals? For ecosystem sustainability? It's often hard to discern. Additionally, management scope is changeable. The focus of fishery management emphasizes different components at different points in time in response to emerging pressures. The result is that overall, the scope of U.S. fishery management has lacked clarity and consistency, leaving decision makers free to emphasize different objectives in response to changing political conditions. In the short run, a flexible scope provides managers with political flexibility, but at the cost of long-run stability (Hanna 1998a).

Management Scale

The optimal scale of management—the size of the management enterprise—is a matter of continuing contention. This contention includes the relationship of scale not only to management costs but also to management performance. At one end of the spectrum is large-scale, centrally controlled management. At the other extreme is small-scale management, with control devolved through comanagement or community-based management arrangements.

U.S. fishery management is at present a hybrid of these two extremes. The regional fishery management council system is at once the regional face of centralized federal authority over fishery resources in the EEZ and the coordination of decentralized decision making among collections of states.

The costs of management are distributed differently in management at different scales. Large-scale centralized arrangements may lower the information and design costs of management but may increase implementation costs. Similarly, small-scale decentralized arrangements may lower postimplementation costs of regulation through better monitoring and compliance but may increase costs of coordination and information (Hanna 1995).

The scale question is tightly linked to management scope. The scope of fishery management has in recent years expanded beyond seafood production and recreational fishing to include habitats, fishing communities, and ecosystems. This expansion reflects an increasing number of interests and increasingly difficult trade-offs among those interests. It also encompasses different points of view as to the appropriate scale of management and the locus of management authority. These too are in transition as traditional approaches to management are forced to adapt an increasing number of claimants to management services.

Origins of the Transition

World fisheries developed rapidly after World War II under national policies to expand territorial control of the oceans and develop shoreside economic growth. Access was open to most fisheries, vessel construction was subsidized, and seafood markets were developed. As a result, fisheries were transformed from being about 60 percent underexploited in the early 1950s to about 60 percent overexploited by the early 1990s (FAO 1997). The Food and Agriculture Organization (FAO) of the United Nations reports that of the fish stocks accounting for the majority of the world's marine landings, about two thirds are in urgent need of management (FAO 1997). The degree of overfishing varies widely among geographic areas and even within geographic areas, but where overfishing exists, species abundance is reduced, and ecological resilience is compromised (Botsford, Castilla, and Peterson 1997; Pauly et al. 1998). Ecosystem productivity in coastal areas is also suffering because of habitat degradation and fragmentation (Gray 1997). These declines in biological productivity translate directly into economic losses.

The postwar environment of world fisheries formed the institutional origin of present U.S. fishery management. In the late 1960s, protecting ecosystems and achieving sustainability were far from people's minds. The public concern was with rehabilitating the fishing industry and promoting fishery growth. A quote from Senator Warren Magnuson illustrates the American view of fisheries and the role of fishery managers at that time. Speaking to a meeting of state and federal agency scientists and managers in 1968, Magnuson said: "You have no time to form study committees. You have no time for biologically researching the animal. . . . Your time must be devoted to determining how we can get out and catch fish. Every activity . . . whether by the federal or state governments, should be primarily programmed to that goal. Let us not study our resources to death, let us harvest them" (Magnuson 1968, 8).

Magnuson was speaking in a time of rising concern over increasing seafood imports and high levels of foreign fishing off the East and West Coasts of the United States. The U.S. fishing fleet was in disrepair and could not compete successfully in world seafood markets. Congress was united on the urgent need to invest in the renovation and expansion of the American fisheries. The 1969 Commission on Marine Science, Engineering and Resources (the "Stratton Commission") reinforced this view. In the commission's review of policy for the nation's coasts and oceans, it produced a report on living marine resources that emphasized the national interest in producing wealth and food from marine fisheries. The cold war further strengthened the desire to reinvigorate U.S. fleets and exert control over ocean territory (Commission on Marine Science, Engineering and Resources 1969; Hanna et al. 2000).

The FCMA was passed in this climate of domestic expansion and foreign exclusion. It provided the fishery management council system with incentives to expand fishing but gave it little direction and few tools with which to accomplish its most difficult task: the allocation of fish among competing and expanding domestic interests. The FCMA was successful in eliminating open-access fishing by foreign fleets but did little to resolve the problem of open-access fishing by domestic fleets. Some councils, determined to provide fishing opportunities for all, left fishing access open, to the detriment of fishery conservation. Fleet capacity expanded. Federal programs designed to assist in fishing industry renovation and expansion—low-interest loans and tax-deferred vessel construction programs—remained in place, exacerbating the capacity problem.

The Institutional Path

The path followed by American fishery management under the FCMA has been true to its origins. It has included a race for fish, an overinvestment in fishing capital, an underinvestment in management capital, and shortened time horizons. The path illustrates the importance of historical context in understanding the evolution of institutions.

The race for fish had commonsense origins and destructive results. With open access to fishing, fishermen could own fish only by capture. Fishermen raced to achieve a competitive edge in fishing by investing in bigger and more powerful fishing vessels. Seafood processors expanded their plants to accommodate the increased volumes of fish being landed. Investments in fishing and processing capacity far exceeded levels that could be supported by the fishery resource over time. Concurrent with the overinvestment in fishing and processing capital was an underinvestment in the management capital needed to stay abreast of changing conditions. Management required decision-making skills, adaptive processes, knowledge about alternative management tools, and systems of monitoring and evaluation (Hanna et al. 2000).

The race for fish combined with fishing overcapacity and management undercapacity to progressively shorten the time horizons of fishery managers and user groups. Short-term actions crowded out long-term strategies. Reactions to crisis overwhelmed planned management. Assurance about the future declined, and conflict among competing interests increased.

From the perspective of rehabilitation and expansion of U.S. fisheries, the path followed by postwar fishery management led to limited short-term success. From the perspective of building an institutional structure for long-term fishery sustainability and economic productivity, the path led to a fishery management system plagued by institutional dysfunction (Hanna 1998a).

Pressures for Change

By the 1990s, pressures for change were growing. The 1996 SFA sent a powerful signal of the public's desire for stabilization and ecosystem protection. It added several important strictures to regional fishery management: councils must eliminate overfishing, rebuild overfished stocks, minimize bycatch (unintended catch), document and protect essential fish habitat, and account for the effects of fishery regulations on fishing communities.

These national changes reflected international trends toward more conservative fishery management. Several international ocean agreements for the protection of marine ecosystems were ratified during the 1990s, including the Rome Consensus on World Fisheries, the Code of Conduct on Responsible Fishing, and the Kyoto Declaration adopted at the Conference on the Sustainable Contribution of Fisheries to Food Security (FAO 1997). These agreements contain guidelines for ending overfishing, reducing bycatch and discards, reducing fishing capacity, strengthening governance, and strengthening the scientific basis for ecosystem management.

One problem with both the international agreements and the 1996 amendments to the MSFCMA is that they are, for the most part, incompatible with the old incentives. The new focus on stabilization and the accommodation of a wider range of resource values is layered over incentives designed to promote development and direct use. For example, at the same time the MSFCMA requires an end to overfishing and a rebuilding of all overfished stocks, federal programs such as low-interest loans, tax incentives for vessel construction, and market development assistance remain in place. Some fisheries remain open access.

The Challenges of Reconstruction

The historical origins and current status of American fishery management are complicating its adaptation to a new environment and limiting its transition to the future. Institutional redesign, although needed, is without a clear model for its process or form. Additionally, several blockages are slowing the pace of reconstruction: changing public expectations, scientific uncertainty, and incompatible incentives.

Changing Public Expectations

The fishery management system implemented by the FCMA reflected values widely held by the U.S. public in the 1970s. Strong sentiment supported domestic investment, fishery expansion, and a degree of regional autonomy. But these values changed as the conditions of fisheries changed. By the 1990s, as more fisheries had gone from being fully exploited to overexploited, the public's attention was drawn to questions of fishery sustainability and the protection of endangered species. Additionally, the constituent base of American fisheries management had broadened over time from the traditional mix of commercial fishermen, recreational fishermen, and coastal communities to a wider group that included environmental organizations and public-interest groups. These more heterogeneous interests represented diverse values requiring a different approach to fishery management.

Twenty-first century fisheries are still valued, as they have traditionally been, for their direct production of seafood. These values are reflected in market prices for seafood as a consumption good. But increasingly, fish are also valued for services not represented by markets. These include the contribution of fish to reproduction and to the genetic diversity of a marine ecosystem, the option for future uses of fish, the knowledge that they exist, their availability as a bequest to future generations, and their potential to provide new goods such as pharmaceuticals (Hanna 1998b). These nonmarket values are more difficult to measure and analyze. Estimation methods are still relatively crude and subject to bias, and as values get increasingly hypothetical, they become correspondingly difficult to assess. Nonmarket value estimations are specific to a given context, preventing generalization of results from one area to another.

Nevertheless, the fishery management task is to understand and accommodate the full distribution of public values within the constraints of the law. Public values are diverse, and management must develop the means to analyze trade-offs among competing actions. Reaching management decisions that reflect public values, including nonmarket values, is a difficult challenge that becomes more intractable as the competition for limited resources increases. Because the demand for fishery resources exceeds the sustainable supply, American fishery management is now operating in an environment in which all actions have distributional outcomes. Benefits generated to some may create a cost to others.

Scientific Uncertainty

Science-based decision making is an explicit requirement of federal fishery management law. Management actions must be based on the best available scientific information, and the potential impacts of those actions on the ecological and human environments must be analyzed (NMFS 1996). There are many uncertainties, however, that stand in the way of ideal science-based management. These uncertainties impede progress toward reconstructing fishery management.

First, scientific knowledge about marine fish populations is limited. It is even more limited with respect to marine ecosystems, in particular with regard to their critical functions and threshold effects. Additionally, ocean processes are subject to large-scale variations that create uncertainty about conditions at any point in time. These biological, ecological, and physical uncertainties combine with equally large uncertainties about the human component of fisheries: economies, social structures, and strategic behaviors. Interacting with management goals that are often vague and unclear, these uncertainties create an ambiguous environment for decision making.

A further complication to the climate of uncertainty is that U.S. fishery management is based on public participation and requires a broad-based stock of human capital to maintain the flow of management services. The stock is contained in the education, knowledge, and skills of decision makers, advisors, and other participants. It produces the flow of services required for coordination, negotiation, scientific review, design, monitoring, and enforcement. As the complexity and sophistication of fishery management has increased, the need for human capital has expanded. This demand has generated heavy loads to be borne by the existing human capital and has impeded the acquisition of new skills required to manage complex systems (Hanna 1997).

Uncertainties about human behavior also make it difficult to determine how to constrain human actions at levels that will ensure sustainable marine ecosystem benefits. The knowledge base of fishery management is primarily one of individual species valued for direct use. Little is known about species interactions or human impacts at the ecosystem scale. This uncertainty makes it particularly difficult to redesign the decision process to accommodate the full scope of marine ecosystems.

A fundamental uncertainty concerns the appropriate scale and structure for managing ecosystems. It is logical to try to match management scale with ecosystem scale, but how to do this within existing political and economic structures is a large research question that is only beginning to receive substantial investments (Ecosystems Principles Advisory Panel 1999). Defining an ecosystem is the first large challenge. Designing the appropriate coordination of management authority over the full range of ecosystem components and uses is the next.

Incompatible Incentives

Continuing scientific uncertainties in the knowledge base of fishery management lead to incompatible incentives facing managers and fishery participants alike. A good example is the dueling incentives related to social and private time horizons. Uncertainties about the drivers, interactions, and constraints of ecological and human systems shorten the time horizons of managers and user groups, yet those same uncertainties provide, from the social perspective, an incentive to take a precautionary approach. Similarly, uncertainties about the tenure of access to fishery resources also cause people to focus on the short term. Both types of uncertainties create substantial disincentives for long-term sustainability.

An incentive to race for fish and to emphasize short-term gains is created by high levels of uncertainty combined with a lack of assurance about rights to resources. In the race for fish, levels of fishing capacity far exceed the productive capacities of fish populations, because long-term incentives for stewardship are missing. The result is that both ecological and economic outcomes are suboptimal.

In addition to the basic incentive problem posed by the race for fish, the more sophisticated problem of not understanding enough about human behavior to make strategic use of incentives is a continuing problem in fishery management. The institutional design challenge is to promote incentive structures that are less vulnerable to short-term interests and that meet long-term ecosystem-scale objectives in a cost-effective way. These structures may be independent of the degree of centralization of authority and are almost certainly dependent on the fishery context.

Worldwide, problems associated with centralized management processes have generated interest in finding alternative institutional forms, particularly those that shift authority further away from the center. But decentralized approaches can also create incentives incompatible with long-term sustainability. Finding the distribution of authority and rights that promotes long-term behavior in any given fishery is complicated by the controversy surrounding assigning rights to individuals in fisheries (National Research Council 1999b).

Finding the incentives that perform best in various contexts remains largely a matter of trial and error in the U.S. fishery management system because knowledge of incentive-based management is limited among participants. Fishery management plans routinely ignore many of the basic rules of incentives, for example, aligning responsibilities with rights. The potential of using market-based tools or new forms of property rights to promote social goals has received little management attention. Progress in the design of institutions that privatize some of the rights to use fishery resources came to a halt in 1996 when a moratorium on the development or adoption of new ITQ programs was written into law.

These three blockages to the reconstruction of U.S. fishery management—scientific uncertainties, changing public expectations, and incompatible incentives—are accompanied by many research needs. Research is needed on nonmarket values, the social and economic dynamics of resource exploitation, the role of fisheries in community economies, the costs and benefits of alternative rights structures in fisheries, the design elements of decision processes, mechanisms to align responsibilities with rights, and systems for monitoring and evaluation. Scholars of common-pool resources have much to contribute to the knowledge base.

The Future

U.S. fisheries have come a long way in the past thirty years: from foreign dominated to "Americanized," from undercapitalized to overcapitalized. Fishery management has undergone a corresponding evolution from promoting growth and development to a search for stabilization and sustainability. The specific management task is to articulate in concrete operational terms a vision of sustainable balance between people and ecosystems. The challenges for reconstituting the fishing commons have never been greater.

The links between the biophysical environment of the oceans and the behavior of resource users are still poorly understood. Although it is generally accepted that conservation actions affect economic well-being, it is less well understood that motives to conserve depend in turn on economic well-being. Similarly, although it is acknowledged that management decisions affect social well-being, it is often a surprise that social well-being influences the effectiveness of management.

In conditions of scarcity, fishery management is often driven by crisis, tending to be reactive, rather than strategic, resolving one problem only to create another. A failure to specify measurable management objectives leaves decisions without long-term performance criteria and vulnerable to short-term political pressures. Yet monitoring progress toward achieving objectives is critical to the ability to adapt and improve institutional design.

Where is U.S. fishery management going? It is clear that the path to reconstruction will require institutional redesign to accommodate limited information, establish performance standards, and introduce accountability for meeting long-term goals. The path will require institutions that have the flexibility to adapt to large-scale changes in oceanic regimes, medium-scale fluctuations in marine populations, and small-scale variation in markets. The alternative paths through the transition are limited by the momentum of history, but the need for effective institutional adaptation is critical to future management effectiveness.

Acknowledgments

The work on which this chapter is based was supported by grant no. NA76RG0476 (project no. R/RCF-05) from the National Oceanic and Atmospheric Administration to the Oregon State University Sea Grant College Program and by appropriations made by the Oregon legislature. The views expressed herein are those of the author and do not necessarily reflect the views of NOAA or any of its subagencies.

References

Botsford, L., J. Castilla, and C. Peterson. 1997. "The Management of Fisheries and Marine Ecosystems." *Science* 277 (July 25):509–515.

Commission on Marine Science, Engineering and Resources. 1969. *Our Nation and the Sea: A Plan for National Action.* Washington, DC: U.S. Government Printing Office.

Costanza, R., F. Andrade, P. Antunes, M. v.d. Belt, D. Boersma, D. Boesch, F. Catarino, S. Hanna, K. Limburg, B. Low, M. Molitar, J. Pereira, S. Rayner, R.

Santos, J. Wilson, and M. Young. 1998. "Principles for Sustainable Governance of the Oceans." *Science* 281:198–199.

Ecosystems Principles Advisory Panel. 1999. *Ecosystem-Based Fishery Management. A Report to Congress as Mandated by the Sustainable Fisheries Act Amendments to the Magnuson-Stevens Fishery Conservation and Management Act, 1996.* Washington, DC: Department of U.S. Commerce, National Oceanic and Atmospheric Administration, National Marine Fisheries Service.

Food and Agriculture Organization (FAO). 1997. *The State of World Fisheries and Aquaculture 1996.* Rome: Food and Agriculture Organization of the United Nations.

Gray, J. S. 1997. "Marine Biodiversity: Patterns, Threats and Conservation Needs." *Biodiversity and Conservation* 6:153–175.

Hanna, S. 1995. "Efficiencies of User Participation in Natural Resource Management." In *Property Rights and the Environment: Social and Ecological Issues*, ed. S. Hanna and M. Munasinghe, 59–67. Washington, DC: World Bank.

Hanna, S. 1997. "The New Frontier of American Fishery Governance." *Ecological Economics* 20(3):221–233.

Hanna, S. 1998a. "Parallel Institutional Pathologies in North Atlantic Fisheries Management." In *Northern Waters: Management Issues and Practice*, ed. D. Symes, 25–35. London: Blackwell Science.

Hanna, S. 1998b. "Institutions for Marine Ecosystems: Economic Incentives and Fishery Management." *Ecological Applications* 8(1, suppl.):170–174.

Hanna, S. 1999. "From Single-Species to Biodiversity—Making the Transition in Fisheries Management." *Biodiversity and Conservation* 8:45–54.

Hanna, S., H. Blough, R. Allen, S. Indicello, G. Matlock, and B. McCay. 2000. *Fishing Grounds: Defining a New Era for American Fishery Management.* Washington, DC: Island Press.

Hanna, S., and M. Hall-Arber, eds. 2000. *Change and Resilience in Fishing.* Oregon Sea Grant. Corvallis: Oregon State University.

Hennemuth, R. C., and S. Rockwell. 1987. "History of Fisheries Conservation and Management." In *Georges Bank*, ed. R. H. Backus, 430–446. Cambridge: MIT Press.

Iudicello, S., M. Weber, and R. Wieland. 1999. *Fish, Markets, and Fishermen: The Economics of Overfishing.* Washington, DC: Island Press.

Magnuson, W. G. 1968. "The Opportunity Is Waiting . . . Make the Most of It." In *The Future of the Fishing Industry of the United States*, 7–8. Publications in Fisheries, vol. 4. Seattle: University of Washington.

Miles, E., S. Gibbs, D. Fluharty, C. Dawson, and D. Teeter. 1982. *The Management of Marine Regions: The North Pacific.* Berkeley: University of California Press.

National Marine Fisheries Service (NMFS). 1996. *Magnuson-Stevens Fishery Conservation and Management Act.* NOAA Technical Memorandum NMFS-F/SPO-23, December.

National Marine Fisheries Service (NMFS). 1999. *Our Living Oceans. Report on the Status of U.S. Living Marine Resources.* Technical memorandum no. NMFS-F/SPO-41. Washington, DC: U.S. Department of Commerce, National Oceanic and Atmospheric Administration.

National Research Council. 1999a. *Sustaining Marine Fisheries.* Washington, DC: National Academy Press.

National Research Council. 1999b. *Sharing the Fish: Toward a National Policy on Individual Fishing Quotas.* Washington, DC: National Academy Press.

Ostrom, E. 1995. "Designing Complexity to Govern Complexity." In *Property Rights and the Environment: Social and Ecological Issues*, ed. S. Hanna and M. Munasinghe, 33–46. Washington, DC: Beijer International Institute of Ecological Economics and World Bank.

Pauly, D., V. Christensen, J. Dalsgaard, R. Froese, and F. Torres Jr. 1998. "Fishing Down Marine Food Webs." *Science* 279:860–863.

Wise, J. P. 1991. *Federal Conservation and Management of Marine Fisheries in the United States.* Washington, DC: Center for Marine Conservation.

4

The Relationship between Resource Definition and Scale: Considering the Forest

Martha E. Geores

We can not only visualize, but can actually see, images of the entire globe with the help of satellite technology. Through models of global climate change we can predict what will happen if there is deforestation of the tropical moist rain forests. Armed with the knowledge of the global benefit of forests, individuals, groups, and governments from the Northern Hemisphere have exerted pressure on countries in the Southern Hemisphere to preserve a resource (rain forests) to which the North has no authoritative claim, or does it? Is there now a global-scale authoritative aspect to defining tropical rain forests as global resources of biodiversity and carbon sequestration? I assert that demands for rain forest preservation are a function not just of scientific knowledge, but of the expansion of our knowledge of scale to the global level. From the smallest village where women gather firewood from a grove of trees to the Amazon rain forest, called the "lungs of the world" (Foster 1990, 36), forests exist across all scales. This chapter examines the implications of scale for forest definition and management.

Forests are different from other common-pool resources (CPRs) for reasons related to resource definition and scale. For all resources, CPRs and others, resource definition is a process dependent on the function of the resource for the people who are defining it. There can always be disputes over resource definition, but those definitional conflicts are fewer when defining grazing land, fish, or irrigation systems than when defining forests. Forests stand out in their complexity. They "are appreciated as renewable natural resources, valued for the use of their products and for their role in maintaining watersheds, soil fertility, and air quality, as well as for their importance as cultural resources in both religious and aesthetic ways" (Geores 1996, 1). Forests are resources that also contain resources; they are made up of biosystems of varying complexity, used for a myriad of social and economic functions as part of complex social systems.

Although we often think of forests as a stationary resource belonging to a certain place on the earth's surface, forests are dynamic and are defined as resources on multiple scales, from that of an individual tree to a global-scale tropical rain forest. Scale is a concept that facilitates the organization of information by establishing a point of reference for observation of phenomena, both biophysical and social. It has a major impact on resource definition. According to Ellen (1982, 5), scale "determines the independent and dependent variables that may be involved in a single correlation." On a fine scale, such as a tree and person scale, there are relatively few variables to be considered, and deciding whether a forest is present and what its form of management is presents a relatively simple definitional problem. On a coarse scale, however, such as that of the Amazon rain forest, there are myriad variables to be considered in deciding the functional definition of the tree-covered landscape and the possibilities of who might have an ownership interest, not to mention how it might be managed. The scale determines the variables, both physical and human related, that will be considered in the definitional process. Even though forest definitions differ by scale, they are not necessarily mutually exclusive. When the scale of observation is changed with respect to forests, the definition may go from a specifically bounded, tree-covered parcel considered sacred to an Indian tribe (Geores 1996, 34) to the tropical moist forests as a global reserve for biodiversity (Native Forest Network 2000). Forests can contain sacred spaces (generally at fine scales) and still be carbon sinks (generally at coarse scales). Their management as CPRs is, however, dramatically different, and perhaps conflicting, for both definitions. This change with scale is not a characteristic shared by other CPRs. Fish may be observed on a more coarse scale, but they are still fish. Grazing land remains grazing land whether observed globally or locally.

This chapter explores three research questions about scale and resource definition and management. A description of the resource definition process follows this section and is followed by a discussion of the concept of scale. Then the following research questions are addressed:

1. Why are forests affected more strongly than other CPRs by changes in scale, especially from local to global (fine to coarse), in regard to both allocative and authoritative rights?
2. Does the change to a coarser scale necessarily result in worse management?
3. Is there always a conflict among resource management strategies at different scales, or can there be a symbiotic relationship in some forest functions?

Defining Resources

According to Giddens (1984), there are two kinds of resources: allocative (material) and authoritative (power). Although this is a useful concept, for our purposes it makes more sense to think of every resource as having both an allocative aspect and an authoritative aspect. With respect to forests specifically, the allocative aspect includes material objects such as the trees, mushrooms, fodder, and firewood, and the authoritative aspect relates to matters of who controls access to the forest and who has the power to define appropriate use of it. In some cases, the same group of people have both allocative and authoritative rights, that is, they have control over the resource as well as being the ones to use the benefits of the resource. Often, however, allocative and authoritative aspects of a resource are controlled by different parties, setting the stage for conflict over definition of the forest and its use.

Oftentimes the authoritative aspect of a forest is controlled at a different scale than the allocative aspect is. For instance, a group of indigenous people may have used a forest to meet their subsistence needs from time immemorial, and they may define the forest as the source of goods and environmental services to meet their needs and perhaps as their spiritual home. A state government may declare forests state property, as has been done in India and Thailand, among other nations; then the authoritative aspect of the resource definition is separated from the allocative aspect, and state rules are imposed on how indigenous people use the forest. The government operates at a state scale that is much more coarse than the local scale at which the indigenous people are operating. In this case, coarser scale accompanies greater authority; scale is an added dimension to the definition process.

Forest Definition Examples	
Allocative Aspects	Authoritative Aspects
Species of trees	Ownership
Wildlife	Right to exclude
Medicinal plants	Control over harvest

One of the particular challenges addressed in this chapter is the increasing acceptance of forests as a global resource, with authoritative control over the forests at scales further removed from the locale of the forests than ever before and an increasing appropriation of allocative resources for the global community. In some cases, authoritative control passes not to coarser-scale levels of government, because of the difficulty of establishing international environmental agreements, but seemingly, to private industry, which exercises control through purchase of forest land and patenting forest products. The rightful locus of authoritative control is an issue discussed below.

Scale

The process of resource definition is sensitive to scale in many ways. Scale is a geographic tool. It is a way of framing conceptions of reality (Marston 2000, 221). It is an intellectual, social construct created for the purpose of organizing data and concepts.

Scale is a consideration in both the allocative and authoritative aspects of forest definition. Scales relevant to the resource definition process range from the very fine scale of a tree (for the allocative aspect) or a person (for the authoritative aspect) to the very coarse global scale of all tropical forests (for the allocative aspect) or international alliances (for the authoritative aspect). The scale at which a particular forest question is examined greatly influences both the question and the answer. Local groups who use a forest for subsistence would advocate using a fine scale, whereas climate modelers trying to predict global warming would advocate a coarse scale. No question, however, can be addressed from a single scale point of view.

The most common categories of scale are cartographic, operational, and spatial (also called geographic). Cartographic scale is the easiest category to define. It is the expression of the mathematical ratio between the distance on the map and the distance on the earth of the area mapped. It is as old as mapmaking itself. Mapping is important in resource definition because of the need to visually represent the extent of an allocative resource or the extent of authority.

Operational scale is the scale at which the phenomenon being studied is observed. Keeping in mind that data carry different information when presented at different scales (Bian 1997), the designation of operational scale influences both the research questions asked and the methodology employed in the research. Phenomena do not occur within an isolated scale. In both human and biophysical systems, they are influenced by events and processes occurring at finer and coarser scales. When a researcher chooses a scale for a study because the object of the study is readily apparent at that scale, certain scale dependencies must be considered. Walsh et al. (1997, 27–28) identified three kinds of scale dependence: "(1) representations of spatial patterns may be different when observed at different scales; (2) certain patterns and processes may not be observable at a particular scale or resolution; and (3) methods used to observe causal relationships between variables are affected by the scale of observation." Scale dependencies are a caution to observers that their observations of events or phenomena must be understood in the context of events and phenomena not visible, but still occurring at coarser and finer scales. Patterns of phenomena may look different at different scales. Observers must keep in mind that patterns of bacterial behavior cannot be observed at the global scale, even though the behavior is still occurring when the observer is operating at a global scale. Finally, the observer must use methods appropriate to the scale involved to observe causal relationships; for instance, a global vegetation map would not be helpful in defining the allocative aspects of a forest in the highlands of northern Thailand. These scale dependencies have been noted in CPR work as well (see Ostrom et al. 1999, 280; Wilson et al. 1999, 243).

Within biophysical and human systems, actions take place at multiple scales and have impacts at multiple scales. It is extremely difficult to designate a discrete, isolated operational scale for such systems. The concept of recursive relationships from structuration is helpful in thinking about operational scales. All actions are influenced by other actions at both coarser and finer scales. The phenomenon being analyzed may even be an aggregation of data and processes at a finer scale. Recognizing that phenomena occur at multiple scales, researchers must be especially careful about generalizing their findings. This concept is particularly appropriate for forest definition because of the complexity of the resource and the multiple processes, both human and physical, taking place with respect to forests at all times. When defining a forest, one must make an informed decision about the scale to be used and realize the impact of other scales on the definition.

Geographic, or spatial, scale refers to the area studied or represented. It is especially an intellectual construct. The boundaries of operational scale are more fluid than fixed, because a phenomenon is being observed, not a set area. Spatial scale, however, sets boundaries that are artificially imposed. Social and biophysical systems are seldom exactly delimited, however, a geographic scale is useful for a particular purpose.

Its political ramifications are perhaps the best-recognized aspects of scale. The definition of scale used in political geography "refers to the nested hierarchy of bounded spaces of differing size, such as the local, regional, national, and global" (Delaney and Leitner 1997, 93). As a social construct, geographic scale is often taken to follow jurisdictional lines. When resources, especially CPRs, transcend international jurisdictional boundaries, there is a disjuncture, which causes conflict over management of the resources (Barkin and Shambaugh 1996). When the extent of the resources does not match the scale of observation or regulation, political disputes arise.

Traditionally scholars have spoken of the levels of scale as (from finest to coarsest) local, regional, national (or national, regional), and global, often explicitly referring to them as "nested scales." This concept of a hierarchy of scales impedes research into natural-resource definition. Just as relationships between structures and agents are recursive, so are actions at any particular scale. No actions occur in a vacuum; influences impinge on actions from finer and coarser scales. It is necessary to recognize that multiscalar interactions are taking place all of the time. There is not a neat progression from a fine to a coarse scale; there may be a direct relationship between the global economy and a local tree plantation.

Labeling a scale "regional" is another source of confusion. In geography there is a subfield devoted to the study of regions, and the process of defining a region is a complex one. For instance, regions have been seen as social organisms by some scholars (Archer 1993). Others maintain that regions are formed around issues or interests and do not have clearly demarcated boundaries (Gilbert 1988). All of this is to say that "regional" as an adjective modifying scale is an empty term.

Scale is not only spatial, it is also temporal. Time scale is especially important in defining forests. Some allocative aspects of forests, such as trees, have a much greater life span than other allocative aspects such as mushrooms or palm fronds. It takes longer than a human lifetime for some species of trees to mature, while a mushroom may live one season. Some insects live less than a week. Authoritative aspects of the definition may change as quickly as a palace coup or a national election. One of the basic questions to be addressed when defining a sustainable forest is: sustainable for how long? Time scale is not a linear matter in forest definition because there are myriad time scales operating at the same moment. A loosening of the meaning of time scale to take it away from a linear scale will assist in establishing the context within which social and biophysical interaction takes place.

The final scale issue discussed here is time-space distanciation. This is another intellectual construct that allows us to conceptualize the way social systems work. Through time-space distanciation Giddens describes the stretching of a social system across time and space. This concept is much like hegemony, in which one social system is imposed on another. The concept of a global system's being applied to local resources is a form of time-space distanciation.

Research Questions: The Particular Impact of Scale on Forest Definition and Management

This section discusses the research questions identified earlier in the chapter: How and why does a shift in scale have such a great impact on forests? What is the impact on management? Is there conflict at different scales? Throughout the discussion both allocative and authoritative rights will be discussed separately in relationship to the research questions.

Local-Scale Allocative, Functional Definitions

The most common way of defining a forest is by its functional definition, that is, naming its allocative aspects. The functional definition of a forest is reflective of the culture in which it is situated. Clarence Glacken's (1967) treatise on natural history in Western thought notes that rights to use the forest in the Middle Ages were extensively codified. In England, Henry III signed the Forest Charter, which became the foundation of the English social scene, with its regulation of forest use according to social class (322–323). Rights to specific trees, such as oaks and honey trees, were also regulated, although their ownership was tree tenure, not forest tenure (see Fortmann and Bruce 1988). Regulations of the forest product use in France and Germany were also codified during this time. Royalty and noblemen were given extensive rights, including the right to hunt in forests. This reservation of the forests for the upper classes was even an issue in the French Revolution. Peasants lived near, but not in, the forest in medieval times. The fact that the crown granted rights to use the forest meant that the authoritative aspects of the resource definition rested with a figure at the top of the hierarchical society. Noblemen who lived closer to the forests than the crown, but further than the peasants, received permission to use them from the crown. The tension invoked by the partition of a resource into its allocative aspects and its authoritative aspects is certainly not a new state of affairs.

Forests can also be cultural artifacts. Jackson (1994, 96) describes the forests of Kazakhstan as immense expanses of nothing but apple trees, pear trees, and apricot trees, imported from elsewhere, cultural artifacts left on the landscape by Indo-European migrants who settled in the area. Plantations of eucalyptus are no less a cultural artifact than the apple forests of Central Asia. The partners of Nova Scotia's Model Forest also define the allocative aspects of forests broadly from the standpoint of a culturally diverse constituency: "Forests vary around the world, across Canada and within Nova Scotia. Likewise, forests mean different things to different people. A hiker sees the forest as a place for recreation, a biologist sees the forest as a provider of habitat for plants and animals, a forester sees it as a dynamic natural resource that can provide multiple benefits to society and First Nations Communities value the Forest as an essential part of their heritage" (Nova Forest Alliance 2001).

Changes in Definition According to Scale

Conflict related to scale becomes apparent when two groups, often on different authoritative scales, lay claim to one area of forest. Traditional users often have complex institutions in place to govern the use of the forests. In both of the following cases, one from northern Thailand, and the other from the Philippines, the local groups exercised both allocative and authoritative rights with integrated management systems. Higher-scale authorities intervened by redefining the resources as national industrial resources (Thailand) and timber lease holds controlled by the state (Philippines). Both the allocative and authoritative rights were redefined from the fine (village) scale to the coarser (national) scale. The conflict was as much about definition as it was about authority.

According to a study of traditional use of the forests in the highlands of northern Thailand, the indigenous people of Thailand do have institutions to govern use of the forests (Ganjanapan 1998). Their system involves three categories of community forests: sacred, watershed, and communal woodland. The sacred forest is "reserved mainly for ceremonial purposes as a shrine for its guardian spirits, a cremation ground, or a pagoda containing Buddha's relics" (78). No utilization of forest products is allowed in a sacred forest. A watershed forest is located at the head of the watershed for the village's water supply. Again, little use of forest products is permitted in this type of forest, and it is sometimes a sacred forest. In the communal forest, harvesting of forest products and grazing of animals is permitted. Villagers protect their forests from excessive use and from outside groups. Ethnic groups in Thailand have traditionally practiced swidden agriculture, over a long period of time on a sustainable basis.

Conflict arose in 1989 between ethnic groups and Thailand's Royal Forestry Department centering on differences in the allocative aspects of the definition of the forests. The Thai government had recently declared national parks and wildlife sanctuaries in areas traditionally used by ethnic groups for subsistence activities and threatened the people with relocation if they did not stop their swidden practices (Ganjanapan 1998, 72). "In 1992 45.9 percent . . . of the national territory was classified as national reserved forest" (73), and in 1993, 27.5 percent was targeted for conservation forest and 16.2 percent for economic forest (73). Use was restricted in conservation forests, but economic forests were open for development. Traditional uses of the forest were not protected by the government, and in fact, tribal people were forced from their swidden sites so that the land could be awarded to a state-owned company to develop a eucalyptus plantation (73).

The state's view of forests in Thailand was and is more oriented toward industrial development than subsistence use. The land that poor people traditionally farmed is now leased to developers for fast-growing tree plantations, commercial orchards, and tourist resorts (Ganjanapan 1998, 79).

Highland villagers are evicted from their traditional lands, while the government allows lowlanders and investors to utilize the land in the name of national development. Local communities have begun, however, to establish formal institutions to assert their rights to forests against the state.

The situation in the Philippines presents another example of the problems that arise when the allocative aspects of a resource become separated from the authoritative aspects. In the Philippines, the state is unable to effectively establish or enforce a forest protection policy. The colonial history of the Philippines still has an impact on forest definition. Under Spanish rule in the sixteenth century, all land not covered by official titles became state property. This edict included ancestral lands. Forests were cleared for sugar plantations and timber for shipbuilding materials (Magno 1998, 64). Under American colonial rule, modern logging was introduced, and the forest was valued solely for timber. Logging was valued above all other allocative aspects of the forest, including non-timber forest products and environmental services. With Philippine independence, the logging policy did not change. Timber lease agreements allowed private individuals and corporations to cut trees on forest lands. Although the leases carried provisions requiring that the lease areas be replanted, state enforcement was ineffective, and substantial forest areas were degraded.

Throughout this time indigenous people were engaging in subsistence activities in the forest areas. In the 1970s, they started to organize. In 1974, the Ikalahan/Kalanguya people, through a local community organization (KEF), negotiated with the Philippine Bureau of Forest Development for a renewable lease to their ancestral lands. They proposed to exercise forest protection and engage in development activities (Magno

1998, 68). KEF set a new standard for forest management by establishing rules at the local level, instead of at the state level. It proposed forest rules that were presented to village meetings, modified according to feedback, and then passed. Both KEF and the village council had forest guards to enforce the rules. This system of control over both allocative and authoritative aspects of the resource improved the condition of the forest lands substantially.

The Special Case of Tree Plantations

No issue is more likely to create a heated argument among forest advocates than tree plantations. An examination of the issue shows that it has two distinct elements, one relating to the definition of a forest as an allocative resource, and the other relating to the authoritative aspects of the resource. Both are related to scale.

Tree plantations are single-purpose entities. They have one product. Generally, the consumers of the product are not the local people; they are seen as a "global" market, operating at a global scale. (Hold in abeyance the question of whether there are any local benefits to tree plantations.) On the authoritative side, ownership, either direct or through leasehold, is often in the hands of nonlocal corporations, again seen as operating at a coarser scale than local. If the products and the authority are on a different scale than local, the issue becomes a political one, over local (often indigenous) rights.

Many tree plantations are specifically eucalyptus plantations. (Other examples of plantation trees are teak, palm [grown for oil], and various other trees grown for fuelwood.) Eucalyptus trees are ideal plantation trees in that they grow quickly and their fiber is usable, but like all trees and other crops, they draw nutrients and water from the soil in which they are grown. Because they are harvested and removed, they give little or nothing back to the soil in the form of litter. They are native to Australia and are an exotic species in most areas where they are used for plantations, especially in Latin America and Southeast Asia. The argument among forest advocates is not over whether eucalyptus plants are trees, but instead over whether a plantation of eucalyptus trees can be a forest.

Consider the following two challenges to defining plantations as forests:

> Plantations are not forests. Plantations do not provide for most of the services provided by forests. Plantations do not help to conserve biological diversity. Plantations are not a durable reservoir of carbon. (Diamond 1998)

> Forests are more than trees, they contain a community of species: fungi, flowers, insects, understory plants, and a host of wildlife. They are reserves for biodiversity. Native forests contain indigenous species in varying degrees [o]f succession, and various states of health. Native forests should be self sustaining by maintaining biological diversity, ecosystem resiliency, and ecological processes. A plantation is not a native forest. Plantations or tree farms vary in their ability to regenerate native forest ecosystems and are established to meet human demands. (Native Forest Network 2000)

In a paper presented to the Eleventh Forestry Congress in October 1997, Kanowski (1997) discussed the plantation forestry phenomenon. He noted that plantations are created primarily to meet global industrial wood needs, that they are monocultures, and that they do not meet any of the nontimber needs forests generally meet. Plantation forests have grown through international and regional cooperation with little attention to the local scale. They use sophisticated technology, including genetic alteration of the trees and chemical inputs. Although they are a success for their owners economically, they fail to fulfil any of the social dimensions of a native forest. Even fuelwood plantations have not been successful in this regard. Kanowski calls for more complex plantation forestry with a greater association between forests (meaning trees) and other land uses, more direct involvement of the local people, and more diverse species composition on plantations.

Plantation forestry is a mode of foreign investment, especially in Latin America and Southeast Asia. In the state of Chiapas, Mexico, multinational timber companies, including International Paper, a U.S.-based company, have acquired large parcels of land for eucalyptus plantations. One of the charges levied against these multinational companies by indigenous groups and forest advocates is that they rent the land for a few years from campesinos to grow a crop of eucalyptus and then return the degraded land to its owners (ACERCA 2000). Planfosur, a Texas-based multinational, grows eucalyptus on plantations in Tabasco and Veracruz, Mexico. At the end of their growing cycle, the trees are harvested and converted into wood chips for pulp and particle board. The plantations supply wood fiber for Planfosur's partners' paper plants (ACERCA 2000). Members of Action for Community and Ecology in the Rainforests of Central America (ACERCA) are concerned about the use of tropical rain forest land for these plantations, as well as the resulting shift in the scale of the economy from local to global.

Defining tree plantations as forests does have its proponents. "One of the most significant developments in contemporary forestry is the expanding role that industrial forest plantations—forests planted, managed, and harvested for industrial wood values—are assuming in meeting the *world's* growing requirements for wood" (Sedjo 1983, 1, emphasis added).

Plantation forestry may offer one of the clearest examples of conflict among people acting on different operational scales. The definition of a forest is changed, with respect to plantation forestry, from that of a multiple-use, biologically diverse entity to that of a monoculture. The allocative resource is also changed in the case of plantations from a variety of forest products to wood for fiber. The entity being considered is still called a forest, but there is a significant redefinition of the allocative resource involved. The authoritative aspects of the resource are also changed in regard to plantation forests. When a forest is turned over for plantation forestry, the local community sees a loss of control over a "traditional forest" that it used as its own but did not legally have rights to. Plantation owners obtain their authoritative rights by buying or leasing the land, perhaps from local people, but more likely from someone on a coarser scale, such as the state government.

The definitional change resulting from inclusion of plantations is not totally without benefit. Plantation owners do hire local people to work on the plantations, so there is some economic benefit for the communities in which the plantations are located. The conflict may come from the lack of sharing of authoritative resources regarding the land that the plantation forestry occupies. Even though the authoritative control the local people traditionally had over forests was not officially sanctioned, it was still de facto control. Problems may be minimized if plantation forest owners include local people in the planning of plantations to address some of the social aspects of forests that plantation forests are perceived as lacking.

U.S. National Forests

One other type of forest that should be examined with respect to the forest definition process is the national forests in the United States. The purpose of these national forests, established in 1898, was to halt the progressive deforestation of the United States by European Americans and to protect watersheds, assure timber for mining, fuel, and lumber, and provide opportunities for grazing and recreation (Geores 1998, 84). Even with statutorily defined allocative aspects, the functional uses of U.S. national forests vary from forest to forest, depending on the composition of the forest and the community surrounding it.

A study of the Black Hills National Forest in South Dakota and Wyoming showed that the allocative aspects of the definition of the forest varied both over time and according to who made the definition. Although the National Forest Service, a part of the U.S. Department of Agriculture, has responsibility for national forests, national-scale definition of both the allocative and authoritative aspects of the Black Hills National Forest did not result in a sustainable forest. When the local community was committed to defining the forest as a multiple-use resource and the regional and local Forest Service personnel were willing to share the authoritative aspects of the definition, however, the forest was maintained in a more sustainable manner. At times single-use interest groups, such as loggers, dominated the resource definition and were supported by the Forest Service. Not until the local community demanded that more uses than logging be deemed acceptable by the Forest Service that the multiple-use definition was accepted.

One of the problems with definition of this forest by single-interest groups was not just the narrowness of their definition of the forest, but the fact that they considered only the parts of the forest most relevant to them; they did not see the forest as a whole. Such single-issue definition, whether by loggers, trout fishermen, recreationists, or miners, resulted in damage to an underlying infrastructure of the forest that everyone took for granted, but no one included in their definition, namely, the watershed of which it is a part. The definition therefore had to broaden for the forest to be sustainable, and the authoritative community had to use the scale of the whole forest in defining it. Technically the authority for operation of the national forests has always rested with the federal government. Only when the authoritative rights were shared at finer scales, however, was the forest managed in a way that enabled it to approach sustainability.

National forest definition and authoritative rights have another aspect: dependence on the national forests by communities within and near the forests. The economies of communities within and near forests in the national forests system in the United States are often dependent on forest products. This dependence often takes the form of employment of workers in communities in or near the forests by timber companies that have contracts with the Forest Service. This kind of passive dependence often has negative consequences for these communities, and in particular, poverty is a long-standing and persistent problem: "In 1989, nearly a fifth of California forest counties had poverty rates equal to or greater than inner city rates. In the decade between 1979 and 1989 counties that had increases in the timber cut did not experience reduced poverty rates" (Fortmann 1993, 189). Fortmann made the case that if communities in or near national forests were involved in the planning process for these forests, local people could develop and implement policies that would take into account the social and ecological diversity of the communities, to the benefit of the local people. State and federal policies would be necessary to facilitate the process, and certainly both timber corporations (which usually have out-of-state corporate headquarters) and urban-based staffs of national environmental organizations could participate in discussions, but good forest policy is community based (189).

The Impact of Scale on Management

How do changes in scale affect management? Acheson and Brewer (chapter 2) provided one example in their discussion of the change of the exercise of authoritative rights over Maine lobsters from the local scale of harbor gangs to the coarser scale of state regulation. This change, as Acheson and Brewer point out, was necessary because the scale of the lobstering industry had outgrown the system through which the industry had traditionally been managed. In the case of Maine lobsters, the exercise of authoritative rights under the new system was shared and long-held local allocative rights were recognized, which prevented the change in management from being abrupt and total and enabled a system for shared management to be set up.

The case studies on forests discussed in this chapter show mixed results on the surface, but there is a common theme among them. Exclusivity in the exercise of authoritative rights over resources when there is an operational-scale change from fine to coarse results in deterioration of management from the point of view of the people who have formerly had authoritative rights. At least in the cases of forests in northern Thailand and the Philippines, the change in operational scale was accompanied by a definitional change that compounded the problem. The same thing happened with the forest plantations discussed earlier. When a scale change did not sever existing authoritative rights but instead included the finer scale in the coarser-scale management scheme, the change was much more positive. When authoritative rights held at a national scale were shared with people at a local scale, management of the resources improved.

The global scale remains a problem, however. Satellite imagery allows scientists to map rain forests on a global basis, without regard to country boundaries. This absence of geographic scale below the global level fosters the concept that the rain forests and indeed all vegetation are a global resource. Climate models and carbon sequestration models plot changes that the earth would undergo if the tropical rain forests were to disappear. There has been a realization and a recognition of the importance of tropical rain forests to climate stability across the globe. This is an allocative aspect of forest definition. A political problem exists, however, because of the operational scale of tropical rain forests; they occur in the tropics, which means that their spatial scale is the Southern Hemisphere. Some of the allocative benefits of the rain forests are global in scope—certainly climate regulation and biodiversity. Other allocative aspects of tropical rainforests, however, are not necessarily global in scale—for example, rubber trees, subsistence hunting and gathering, or swidden agriculture.

If some of the allocative aspects of a forest are global, on what scale should the authoritative aspects of the resource be considered? This question is the subject of much controversy between countries of the Southern Hemisphere and those of the Northern Hemisphere. Southern Hemisphere countries would ideally like at least country-level control over forests within their boundaries. Northern Hemisphere countries recognize their need for carbon sinks and biodiversity and would like to exercise some authoritative control over these resources at their source.

Having Southern Hemisphere countries operate under different environmental rules than Northern Hemisphere countries did when they were first industrializing, however, seems unfair to the Southern Hemisphere countries.

On issues of demonstrated global importance, there has been an attempt to reach global agreements on country-level regulation of behavior that has an environmental impact. The Earth Summit in Rio de Janeiro in 1992 was one such effort (Panjabi 1997). An international agreement on the environment that has been discussed and negotiated among all participating countries is much less hegemonic than if the United States or the European Community imposed trade sanctions on countries that did not meet their environmental standards.

Must There Be Conflict?

There is conflict over forests around the world. Conflict, however, does not have to be the norm. One way to reduce the amount of conflict over forest use would be to increase understanding of the difference between allocative and authoritative aspects of forest definition. Governments and landowners with authoritative rights need to recognize the allocative rights that have traditionally been exercised forests. Negotiations over forest use would be more productive if both allocative and authoritative aspects of forest definition were considered.

Take, for instance, the conflict between local people and multinational corporations over forest plantations. Forest plantations are allocative resources. The land on which they are planted is an allocative resource, and the products of the plantations are allocative resources. Multinational corporations purchase or lease the land with another allocative resource, money; they have no inherent authoritative resources. They can obtain authoritative resources only from the owner of the land, be it a person or the state. If the authoritative resources rest in the local community or in their government, then there has to be some negotiation over management of the resource, since management of a resource requires the exercise of both allocative and authoritative resources. For management to work effectively, it must be shared. Corporations are realizing this and are including local people in the planning process for plantation development. The bargaining positions of the two sides may not seem equal, but there are nonetheless bargaining rights to be exercised.

In some of the case studies discussed in this chapter, the forest was defined in terms of boundaries set by the state or a colonial power. In these cases, in which the geographic scale is set, it is important to see whether the authoritative aspects of resource definition conform to the set boundaries. In the cases of the northern Thai highland groups and the indigenous people in the Philippines, it was clear that traditional leaders exercised another layer of authoritative rights over the forests, in addition to that of the state authority. Although the authority of these leaders was in conflict with the state authority, they clearly had traditional authority over the use of the forests. It was the source of much frustration to the Thai villagers to have their state government lease land to foreign corporations to establish industrial forest plantations. One of the ways that state governments get revenue from local resources is through royalties, licenses, and taxes. As long as the state defines the allocative resources and claims them for itself, conflict will exist.

It is interesting to note that in the Man and the Biosphere Program, sponsored by the United Nations Education, Scientific, and Cultural Organization, which establishes biosphere reserves in areas that need special protection to preserve fragile environments or biodiversity, local people are involved in the planning of the reserves from the beginning. All parties potentially interested in a particular reserve, from local through global scale, help formulate the plans for that reserve. In establishing a reserve, the needs of the local population are considered, making success of the reserve more likely than if they were not considered.

Conclusion

There is a progression of operational scales for management of resources with a static definition. Forests, however, are dynamic resources, and their definition changes with scale. This definitional change is the source of much conflict among those with authority over forests. Imposition of specific definitions of forests that are linked exclusively to one scale will certainly result in conflict. Witness the early attempts at establishing protected areas for global biodiversity by excluding all human activity under the authority of the state versus the current practices of shared management of these areas between local, state, and nongovernmental organizations.

Scale defines the variables to be considered. It seems important, given the present state of our knowledge about the global role of forests in maintaining the biophysical systems of the earth, that global-scale variables always be considered in the definitional process. We must also recognize, however, that forest definitions change as the scale becomes finer, and that at each point of definition, the recursive relationships between definitions at other scales need to be taken into account. People exercising both authoritative and allocative rights should see themselves as occupying a place on a continuous scale from fine (tree/person) to coarse (continental forest/international group) and be mindful of the recursive relationships that exist when they exercise their rights. Authoritative rights are not always exercised at the global level for forests, but the exercise of authoritative rights at any scale may have global implications. We can observe forests on a global scale, without boundaries. But boundaries exist at finer scales than global, and forests also exist within those boundaries.

In thinking about effective management of forests, operational scale is a crucial concept. Who has authoritative rights over the forest as it is being used, and who is reaping its benefits? Representatives of all people claiming allocative and authoritative rights over a particular forest from various scales should be reflected in the management group for that forest, thereby attempting to accommodate the multiple definitions of the forest. Including scale in the definitional process for forests allows for a true working definition of a dynamic resource.

References

Action for Community and Ecology in the Rainforests of Central America (ACERCA). 2000. "Eucalyptus, Neoliberalism and NAFTA in Southeastern Mexico." Available at <http://www.acerca.org/eucalyptus_mexico1.html>.

Archer, K. 1993. "Regions as Social Organisms." *Annals of the Association of American Geographers* 83(3):498–514.

Barkin, J. S., and G. E. Shambaugh. 1996. "Common-Pool Resources and International Environmental Politics." *Environmental Politics* 5(3):429–447.

Bian, Ling. 1997. “Multiscale Nature of Spatial Data in Scaling Up Environmental Models.” In *Scale in Remote Sensing and GIS*, ed. D. A. Quattrochi and M. F. Goodchild, 13–26. Boca Raton, FL: Lewis Publishers.

Delaney, D., and H. Leitner. 1997. “The Political Construction of Scale.” *Political Geography* 16(2):93–97.

Diamond, B. 1998. “Critique of Plantation Forestry.” Available at <http://www.metla.fi/archive/forest/1998/08/msg00065.html>.

Ellen, R. 1982. *Environment, Subsistence and System: The Ecology of Small-Scale Social Formations.* Cambridge: Cambridge University Press.

Fortmann, L. 1993. “Commentary: Testimony at the White House Forest Conference.” *Society and Natural Resources* 7:189–190.

Fortmann, L., and J. W. Bruce. 1988. *Whose Trees: Proprietary Dimensions of Forestry.* Boulder, CO: Westview Press.

Foster, D. 1990. “No Road to Tahuanti.” *Mother Jones* 15(5):36.

Ganjanapan, A. 1998. “The Politics of Conservation and the Complexity of Local Control of Forests in the Northern Thai Highlands.” *Mountain Research and Development* 18(1):71–82.

Geores, M. E. 1996. *Common Ground: The Struggle for Ownership of the Black Hills National Forest.* Lanham, MD: Rowman & Littlefield.

Geores, M. E. 1998. “The Historic Role of the Forest Community in Sustaining the Black Hills National Forest as a Complex Multiple Use Resource.” *Mountain Research and Development* 18(1):83–94.

Giddens, A. 1984. *The Constitution of Society.* Cambridge, UK: Polity Press.

Gilbert, A. 1988. “The New Regional Geography in English and French-Speaking Countries.” *Progress in Human Geography* 12:208–228.

Glacken, C. J. 1967. *Trace on the Rhodian Shore.* Berkeley and Los Angeles: University of California Press.

Jackson, J. B. 1994. *A Sense of Place, A Sense of Time.* New Haven, CT: Yale University Press.

Kanowski, P. J. 1997. “Afforestation and Plantation Forestry.” Special paper for Eleventh World Forestry Congress, October 13–22. Available at <http://www.coombs.anu.edu.au/Depts/RSAS/RMAP/kanow.htm>.

Magno, F. A. 1998. “Forest Protection in the Caraballo Sur, Northern Philippines.” *Mountain Research and Development* 18(1):63–70.

Marston, S. A. 2000. “The Social Construction of Scale.” *Progress in Human Geography* 24(2):219–242.

Native Forest Network. 2000. “Working Definition of a Native Forest.” Available at <http://www.nfn.org.au>.

Nova Forest Alliance. 2001. “Sustainable Forest Management Community Vision.” Available at <http://www.novaforestalliance.com>.

Ostrom, E., J. Burger, C. Field, R. B. Norgaard, and D. Policansky. 1999. "Revisiting the Commons: Local Lessons, Global Challenges." *Science* 284(5412) (April 9):278–282.

Panjabi, R. K. L. 1997. *The Earth Summit at Rio.* Boston: Northeastern University Press.

Sedjo, R. A. 1983. *The Comparative Economics of Plantation Forestry.* Washington, DC: Resources for the Future.

Walsh, S. J., A. Moody, T. R. Allen, and D. G. Brown. 1997. "Scale Dependence of NDVI and Its Relationship to Mountainous Terrain." In *Scale in Remote Sensing and GIS*, ed. D. A. Quattrochi and M. F. Goodchild, 27–56. Boca Raton, FL: Lewis Publishers.

Wilson, J., B. Low, R. Costanza, and E. Ostrom. 1999. "Scale Misperceptions and the Spatial Dynamics of a Social-Ecological System." *Ecological Economics* 31(2) (November):243–257.

III

Privatization

5

Privatizing the Commons . . . Twelve Years Later: Fishers' Experiences with New Zealand's Market-Based Fisheries Management

Tracy Yandle and Christopher M. Dewees

Introduction

Among common-pool resource (CPR) issues, fisheries management receives considerable attention, perhaps because of the impact of fisheries on people and economies. For example, the FAO estimates that the world catch is worth $80 billion annually (Carr 1998), and Garcia and Newton (1997) estimate that 200 million people worldwide receive their income from fishing. The scale of fishing also has increased dramatically. "From 1952–1992 marine fishery catches increased 300% from 18.5 to 82.5 million metric tons" (Garcia and Newton 1997, 4). As the size and importance of fishing increases, so do concerns over the state of fisheries worldwide. At the global level, "of the top 200 of the world's marine fisheries . . . 60% were fully exploited or over-exploited" at the close of the twentieth century (Mace 1999, 30). Similarly, "70% of the fish resources for which data are available are either heavily or fully fished, overexploited, overfished, depleted, or recovering from depletion" (Garcia and Newton 1997, 23).

As these fisheries decline, harvesters have begun to place increased effort into exploiting the resources of other fisheries. Furthermore, fishing power and technology have advanced significantly, further threatening fisheries sustainability. Fisheries management research has accelerated as the scope of management problems has become apparent.[1] As a result, a large body of literature in economics, anthropology, political science, and other social sciences addresses problems and innovations in fisheries management. Much of this literature focuses on fisheries as an example of the management problems surrounding CPRs.

A central concern of literatures addressing CPR management is the characteristics of institutions used to manage CPRs. Indeed, chapter 1 offers a synthesis of the factors affecting CPR use and extensively discusses institutional design issues. Within the fisheries literature, Charles (1992) describes three distinct schools of thought that dominate the fisheries management literature: conservation, rationalization, and community. Thinking of these in terms of the wider policy literature, three different arrangements of management institutions for fisheries can be identified: bureaucracy-based, market-based, and community-based regulation. Charles describes each approach as occupying one point of a triangle (see figure 5.1).

New Zealand's quota management system (QMS), with its emphasis on ITQs, removal of subsidies, and promotion of international export, is a long-standing program viewed as an example of the market-based approach. As the earliest (1986) nation to introduce a market-based QMS for most of its marine fisheries, New Zealand also presents a useful opportunity to study the influence that such an institutional arrangement has on the fishing industry and community. Because of the length of time New Zealand's QMS has been in place, it is possible to conduct a long-term analysis of both the strengths and weaknesses of this regulatory approach.

Specifically, this chapter examines the changes in fishers' perceptions of the QMS and the degree to which these perceptions are consistent with the predictions made in the literature. For example, has ITQ management led to a perceived increase or a perceived decrease in violations

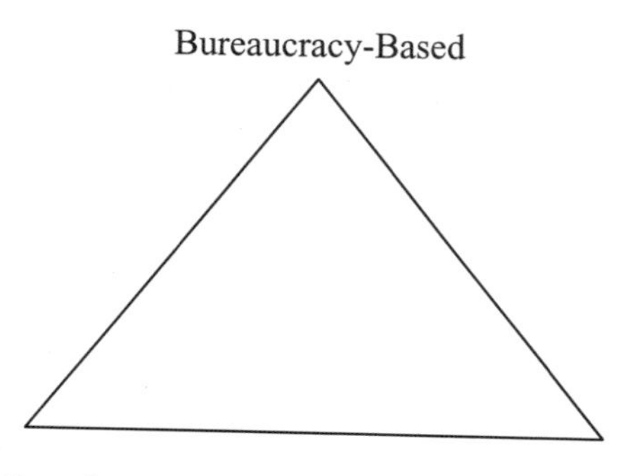

Figure 5.1
Approaches to fisheries management.
Source: Charles 1992.

of fisheries law? Has the structure of the fishing industry changed since the introduction of ITQ management? Is ITQ management perceived as improving or hurting the fishing industry? Do these perceptions vary depending on the amount of quota a participant owns?

After a review of the literature surrounding ITQs and an overview of New Zealand's QMS, the chapter examines changes that have occurred in the Auckland region's fishing community since ITQ management was introduced in 1986. This study primarily relies upon data from two sources: a panel survey of fishers in the Auckland region (conducted in 1987, 1995, and 1999) and a 1999 nationwide survey of fishing industry companies.

Background

Although discussion of the market-based approach has been ongoing within the fisheries and CPR management community for a considerable period of time (e.g., Gordon 1954; Scott 1955; Kneese and Schultze 1975), the approach emerged as an important policy tool only during the 1980s and 1990s. Several factors contributed to the emergence of this approach as a policy tool. Economic principles and concepts spread to a variety of academic disciplines (including natural-resource management), and criticism of the bureaucracy-based approach helped to create an environment more favorable to a market-based approach. As a result, a growing number of policy analysts began to explore or advocate the use of a market-based approach rather than a bureaucracy-based approach.[2] The two countries that have the most experience with the market-based approach are New Zealand and Iceland. Of the two, New Zealand is usually presented as the success story, whereas Iceland is often subject to more critical evaluations.[3] Other countries are also experimenting with ITQs, including the United States (Buck 1995), Canada (Grafton 1996), the United Kingdom (Hatcher 1997), and Australia (Sanders and Beinssen 1997).

Based on the experiences in these countries and the research generated by their experiences, it is possible to make some generalizations about the market-based approach (see tables 5.1 and 5.2). The primary emphasis in the market-based approach is on increasing the economic efficiency and productivity of the fishing industry while maintaining fish stocks at

Table 5.1
Characteristics of approaches to fisheries management

	Bureaucracy-based approach	Market-based approach	Community-based approach
Primary emphasis	Stock protection and maintaining fisheries at sustainable levels	Wealth generation for the fishing industry	Community control over the fishery
Competing objectives	Conservation Resource maintenance Administrative efficiency Accountability	Market efficiency Productivity Resource maintenance Accountability	Fisher control Community welfare Distributional equity Other social/cultural benefits Resource conservation
Resource ownership	Government: Property rights held by state	Fishers: Property rights allocated through ITQs to boat owners/fishers	Community: Property rights held by community or group of individuals within community
Vision of fishers	Components of predatory fleet	Individual fishing firms acting in economically rational manner	Members of cohesive community
Policy tools	Focus is on regulating inputs: Licenses Gear restrictions Seasonal restrictions Closures	Focus is on regulating the outputs using primarily ITQs: Percentage of total allowable catch Tonnage	Mixed inputs and outputs selected by self-regulation or comanagement: Gear limits Seasonal restrictions Location restrictions Rotating pressure Ownership of fishing grounds

Table 5.1
(continued)

	Bureaucracy-based approach	Market-based approach	Community-based approach
Cheating behavior	Illegal gear Fishing during closures or in closed areas Violating catch limitations Reporting false catch information	Quota busting (high grading and discarding) Offloading Leakage from monitoring system (e.g., reporting false catch information)	Violating communal rules (e.g., gear limits) Outsiders violating fishing rules
Enforcement focus	Fines or license revocation for violating rules of gear, closure, etc.	Fines or forfeiture of quota for reporting violations or quota-busting activities	Social sanctions and agreed-upon penalties

Source: Imperial and Yandle 1998.

a sustainable level. The primary policy instrument is a system of tradable permits: ITQs. Although variations on ITQs exist (e.g., leasing, measuring by tonnage vs. proportion of catch, use of ITQs as loan collateral), as a general matter ITQs can be defined as

> a specific portion of the total annual catch in the form of quota shares. . . . ITQs divide the total allowable catch quota into smaller individual portions. ITQs are generally transferable, which means fishing vessel owners can sell their ITQ certificates or buy others' certificates. (Buck 1995, 1)

The theory of ITQ management in fisheries has been supported by modeling and theoretical analysis.[4] Essentially, the market-based approach views fishers as individual fishing firms that wish to maximize their returns on their investment. Thus, whereas the bureaucracy-based approach focuses on inputs, the market-based approach focuses on output (the amount of fish removed). The latter approach has been embraced within the resource economics community and is gaining support in the corporate fishing industry. Its biggest critics tend to be supporters of the community-based model (Charles 1992).

Table 5.2
Perceived outcomes of fisheries management approaches

	Bureaucracy-based approach	Market-based approach	Community-based approach
Definition of success	Rules limit total catch so that MSY is not exceeded	Quota is set so that MSY is not exceeded and market is able to operate efficiently	Community is able to maintain the fishery at a socially and biologically viable level
Potential positive outcomes	Centralized government control over resource allocation Resource protection Stability of the rules governing the fishery Low administrative costs Accountability Equitability Preservation of small fishers	Economic efficiency and higher incomes for fishermen Elimination of capital stuffing[1] and derbies Stock conservation through allocating quotas Accountability with respect to quotas Fleet/industry modernization Stability for fishermen and producers Rapid response to environmental change through annual TAC setting	Local management Preservation of community culture and values Preservation of small-scale fishers/producers Often minimal environmental impacts Viewing of rent-seeking behavior with respect to negotiating fishing rights in positive terms Greater robustness to scientific uncertainty through use of local knowledge
Potential negative outcomes	Rent-seeking behavior with respect to regulations Agency capture by fishers, industry, or conservation groups Inefficiency	Rent-seeking behavior with respect to quotas Agency capture by fishing industry Equity problems Loss of small fishers/producers	Subject to capture by community leaders No external accountability Economically inefficient Unsafe fishing practices Lack of adaptability to dramatic

Table 5.2
(continued)

	Bureaucracy-based approach	Market-based approach	Community-based approach
	Capital stuffing and derbies Lack of adaptability Administrative costs of monitoring and enforcement Slow response to environmental change because of administrative process Susceptibility to scientific uncertainty concerning whether regulations will prevent overuse	Industry consolidation Administrative costs of tracking quota allocations and setting new quotas Loss of community Susceptibility to scientific uncertainty concerning whether quota has been set correctly	changes in practices, technology, stock, culture or environmental conditions Susceptibility, through lack of scientific knowledge, to environmental changes or resource overuse not detected through local knowledge

Source: Imperial and Yandle 1998.
[1] Capital stuffing is the practice of loading as much gear or technology as possible onto a vessel, while staying within the technical limits of the laws regarding vessel size.

Analysts suggest that there are several positive outcomes associated with the market-based approach. The first is economic efficiency and higher incomes for fishers and the fishing industry (e.g., Beckerman 1990; Clark 1993, 1994; Clark, Major, and Mollett 1988; Grafton 1996). This can help modernize the industry (e.g., Clark, Major, and Mollett 1988), help prevent overcapitalization (e.g., Buck 1995; Grafton 1996),[5] and help eliminate fishing derbies (e.g., Grafton 1996).[6] Second, the market-based approach is also perceived as an effective means of stock conservation, since it sets a limit (or a total allowable catch, or TAC) on the total harvest allowed from a particular resource in a specified time period (e.g., Boyd and Dewees 1992; Clark 1994). The market-based approach can also be adaptable, since the TAC can be set yearly or seasonally, allowing adjustments for stock changes (Squires, Kirkley, and Tisdell 1995). Finally, since the TAC and ITQs are set, fishers and processors are able to make better operational decisions and investments (Clark 1994).

However, the market-based approach also has potential negative outcomes.[7] Industry consolidation and loss of small fishers are often viewed as negative results (e.g., Young and McCay 1995; Palsson and Helgason 1995). Some researchers describe a variety of social problems arising from this approach, such as unemployment (e.g., Squires, Kirkley, and Tisdell 1995; Palsson and Helgason 1995) coupled with loss of community and damage to existing local institutions (Schlager 1990; Palsson and Helgason 1995). Also, equity problems are created as ITQs are consolidated among the largest fishers and new entries (fishers) to the system are restricted (Palsson and Helgason 1995).

The market-based approach also is vulnerable to some problems similar to those found under the bureaucracy approach. For example, the form of cheating changes from behaviors typically seen under the bureaucratic approach (such as using illegal gear, violating catch limits, or fishing during closures) to different forms such as high-grading (discarding fish that sell for less value and replacing them with more "expensive" fish), dumping bycatch (throwing back fish for which one does not have quota), and other forms of quota busting (e.g., Copes 1996a, 1996b; Turner 1997).[8] These forms of cheating can hurt a fishery, since many fish will not survive the catching process (such as fish brought up from a deepwater trawl). Thus, both high-grading and dumping bycatch result in underreported catch and an inaccurately set TAC. The market-based approach also relies on an accurate understanding of the fisheries population dynamics so that the TAC may be set at a level appropriate to achieving the desired goals of managing the fishery. If the TAC is inaccurately set, a fishery can be decimated, possibly before scientists have an opportunity to discover and correct the error.[9]

New Zealand's Quota Management System

As one of the oldest tradable quota systems in fisheries management, New Zealand's QMS provides an excellent opportunity to evaluate the market-based approach that has attracted so much positive and negative attention in the fisheries management literature. A broad overview of the New Zealand fisheries management system shows that the 200-mile EEZ around New Zealand covers an area of 1.2 million square nautical miles, or approximately fifteen times New Zealand's land mass, and 185 separate fish stocks (as of the 1996–1997 season) totaling approximately 531,000 tons of quota-managed species and 79,000 tons of species not under quota management (Clement & Associates 1998). In 1997, there were 2,170 domestic vessels, 59 foreign-charter vessels, and 16 foreign-licensed vessels (Statistics New Zealand 1999, 927) fishing New Zealand's waters commerically. In 2000, seafood exports accounted for NZ$1.43 billion, making it the fourth-largest source of foreign currency in the New Zealand economy (New Zealand Seafood Industry Council 2001).

Historically, New Zealand has had little widespread interest or concern with fishing or fisheries management. A small proportion of the country has continued to fish professionally over the years with a deep commitment to the industry (Makarios 1996; Martin 1969; Slack 1969), and a few fishing companies such as Sanfords have long histories (Titchener 1981). But New Zealand as a whole has been more focused on other primary industries, such as farming and forestry.[10] In fact, until the declaration of New Zealand's EEZ in 1978, the nation's fishing industry was small and confined to a domestic inshore industry. There was no New Zealand deepwater fishing. Instead, prior to 1978, the country's deepwaters were fished by other nations' trawlers, primarily those from Korea, Japan, and the former Soviet Union. Within the inshore fisheries, a variety of management approaches were used. Clark characterized New Zealand's changing fisheries management regimes as follows: "Management of fisheries during this time was . . . characterized by fundamental changes. From 1938 to 1963 the fishery was managed under a restrictive licensing system with very tight controls. In 1963 the fishery was completely deregulated and remained that way, by and large, until 1980 when a moratorium on issuing further wet fish permits was introduced" (Clark 1993, 340). This minimalist approach to management of the New Zealand fishing industry began to change in 1978 when New Zealand claimed its EEZ. Then, with clear warnings of an imminent collapse of the inshore fisheries at hand, the Fisheries Act of 1983 was passed, introducing property rights as tradable quotas and incorporating biological preservation and economic development into the management of New Zealand's inshore fisheries (Clark, Major, and Mollett 1988). In 1986 legislation, New Zealand's QMS was expanded to include its deepwater fisheries.

Over the last sixteen years, the scope of the QMS in New Zealand has expanded and changed in respect to some details (e.g., a switch from tonnage-based quota to proportion-based quota and from resource rental funding to cost recovery funding; introduction of Maori rights), but the fundamental principles of the system have remained constant. The QMS can be seen as having two primary goals: maintaining (or building) healthy fisheries and doing so in a manner that encourages an economically efficient industry. Within these two broad goals, wide ranges of more specific objectives have been articulated (e.g., Clark 1993; Clement & Associates 1997).

As might be expected, New Zealand's QMS has attracted considerable attention from both the policy community and the CPR community.[11] Literature on the QMS primarily presents descriptions of how the QMS works, or economic analyses that highlight the success of the QMS in conserving resources and encouraging economic performance. Other articles focus on specialized issues such as enforcement (McClurg 1994), stock assessment (Mace 1993; Annala 1993), and more recently the development of comanagement organizations (e.g., Hughey, Cullen, and Kerr 2000).

Literature on the socioeconomic effects of QMS as an institution has been limited primarily to assessment of effects on the rural Northland and/or Maori communities (e.g., Fairgray 1986; Cassidy 1995). The exception to this is Dewees's continuing research on the social consequences of QMS.[12] This chapter is a continuation of that effort, extending the panel data survey for Auckland region fishers to include 1986–1987, 1995, and 1999 as well as incorporating national data on the characteristics and composition of the larger company segment of the fishing industry in 1999.

Changes in the New Zealand Fishing Industry

As one of the first nations to adopt an ITQ system, New Zealand provides an important case for studying the effects of this management approach, not only from a biological and industrial perspective, but also from a more social perspective. For example: Has the character of the industry changed since the adoption of the ITQ system? Have attitudes toward this management approach changed over time? Is the approach perceived as encouraging sustainability of the fishery or as beneficial to the industry? The following is a preliminary effort to explore the changes these questions address.

Historic Auckland Region Surveys

Assessment of the effects of the implementation of QMS has been an ongoing project. The first survey of such effects was conducted over a nine-month period in 1986–1987. Subsequent surveys were conducted in 1995 and 1999. The initial list of commercial fishers and fishing company managers to be surveyed represented an unstratified random sample of 100 fishers and companies in the Auckland region. These 100 were randomly selected from the 400 provisional quota holders (those who received quotas in the initial distribution of quota rights) in the Auckland region.

Subsequent surveys (conducted in 1995 and 1999) were based upon the list of sixty-two fishers and company managers who participated in the initial survey. Table 5.3 shows the number of participants in each round of surveys. Between 1986–1987 and 1999, the number of survey participants dropped from sixty-two to thirty-nine. The decrease in the number interviewed in each subsequent survey is not surprising. Indeed, by 1999, of the original sixty-two, five had died or were too sick to be interviewed and twelve could not be located, even on regional voter rolls.

Table 5.3
Summary of historic Auckland region survey

	Wave 1[a]	Wave 2	Wave 3
Year	1986–87	1995	1999
Number of respondents	62	48	39
Percentage of Wave 1		77	63
Percentage of Wave 2			81

[a]Survey based on random sample of 100 quota owners.

Thus, the decline in participation can be viewed as typical decay in a panel survey over a twelve-year period.

Small-scale fishers (those with one or two boats under 20 meters) dominate the sample, with a few large vertically integrated companies based in the Auckland region also included in the survey. Because this survey is focused on the Auckland region, it should not be viewed as representative of larger New Zealand fishing industry, or the smaller fishers throughout New Zealand. The Auckland region's fisheries are considered by many to be under greater stress than other fishing regions, so the opinions expressed in this survey may be more pessimistic than those in a national sample.

Because of the small sample size in this survey, the exact Pearson's test (a variant of a chi-square test) was used for statistical analysis (results reported in table 5.4).[13] The exact Pearson's test examines, for each survey item, the null hypothesis that the same proportion of respondents mentioned or agreed with that item in all three years. The Pearson test was selected because it is not based on the asymptotic chi-square distribution, and thus it can be used even for data with low numbers or zeros in certain cells.

Furthermore, when analyzing the results obtained, we focus on the patterns of opinions across multiple questions, rather than a detailed analysis of any single question. Examining these broad patterns helps ensure that the most accurate conclusions are drawn (Yin 1993).

Focusing on large differences in responses across the three waves of the panel study (presented in table 5.4), some noteworthy trends can be observed over the lifetime of the QMS in New Zealand. Examining the responses that show consistent trends or patterns over time can yield insights into perceptions of the QMS.

First, (as is noted above) there has been a pronounced decrease with each survey wave in the number of participants interviewed. The decrease in participants has been steady (roughly ten every survey wave). Furthermore, the reduction in the number of respondents owning quotas is substantial. In the first wave, forty-nine of the original sixty-two respondents (79 percent) owned quota. In the second wave, only thirty-four out of forty-eight respondents (71 percent) owned quota, and only nineteen out of thirty-nine (49 percent) owned quota by the time of the third wave. This decrease in the number of quota owners is indicative

Table 5.4
Response to Auckland region survey in 1987, 1995, and 1999

	1987	1995	1999	Exact Pearson's test result (*p* value)[a]
Number interviewed	62	48	39	
Number (%) who own quotas	49 (79%)	34 (71%)	19 (49%)	0.0052
Number (%) agreeing with statements				
Fishing industry better off	36 (58%)	36 (75%)	25 (64%)	0.1890
ITQs compatible with beliefs	35 (56%)	35 (73%)	19 (49%)	0.0621
ITQs conserve stocks	35 (56%)	33 (69%)	25 (64%)	0.4204
Fishing is safer with ITQs	24 (39%)	24 (50%)	5 (13%)	0.0010
My economic situation improved	30 (48%)	22 (46%)	20 (51%)	0.8732
More secure about retirement	45 (73%)	31 (65%)	11 (28%)	<0.0001
Difficult for young to enter	59 (95%)	47 (98%)	33 (85%)	0.0368
Number (%) mentioning positive ITQ effects				
Conserve fish stocks	33 (53%)	24 (50%)	23 (59%)	0.7123
Provide asset/security	26 (42%)	13 (27%)	5 (13%)	0.0067
Reduce effort	14 (23%)	6 (13%)	2 (5%)	0.0502
Improve quality	0 (0%)	9 (19%)	1 (3%)	0.0002
Number (%) mentioning ITQ problems				
High-grading	41 (66%)	12 (25%)	1 (3%)	<0.0001
Enforcement	25 (40%)	10 (21%)	11 (28%)	0.0798
Company control	16 (26%)	22 (46%)	12 (31%)	0.0843
Resource allocation	4 (6%)	16 (33%)	21 (54%)	<0.0001
Complexity	0 (0%)	17 (35%)	25 (64%)	<0.0001

[a]For *p* equal to 1.000 (or 0.000) the observed arrangement of responses is one of the most extreme allowed. The results are as much a result of the data structure (0s and 1s) as the data values.

of a consolidation in quota ownership over time, as is suggested in the literature.

Results from the closed-ended questions reflect both negative and positive reactions to the QMS among respondents. Loss of confidence in the QMS are expressed by the proportion of respondents agreeing that "ITQs increase retirement security" dropping from forty-five out of sixty-two (73 percent) in 1987 to thirty-one out of forty-eight (65 percent) in 1995 to eleven out of thirty-nine (28 percent) in 1999. A similar overall drop in confidence is observed between the 1987 and 1999 waves of the survey where fewer respondents agree that the QMS increases safety. Here (in spite of a temporary increase in confidence in the 1995 survey), the proportion decreases from twenty-four out of sixty-two (39 percent) in 1983 to five out of thirty-nine (13 percent) in 1999.

In spite of these concerns, consistently well over half of the participants in each survey wave are in agreement that the New Zealand fishing industry is better off under the QMS: starting at thirty-six out of sixty-two (58 percent) in 1987, increasing to thirty-six out of forty-eight (75 percent) in 1995, then leveling off at twenty-five out of thirty-nine (64 percent) in 1999. Also, when asked about conservation, thirty-five out of sixty-two (56 percent) agree that ITQs conserve stocks in 1987, compared to thirty-three out of forty-eight (69 percent) in 1995 and twenty-five out of thirty-nine (64 percent) in 1999. Perhaps the best explanation of these results is that although many participants feel that QMS does not help them individually (e.g., retirement, allocation, complexity), it is seen as a positive force for the industry and the resource at a broader level (e.g., stock conservation, condition of industry).

Among open-ended questions, the most positive results are found in the area of enforcement. High-grading (dumping fish with low commercial value) has dropped from being mentioned as a problem by forty-one out of sixty-two respondents (66 percent) in 1987 to twelve out of forty-eight respondents (25 percent) in 1995 to only one out of thirty-nine respondents (3 percent) in 1999.[14]

Most of the results from the open-ended questions, however, suggest a growing discontent with QMS within our Auckland sample. For example, fewer participants mention positive results such as ITQs being an asset, with the proportion dropping from twenty-six out of sixty-two (42 percent) in 1987 to thirteen out of forty-eight (27 percent) in 1995 to five out of thirty-nine (13 percent) in 1999. Similarly, dramatically more respondents mention resource allocation problems in later survey waves than in the earliest one. The number of participants citing such problems rises from four out of sixty-two (6 percent) in 1987 to sixteen out of forty-eight (33 percent) in 1995 to twenty-one out of thirty-nine (53 percent) in 1999. Also, the proportion of respondents mentioning complexity issues increases from none in 1987 to seventeen out of forty-eight (35 percent) in 1995 to twenty-five out of thirty-nine (64 percent) in 1999. Furthermore, the number of respondents agreeing that ITQs reduce effort drops from fourteen out of sixty-two (23 percent) in 1987 to six out of forty-eight (13 percent) in 1995 to two out of thirty-nine (5 percent) in 1999.

Finally, it is worth noting that for many questions, there is a lack of consistent patterns across the three waves of the survey. Of the reported results, over half showed inconsistent trends over time. Most of these initially indicated a growing optimism (in the comparison of the 1995 and 1987 results) then an increasing pessimism (in the comparison of the 1995 and 1999 results). For example, the percentage of participants who agreed that QMS was compatible with their beliefs rose from thirty-five out of sixty-two (56 percent) in 1987 to thirty-five out of forty-eight (73 percent) in 1995, then fell to nineteen out of thirty-nine (49 percent) in 1999. Similarly, the percentage of participants who agreed that fishing is safer with ITQs increased from twenty-four out of sixty-two (39 percent) in 1987 to twenty-four out of forty-eight (50 percent) in 1995, then decreased to five out of thirty-nine (13 percent) in 1999. This pattern can be seen as indicating ambivalence over ITQ management.

One explanation for this pattern is the increased expense involved and complexity encountered in complying with QMS between 1995 and 1999. Examples of this increased expense and complexity include the 1996 Fisheries Act, which included penalties for noncompliance that many fishers described as "draconian," a rapid increase in cost recovery fees[15] (in part brought on by increased environmental requirements in the 1996 law), and a growing frustration with the failure to reach a final distribution of the Treaty of Waitangi settlement (Maori fishing rights settlement) that was signed in 1994. Another possible explanation specific to the Auckland Region surveys is that the sample was dominated by small-scale fishers, who are operating in a difficult business environment, including pressure on the snapper fisheries from recreational fishers (snapper is one of the dominant commercial fisheries in the region, particularly for small-scale fishers), cutbacks in TACs for snapper, and competition with recreational vessels for facilities such as docking space.

Comparison of 1999 Auckland Small-Scale Fishers and North Island Company Surveys

In addition to the 1999 Auckland survey, another survey was conducted to obtain information on the opinions of company managers in large and medium-sized fishing and processing companies on both the North and South Islands of New Zealand. The New Zealand Seafood Industry Council's research librarian identified twenty-nine such companies. Four active members of the fishing industry then identified appropriate people at each of the companies to participate in the interviews and in some cases identified additional companies or suggested removing companies on the original list that were involved in export brokering rather than the fishing industry.

Questions for the corporate survey were based on the Auckland panel survey, with a few questions removed as inappropriate for the corporate setting.[16] Analysis and coding remained consistent between the two surveys. In the analysis presented here, responses from the company managers based on the North Island are compared to responses from the small-scale fishers (all fishers who are not company managers) in the Auckland Region survey.

Comparison of these surveys raises a few issues. The first issue is geographic coverage. Since the North Island company survey covers a larger geographic area than the Auckland fisher survey, comparison of the results of the two surveys cannot be considered to yield a perfect comparison of the differences between companies and small-scale fishers in the Auckland region. However, the number of companies based in the Auckland region is too small to allow an exclusive comparison of small-scale Auckland fishers versus companies based in the Auckland region. Thus, the comparison undertaken was the most focused comparison available. A related issue is the degree to which differences in opinion revealed between the two surveys are due to respondents' economic position (small-scale fisher compared to company manager) or their location (the stressed Auckland region compared to the larger North Island). Undoubtedly, both factors influence the results. But the fact that many of the companies surveyed are based in Auckland and that most rely on the inshore Auckland fishery for part of their revenue stream (and in some cases all of their revenue stream) increases our confidence that most of the difference in response is due to respondents' economic position.

Table 5.5 presents a comparison between the responses of small-scale fishers in the Auckland survey and those of companies based on the North Island. Fisher's Exact test is employed to examine statistically the results of this analysis. It tests, for each item, the null hypothesis that

Table 5.5
Responses by 1999 Auckland region small-scale fishers and North Island fishing company managers

	Small-scale fishers	Company managers	Fisher's Exact test result (*p* value)
Number interviewed	23	17	
Number (%) agreeing with statements			
Fishing industry better off	14 (61%)	16 (94%)	0.0257
ITQs compatible with beliefs	9 (39%)	16 (94%)	0.0006
ITQs conserve stocks	13 (57%)	16 (94%)	0.0119
Fishing is safer with ITQs	4 (17%)	6 (35%)	0.2743
Difficult for young to enter	18 (78%)	16 (94%)	0.2156
Number (%) mentioning positive ITQ effects			
Conserve fish stocks	16 (70%)	8 (47%)	0.1991
Provide asset/security	2 (9%)	3 (18%)	0.6340
Reduce effort	1 (4%)	0 (0%)	1.0000
Improve quality	1 (4%)	1 (6%)	1.0000
Number (%) mentioning ITQ problems			
Highgrading	1 (4%)	0 (0%)	1.0000
Enforcement	9 (39%)	2 (12%)	0.0786
Company control	7 (30%)	1 (6%)	0.1074
Resource allocation	16 (70%)	3 (18%)	0.0016
Complexity	16 (70%)	9 (53%)	0.3355

the same proportion of fishers and company managers mentioned or agreed with that item. Fisher's Exact test is used here because it is a powerful test when the number of observations is small.[17]

Among the statistically significant results presented in table 5.5, the company managers are found to be consistently more satisfied or optimistic about the QMS than the small-scale fishers. These managers show this particularly in their responses to the broader, more philosophical questions the survey posed. For example, dramatically more of the company managers than the small-scale fishers—sixteen out of seventeen (94 percent) versus fourteen out of twenty-three (61 percent)—agree that the fishing industry is better off under the QMS. Similarly, sixteen out of seventeen company managers (94 percent) versus nine out of twenty-three small-scale fishers (39 percent) agree that the QMS is compatible with their own beliefs.

Company managers also consistently show lower levels of concern over problems with the QMS. This is particularly evident in the area of resource allocation, where only three out of seventeen (18 percent) company managers mention a problem, compared to sixteen out of twenty-three (70 percent) of small-scale fishers. Similar results are seen in the area of stock conservation, where sixteen out of seventeen (94 percent) of company managers agree that ITQs conserve fish stocks, compared to thirteen out of twenty-three (57 percent) of small fishers.[18]

This broad trend of company managers' being more optimistic than the fishers continues in the results that are not statistically significant. Except for two cases in which the fishers' and managers' responses were essentially the same,[19] the company managers consistently expressed more optimistic sentiments than the small fishers about the desirability and effects of ITQs.

Comparing quotes from the Auckland region small-scale fishers and the managers of the North Island companies also illustrates differences between the two groups in their perception of ITQs as a management approach. For example, when asked about the effects of the QMS on fish stock, company managers gave responses such as "ITQs protect fish from the ravages of man. It does protect the stock" (Company Survey #3) and "ITQs give us a certain amount of sustainability, an assurance that primary species like orange roughy won't get decimated" (Company Survey #22). In contrast, fishers in the Auckland region survey expressed more mixed opinions about these effects: "It depends on the species. Different species, yes. Mullet and flounder are still being hammered down" (Auckland Survey #4) or "None of the species has recovered enough to increase TAC. Instead, it reduces my catch" (Auckland Survey #18).

Similar results are seen when cheating behavior is discussed. Most company managers were positive, making statements like "Cheating is reduced under ITQs. Once people have ownership they have a drive to protect it. Owners are the best patrollers. There's peer pressure. Its very unfashionable to be a pirate in the industry" (Company Survey #2). In contrast, many fishers made statements such as "Cheating has gotten more hidden and more underground. You have to make a living and if you went above board between taxes and quota you couldn't make any money" (Auckland Survey #4).

Finally the comments of company managers and Auckland region fishers reveal very different perspectives on the current state of the industry and the loss of smaller fishers. Managers are more likely to see the process as benign or as a necessary evil, as is illustrated in the following quotes: "The reality is that there's been a restructuring in the industry. . . . There's pain in the smaller communities. But there's also more value added now either on ships or on shore" (Company Survey #5). "Smaller fishers had their chance. They were allocated quota. Where small fishers went wrong was selling quota with an agreement to continue catching. They sold themselves out for cash" (Company Survey #15). In contrast, many (but not all) fishers feel exploited or let down by the QMS: "Combined with cuts, the small fishers couldn't get the money to compete. . . . The guy leasing quota to me makes five times more than me. It pushes the smaller fishers to the wall" (Auckland Survey #19). "Conditions are quite shocking. Only reason people are in it is the lifestyle. . . . People nearly go broke before they get out. We're fishing on our pride" (Auckland Survey #35).

Both the results from examining the aggregate survey responses and those from looking at individual comments demonstrate that North Island company managers have a greater satisfaction with the QMS than Auckland region small-scale fishers. This suggests that the QMS, as implemented in New Zealand, has created a setting in which the larger-scale companies are favored over the small-scale fishers.

Analysis and Conclusions

The analysis presented in this chapter focuses on the influence of New Zealand's QMS on the fishers and companies fishing in the Auckland region. More specifically, it examines the evolution in the opinions of Auckland region fishers and company managers toward QMS as a management approach. The results show mixed perceptions of the performance for QMS. Although many positive aspects of ITQ management are apparent to interviewees, so are many negatives.

Attrition from the survey and the proportion of remaining participants who own quota suggest that ITQ-based systems like New Zealand's QMS with high quota ownership aggregation limits do indeed encourage an industry consolidation, as is predicted by the literature. As an institutional arrangement, the QMS appears to result in a variety of effects, both positive and negative. There is broad agreement that the New Zealand fishing industry as a whole is better off under the QMS and that the QMS preserves the country's fish stocks. Since these are the two main goals of the QMS, it does speak positively for the approach.

Survey participants also expressed a body of concerns, however, that warrant careful attention. Over time, there is a decline in the proportion of survey participants agreeing that the QMS provides retirement security and that the QMS is compatible with their belief systems. There is also a decline in the percentage mentioning that the QMS provides an asset or security. Since one of the fundamental principles of an ITQ-based system is that quota is seen as a long-term asset worth investing in, this suggests that there may be a weakening in the "currency" of QMS and a possible degradation in the perception of ITQs as a property right. Furthermore, survey participants are consistent in expressing concerns over the complexity of QMS and the barriers young people face entering a QMS-managed fishery. It is unclear whether these concerns show a coming weakness in the industry or a growing professionalization in the industry.

Finally, both aggregate survey results and individual comments provide evidence for the split between "classes" of fishers as described by some of the literature critical of ITQ-based management. Broadly speaking, our results identify a split between companies and small-scale fishers. This split shows the companies and quota owners being broadly more optimistic than their small-scale or nonowning counterparts under New Zealand's QMS. These results suggest a disenfranchisement of the smaller, more traditional Auckland region fishers and raise questions about how the characteristics of the New Zealand fishing industry will change in the long term.

In comparing these results with others, it is interesting to note that Einar Eythórsson's chapter on Iceland's ITQ system (chapter 6) also notes many of the changes and industry characteristics described here. For example, Eythórsson provides extensive documentation of the development and consequences of quota leasing and contract fishing, as well as increased barriers to entry, and the relative success of vertically integrated companies. Although other issues discussed in this chapter are not discussed in Eythórsson's analysis of Iceland, the parallels between these two cases suggests that some of these results are not solely due to case selection but instead are associated with the ITQ management approach itself.

Based on these results, we propose that the market-based approach is neither the panacea nor the curse that some characterizations suggest. Instead, it is an institutional arrangement with an important mixture of strengths and weaknesses that creates important and long-lasting changes in the fishing industry, fishing community, and regulatory community associated with it. Nations or fisheries considering ITQ-based management regimes need to recognize that they are embarking on a major effort of institutional design. This can influence not only the behavior but also the characteristics of the fishing industry and management agency. Thus, when considering such an approach, decision makers need to set goals carefully—not only for catch limits, but also for fishing industry characteristics—and reflect on the set of changes and challenges they are likely to face if they adopt a market-based approach.

Acknowledgments

The authors would like to thank Lance Waller and Traci Leong at Emory University's Rollins School of Public Health for their statistical advice and generous sharing of expertise.

Notes

1. It is also argued that there are significant changes regionally in biological productivity based on cyclical changes in marine climate.

2. Examples of this approach include Maloney and Pearse 1979; Clark, Major, and Mollett 1988; Green and Nayar 1988; Schlager 1990; Pearse and Walters 1992; and Squires, Kirkley, and Tisdell 1995.

3. For discussion of New Zealand's ITQ system, see Clark 1994; Sharp 1997, 1998; Mace 1993; Boyd and Dewees 1992; Dewees 1989; and McClurg 1994. For discussion of Iceland's ITQ system, see Eythórsson 1996a, 1996b; Palsson and Helgason 1995; Eggertsson 1996; and Matthiasson 1997.

4. For example, Terrebonne modeled entrepreneurial fishers with heterogeneous production and employment opportunities outside of the fishery. He found that under an open-access model, fishers' income is proportional to the price that they receive for their catches. He also found a reason for fishers to support the use of ITQs, because in his model, fishers received more income under the ITQ model

than the open-access model. See Terrebonne 1995; Arnason 1991; Grafton 1995; and Charles 1988, 1992, for examples.

5. Others, however (e.g., Schlager 1990), argue that ITQs encourage overcapitalization.

6. A "derby fishery" is one where the fishermen race each other to catch the fish before the other fishermen beat them to the resource.

7. It is also interesting to note that many of these problems are social issues not addressed in the economics literature that supports this approach or are the "flip side" of what is described by supporters of the approach as a positive outcome. See tables 5.1 and 5.2 for further illustration of this point.

8. Other forms of quota busting include false reporting of catch information and diverting catch to a gray or black market so that it is outside of the monitoring system.

9. Works that advance this critique include Loayza 1994; Mace 1993; and Sissenwine and Mace 1992. An excellent example of this type of problem with improperly set TACs (described in detail by Mace) is the ongoing controversy over what constitutes a sustainable fishing level for New Zealand's orange roughy. But one should also ask: Would the situation before have been different if the orange roughy catch were managed without ITQs? Bad stewardship (or inaccurate estimates) is bad stewardship under any system!

10. Indeed, between two prominent New Zealand history books there is no discussion of fishing or the fishing industry, but over thirty index references to farming (Barber 1989; Sinclair 1997). This ignores, however, the native Maori population, which has a long fishing tradition but until recently was not actively considered in national fishing policy.

11. Examples of descriptive works include Clark, Major, and Mollett 1988; Sissenwine and Mace 1992; Batkin 1996; and Annala 1996. Examples of economic analyses include Clark 1993; Sharp 1997; and Batstone and Sharp 1999.

12. See for examples Dewees 1989, 1996a, 1996b, 1998; and Boyd and Dewees 1992.

13. In tables 5.4 and 5.5 we first report the number of respondents (and the percentage of respondents in a particular wave) who agreed or strongly agreed with a series of structured statements about the status of the industry. In addition to these structured statements, all respondents were asked a series of open-ended questions about ITQs. The responses in all three survey waves were coded with a consistent set of categories. Thus, it is possible to report the number (and percentage) of the respondents who volunteered a particular type of response to all of these open-ended questions. The survey instrument is available from Tracy Yandle. See also Yandle 2001.

14. However, this may also represent an acceptance among fishers of high grading as a standard practice. Similarly, the proportion of respondents mentioning enforcement problems dropped between 1987 and 1999—hitting a low point in 1995. In 1987, twenty-five out of sixty-one (40 percent) mentioned prob-

lems, compared to ten out of forty-eight (21 percent) in 1995 and eleven out of thirty-nine (28 percent) in 1999.

15. Cost recovery refers to the policy of charging quota owners for the expenses the government incurs from regulating commercial fishing activities. This policy was introduced in 1994.

16. For example, agreeing or disagreeing with "my retirement is more secure" under the QMS was viewed as nonapplicable to respondents who were company managers.

17. Pearson's Exact test is used for the analysis in table 5.4 because it is able to test for differences across three time periods. Fisher's Exact test is used for the analysis in table 5.5, however, because Fisher's exact test is specially designed for 2×2 tables with small sample sizes, such as those in table 5.5, but does not extend directly to multiway tables, such as those in table 5.4.

18. Curiously, however, in the open-ended questions, fewer company managers mentioned stock conservation as an advantage of ITQs than small fishers (eight out of seventeen or 47 percent of company managers compared to sixteen out of twenty-three or 70 percent of small fishers). This difference may be due to the different priorities of the two groups. Fishers (who more closely observe the stock daily) may be more likely to mention this issue than managers (whose daily concerns focus more on administrative matters). In spite of this, the difference in the response when the two groups were directly asked about stock conservation (94 percent positive for managers versus 57 percent positive for small fishers) confirms the pattern of relative optimism among company managers on this issue.

19. "Reducing effort," which one fisher and no managers mentioned as a benefit of ITQs, and "improved quality," which one fisher and one manager mentioned as a benefit.

References

Annala, John H. 1993. "Fisheries Assessment Approaches in New Zealand's ITQ System." In *Proceedings of the International Symposium on Management Strategies for Exploited Fish Populations*, ed. G. Kruse et al., 791–805. Fairbanks: University of Alaska Press.

Annala, John H. 1996. "New Zealand's ITQ System: Have the First Eight Years Been a Success or Failure?" *Reviews in Fish Biology and Fisheries* 6:43–62.

Arnason, Ragnar. 1991. "Efficient Management of Ocean Fisheries." *European Economic Review* 35:408–417.

Barber, Laurie. 1989. *New Zealand: A Short History*. Auckland, New Zealand: Century Hutchinson.

Batkin, Kirsten M. 1996. "New Zealand's Quota Management System: A Solution to the United States' Federal Fisheries Management Crisis?" *Natural Resources Journal* 36:855–880.

Batstone, C. J., and B. M. H. Sharp. 1999. "New Zealand's Quota Management System: The First Ten Years." *Marine Policy* 23(2):177–190.

Beckerman, Wilfred. 1990. *Pricing for Pollution: Market Pricing, Government Regulation, Environmental Policy*. London: Institute for Economic Affairs.

Boyd, R. O., and C. M. Dewees. 1992. "Putting Theory into Practice: Individual Transferable Quotas in New Zealand's Fisheries." *Society and Natural Resources* 5:179–198.

Buck, Eugene H. 1995. *Individual Tradable Quotas in Fishery Management*. Washington, DC: Congressional Research Service.

Carr, Edward. 1998. "The Deep Green Sea." *Economist* 347 (May 23, suppl.): 17–18.

Cassidy, M. 1995. "Providing for Maori Rights and Interests in New Zealand's Fisheries." Paper presented at Third Australian Fisheries Managers' Conference.

Charles, Anthony T. 1988. "Fisheries Socioeconomics: A Survey." *Land Economics* 64(3):276–295.

Charles, Anthony T. 1992. "Fisheries Conflict: A Unified Framework." *Marine Policy* 16(5):379–393.

Clark, Ian. 1993. "Individual Transferable Quotas: The New Zealand Experience." *Marine Policy* 17(5):340–352.

Clark, Ian. 1994. "Fisheries Management in New Zealand." In *Managing Fisheries Resources: Proceedings of a Symposium Co-sponsored by the World Bank and Peruvian Ministry of Fisheries Held in Lima Peru*, ed. Eduardo A. Loyaza, 49–58. World Bank Discussion Papers Fisheries Series no. 217. Washington, DC: World Bank.

Clark, Ian, Philip J. Major, and Nina Mollett. 1988. "Development and Implementation of New Zealand's ITQ Management System." *Marine Resource Economics* 5:325–349.

Clement & Associates. 1997. *New Zealand Commercial Fisheries: The Guide to the Quota Management System*. Tauranga, New Zealand: Author.

Clement & Associates. 1998. *New Zealand Commercial Fisheries: The Atlas of Area Codes and TACCs 1998/99*. Tauranga, New Zealand: Author.

Copes, Parzival. 1996a. "Adverse Impacts of Individual Quota Systems on Conservation and Fish Harvest Productivity," Discussion paper 96–1. Burnaby, British Columbia: Institute of Fisheries Analysis, Simon Fraser University.

Copes, Parzival. 1996b. "Social Impacts of Fisheries Management Regimes Based on Individual Quotas." Discussion paper 96–2. Burnaby, British Columbia: Institute of Fisheries Analysis, Simon Fraser University.

Dewees, Christopher M. 1989. "Assessment of the Implementation of Individual Transferable Quotas in New Zealand's Inshore Fishery." *North American Journal of Fisheries Management* 9(2):131–139.

Dewees, Christopher M. 1996a. "Fishing for Profits: New Zealand Fishing Industry Changes for 'Pakeha' and Maori with Individual Transferable Quotas."

Paper presented at Social Implications of Quota Systems in Fisheries Workshop, Vestman Island, Iceland, May 25–26.

Dewees, Christopher M. 1996b. "Industry and Government Negotiation: Communication and Change in New Zealand's Individual Transferable Quota System." In *Fisheries Resource Utilization and Policy: Proceedings of the World Fisheries Congress, Theme* 2, ed. R. M. Meyers et al., 333–341. New Delhi: Oxford and IBH Publishing.

Dewees, Christopher M. 1998. "Effects of Individual Quota Systems on New Zealand and British Columbia Fisheries." *Ecological Applications* 8(1):S133–S138.

Eggertsson, Thrainn. 1996. "No Experiments, Monumental Disasters: Why It Took a Thousand Years to Develop a Specialized Fishing Industry in Iceland." *Journal of Economic Behavior and Organization* 30:1–23.

Eythórsson, Einar. 1996a. "Theory and Practice of ITQs in Iceland: Privatization of Common Fishing Rights." *Marine Policy* 20(3):269–281.

Eythórsson, Einar. 1996b. "Coastal Communities and ITQ Management. The Case of Icelandic Fisheries." *Sociologia Ruralis* 36(2):212–223.

Fairgray, J. D. M. 1986. *Individual Transferable Quotas Implications Study. Second Report: Community Issues*. Fisheries Management Series no. 20. Wellington, New Zealand: Ministry of Agriculture and Fisheries.

Garcia, S. M., and C. Newton. 1997. "Current Situation, Trends, and Prospects in World Capture Fisheries." In *Global Trends: Fisheries Management*, ed. E. L. Pikitch, D. D. Huppert, and M. P. Sissenwine, 3–27. American Fisheries Society Symposium no. 20. Bethesda, MD: American Fisheries Society.

Gordon, H. Scott. 1954. "The Economic Theory of a Common-Property Resource: The Fishery." *Journal of Political Economy* 62:124–142.

Grafton, R. Quentin. 1995. "Rent Capture in a Rights-Based Fishery." *Journal of Environmental Economics and Management* 28:48–67.

Grafton, R. Quentin. 1996. "Experiences with Individual Transferable Quotas: An Overview." *Canadian Journal of Economics* 24(special issue):S135–S138.

Green, G., and M. Nayar. 1988. "Individual Quotas in the Southern Blue Fin Tuna Fishery: An Economic Appraisal." *Marine Resource Economics* 5:365–387.

Hatcher, Aaron C. 1997. "Producers' Organizations and Devolved Fisheries Management in the United Kingdom: Collective and Individual Quota Systems." *Marine Policy* 21(6):519–533.

Hughey, Kenneth F. D., Ross Cullen, and Geoffrey N. Kerr 2000. "Stakeholder Groups in Fisheries Management." *Marine Policy* 24:119–127.

Imperial, Mark T., and Tracy Yandle. 1998. "Marching towards Leviathan, Embracing the Market, or Romancing the Commons: An Examination of Three Approaches to Fisheries Management." Paper presented at the Association for Public Policy Analysis and Management Twentieth Annual Research Conference, New York, October 29–31.

Kneese, Allen V., and Charles L. Schultze. 1975. *Pollution, Prices, and Public Policy.* Washington, DC: Brookings Institution.

Loayza, Eduardo A., ed. 1994. *Managing Fisheries Resources: Proceedings of a Symposium Co-sponsored by the World Bank and Peruvian Ministry of Fisheries held in Lima, Peru, June 1992.* World Bank Discussion Paper Fisheries Series no. 217. Washington, DC: World Bank.

Mace, Pamela M. 1993. "Will Private Owners Practice Prudent Resource Management?" *Fisheries* 18(9):29–31.

Mace, Pamela M. 1999. "Current Status and Prognosis for Marine Capture Fisheries." *Fisheries* 24(3):30.

Makarios, Emmanuel. 1996. *Nets, Lines and Pots: A History of New Zealand Fishing Vessels.* Vol. 1. Wellington, New Zealand: IPL Books.

Maloney, D. G., and P. H. Pearse. 1979. "Quantitative Rights as an Instrument for Regulating Commercial Fisheries." *Journal of the Fisheries Research Board of Canada* 36:859–866.

Martin, E. R. 1969. *Marine Department Centennial History: 1866–1966.* Wellington: New Zealand Marine Department.

Matthiasson, Thorolfur. 1997. "Consequences of Local Government Involvement in the Icelandic ITQ Market." *Marine Resource Economics* 12:107–126.

McClurg, T. 1994. "Two Fisheries Enforcement Paradigms: New Zealand before and after ITQs." In *OECD Documents: Fisheries Enforcement Issues*, 123–139. Paris: Organization for Economic Cooperation and Development.

New Zealand Seafood Industry Council. 2001. "New Zealand's Seafood Business: Economics and Trade." Available at <http://www.seafood.co.nz/nzseabus.cfm?SEC_ID=67&DOC_ID=95>.

Palsson, Gisli, and Agnar Helgason. 1995. "Figuring Fish and Measuring Men: The Individual Transferable Quota System in the Icelandic Cod Fishery." *Ocean and Coastal Management* 28(1–3):117–146.

Pearse, P. H., and C. J. Walters. 1992. "Harvesting Regulations under Quota Management Systems for Ocean Fisheries." *Marine Policy* 16:167–182.

Sanders, M. J., and K. H. H. Beinssen. 1997. "Uncertainty Analysis of a Fishery under Individual Tradable Quota Management: Applied to the Fishery for Blacklip Abalone Haliotis Rubra in the Western Zone of Victoria (Australia)." *Fisheries Research* 31:215–228.

Schlager, Edella. 1990. "Model Specification and Policy Analysis: The Governance of Coastal Fisheries." Ph.D. diss. Bloomington: Indiana University.

Scott, Anthony. 1955. "The Fishery: The Objective of Sole Ownership." *Journal of Political Economy* 63:116–124.

Sharp, Basil M. H. 1997. "From Regulated Access to Transferable Harvesting Rights: Policy Insights from New Zealand." *Marine Policy* 21(6):501–517.

Sharp, Basil M. H. 1998. "Fishing." In *The Structure and Dynamics of New Zealand Industries*, ed. Michael Pickford and Alan Bollard, 53–86. Palmerston North, New Zealand: Dunmore Press.

Sinclair, Keith, ed. 1997. *The Oxford Illustrated History of New Zealand*. New York: Oxford University Press.

Sissenwine, M. P., and P. M. Mace. 1992. "ITQs in New Zealand: The Era of Fixed Quota in Perpetuity." *Fishery Bulletin* 90:147–160.

Slack, E. B. 1969. "The Fishing Industry in New Zealand: A Short History." In *Fisheries and New Zealand: Proceedings of a Seminar on Fisheries Development in New Zealand*, ed. E. B. Slack, 3–24. Wellington, New Zealand: Victoria University of Wellington.

Squires, Dale, James Kirkley, and Clement A. Tisdell. 1995. "Individual Transferable Quotas as a Fisheries Management Tool." *Reviews in Fisheries Science* 3(2):141–169.

Statistics New Zealand. 1999. *New Zealand Official Yearbook 1998: 101st Edition*. Wellington, New Zealand: GP Publications.

Terrebonne, R. Peter. 1995. "Property Rights and Entrepreneurial Income in Commercial Fisheries." *Journal of Environmental Economics and Management* 28:68–82.

Titchener, Paul. 1981. *The Story of Sanford Ltd: The First 100 Years*. Auckland, New Zealand: Sanford Ltd.

Turner, Matthew A. 1997. "Quota-Induced Discarding in Heterogeneous Fisheries." *Journal of Environmental Economics and Management* 33:186–195.

Yandle, Tracy. 2001. "Market-Based Natural Resource Management: An Institutional Analysis of Individual Tradable Quotas in New Zealand's Commercial Fisheries." Ph.D. diss. Bloomington: Indiana University.

Yin, Robert K. 1993. *Applications of Case Study Research*. Newbury Park, CA: Sage Publications.

Young, Michael D., and Bonnie J. McCay. 1995. "Building Equity, Stewardship, and Resilience into Market-Based Property Rights Systems." In *Property Rights and the Environment: Social and Ecological Issues*, ed. Susan Hanna and Mohan Munasinghe, 87–102. Washington, DC: World Bank.

6

Stakeholders, Courts, and Communities: Individual Transferable Quotas in Icelandic Fisheries, 1991–2001

Einar Eythórsson

Introduction

In chapter 5, Tracy Yandle and Christopher Dewees outline three approaches to fisheries management: the bureaucracy-based approach, the community-based approach, and the market-based approach. A fourth approach, the comanagement or participatory approach, is the point of departure of this chapter. User participation and stakeholder involvement are usually considered as desirable qualities of management institutions, even if there is a need for balancing stakeholder interests and public interest (Mikalsen and Jentoft 2001). This chapter, on the first decade of fisheries management by ITQs in Iceland, focuses on changes in stakeholder involvement during that decade and changes in the mode of policymaking, from a consensus-based policy to open conflict, a process in which the relative power of different stakeholder groups has changed dramatically. An important element of these changes is reflected in litigation processes related to the property rights to fishing quotas, processes that have gradually strengthened the legal status of fishing rights as private property. The economic and distributive aspects of ITQs have also been a great source of conflict. The initial goals of reducing catch capacity and improving efficiency have not been met to the extent anticipated by the proponents of ITQs. The distributive effects of the system, in terms of its impacts on income distribution between owners and crew, have been a source of a prolonged labor unrest in the industry. The analysis is presented within a context of a somewhat detailed empirical account of the formation of and changes in the Icelandic fisheries management institutions during the last two decades.

As an island nation in the North Atlantic, Iceland is heavily dependent upon its fish resources. Fish products are the nation's most important export commodity, and fluctuations in catches or seafood market prices tend to have immediate impacts on the living standard of most Icelanders.

The most important of the Icelandic fisheries is the demersal or groundfish fishery. In recent years, this fishery has usually generated over 80 percent of the country's total wetfish value (Runólfsson 1997). The demersal catches (cod, haddock, saith, redfish, and Greenland halibut) from Icelandic waters fluctuated between 400,000 and 650,000 tons per year during the 1980s and 1990s. The fisheries industry is geographically spread along the coastline, and the majority of the fishing communities consist of relatively small villages with a rather one-sided employment structure. Out of a total of sixty-one fishing harbors in Iceland, thirty-six are located in communities with less than 1,000 inhabitants (*Útvegur* 1993).

Transparency is another important characteristic of Icelandic fisheries. In 1996, there were only 2,000 registered vessels (800 decked vessels) and sixty-one fishing ports. The structure of the industry makes it relatively manageable in terms of control, reliability of catch statistics, and enforcement costs in general.

Considering these conditions, it is no wonder that fisheries issues are under constant public debate in Iceland. The resource situation, the economic performance of the industry, and last but not least the fairness and effectiveness of the resource management system are not internal issues, debated only within closed fisheries circles. They are issues of great concern for the public at large.

For these and other reasons, Iceland is an interesting case for the study of marine resource management. Because of the transparency of fisheries sector, Iceland can be seen as a suitable laboratory for the testing of theoretical management models such as the system of ITQs. The extensive public debates on fisheries issues, on the other hand, can provide documentation on conflicting values and ethical dilemmas surfacing in the wake of new management practices. An outline of the ITQ model is presented by Yandle and Dewees in chapter 5, and a more detailed discussion of ITQs as a management tool is found in Squires, Kirkley, and Tisdell 1995, Copes 1994, Hannesson 1991, and Árnason 1991.

Different aspects of the Icelandic experience during the first years of ITQ management have been the subject of research by social scientists and economists during recent years. Gísli Pálsson and Agnar Helgason have done extensive work on evaluating the redistribution of quotas and the nature of the quota trade, as well as on the public discourse and the moral issues concerning ITQs (Pálsson and Helgason 1996a, 1996b, 1999; Helgason 1995). My own work involves discussions on the distributive effects of ITQs, including the issues of quota leasing, contract fishing, and quota market prices, as well as the effects of ITQs on fisheries communities (Eythórsson 1996a, 1996b). Whereas most social scientists have taken a critical view of the ITQ model, works by fisheries economists like Hannesson (1991) and Árnason (1991) are primarily theoretical and supportive of the ITQ model. Lately, however, Runólfsson (1997, 1999), Runólfsson and Árnason (1999a, 1999b), and Árnason (1995, 1997) have published empirical work on the effects of ITQs, which I will return to later in this chapter. Whether the record of ITQs in Iceland is presented as less of a success story than in New Zealand (chapter 5) is thus usually somewhat dependent on whether the presenter is a fisheries economist or a social scientist.

The Roots of the Present Management System

In looking for the rationale and justification for the present system, it is reasonable to start with Iceland's extension of its EEZ to 50 nautical miles in 1972 and to 200 nautical miles three years later. The rationale behind Iceland's "nationalization" of its fish resources was twofold:

1. An urgent need to protect the resources. As it seemed evident that the North Atlantic cod might fall victim to the tragedy of the commons if it remained outside the jurisdiction of any state that could introduce an effective resource management regime, it was considered too risky to wait for a new international management regime to become workable.
2. Protection of Iceland's national interests. It was argued that the national economy, and indeed the future of Iceland as an independent state, were totally dependent upon the nation's fish resources. Consequently, national control over these resources was a necessity from an economic as well as a political point of view.

The Icelandic policy of assuming national control over the resource base in the surrounding ocean was met with a fierce opposition from other fishing nations, especially Britain, which sent its navy to protect British fishing vessels in Icelandic waters. Iceland's decision to take control of its ocean resources was based on a very broad consensus among political parties and every stakeholder organization within the fisheries sector. The foundation of the Icelandic fisheries management policy, a combination of ecological issues and national interests, was understandably enough an issue of a broad consensus and popular mobilization. Most Icelanders felt that they were fighting together for a just cause against an external foe, the British. There was a common belief that once the resources were under Icelandic control, they would certainly be harvested and managed in the interests of the people. In retrospect, it can be argued that fisheries management has evolved from being an issue of great consensus and national unity during the 1970s to become the most divisive and conflict-laden issue in Icelandic politics and public debates in the 1990s.

The Introduction of Resource Management in Iceland's Fisheries

Britain withdrew from the last "cod war" in 1976, and Iceland could finally harvest the resources within its EEZ without foreign competition. In practice, those resources had been appropriated as national property. As still stated in the fisheries legislation, the fish resources within the EEZ are considered the property of the Icelandic people. During the 1960s, approximately one third of the total catch in Iceland's waters had been taken by foreign (primarily British and German) vessels. In the early 1970s, it therefore seemed reasonable that the departure of foreign vessels from Icelandic waters would allow for a substantial increase in catches by the domestic fleet. These promising prospects triggered a rush of investment in modern fishing trawlers, and the fleet of stern trawlers increased from none in 1970 to an impressive eighty vessels in 1980. The optimism of the early 1970s was, however, deflated by the so-called black report from the Marine Research Institute in October 1975. The report concluded that the condition of the Icelandic cod stock was poor and that it could, in the worst case, suffer the fate of the North Atlantic herring, the stock of which had collapsed dramatically in 1967–1968.

As a reaction to the apparent threat to the cod stock identified in the black report, Iceland's government introduced a TAC for cod and a set of restrictive measures with respect to commercial fishing in 1976. In essence, these measures were aimed at limiting fishing effort, especially in the cod fisheries. Each vessel was obligated to refrain from cod fishing for a certain number of days each year, and measures were taken to restrict the entrance of new vessels into the fisheries. The Marine Research Institute was also authorized to close fishing grounds on short notice if necessary to protect juvenile cod.

With certain variation from one year to another, these measures were in effect from 1977 to 1983, with some improvement in the cod stock as a result. By 1983, however, the cod stock was once again in a poor condition. It seemed evident that the 1976 management measures had not been sufficiently effective. The fishing fleet continued to grow, and the TAC for cod was repeatedly exceeded. This was the background for the introduction of *vessel quotas* in the demersal fisheries in 1984. The debate and the political process prior to the decision to impose these quotas will be discussed further after the stakeholder organizations in the fisheries are introduced.

Stakeholder Organizations and the Fisheries Assembly

In academic debates on fisheries management, key concepts such as "fisherman" and "stakeholder" are often poorly defined. When ITQ theorists speak of allocating quota to fishermen, they do not literally mean distribution of these quotas among fishing men. Usually, quotas are distributed among owners of fishing vessels, who are in many cases fishing enterprises of different shapes and sizes. Even when a vessel is owned by a fisherman, the quota is allocated to him, not to the hired crew.

Stakeholder (or user group) involvement in fisheries management is generally seen as desirable, but the stakeholder concept is often applied as referring only to organized interest groups in the harvesting sector, primarily fishermen (that is, boat owners). It is far from obvious, however, who are or should be considered as stakeholders when it comes to a distribution of fishing rights through a quota system. Mikalsen and Jentoft (2001) have grouped fisheries stakeholders in Norway according to their positions as definitive, expectant, or latent stakeholders. The definitive stakeholders, according to Mikalsen and Jentoft, are fishers, fish processors, bureaucrats, enforcement agencies, scientists, and fish workers. These authors consider local communities to be expectant stakeholders and citizens in general to be latent stakeholders. Further, they evaluate each of these groups on three dimensions of salience: legitimacy, power, and urgency. Communities, for instance, score high on legitimacy and urgency, but low on power.

In Iceland, where the ups and downs of the fishing industry affect the livelihoods of most of the nation's people and where the fish resources are defined as the common property of the nation, every citizen can with a certain legitimacy be considered a stakeholder. Those whose stakes have the highest urgency are certainly the men and women who depend directly upon the fisheries, as boat owners, fishing crew, owners, and managers and workers in the fish-processing industry and related industries. In particular, all those who have risked their lifetime savings by investing in a fishing community certainly have high stakes in the distribution of fishing rights. The power of the different groups of stakeholders, however, depends on their economic and political position, reflected in the strength of their organizations and unions. The experience in Iceland is that the balance of power among different groups of definitive stakeholders has radically changed in favor of the quota owners. As a result, it has become impossible to continue with a mode of policymaking based on deliberation and consensus among the stakeholder groups.

In essence, Icelandic fishermen belong to different unions and associations depending upon their employment and professional status. Hired deckhands are organized in the Icelandic Seamens' Federation (Sjómannasamband Íslands, or SSÍ), officers (skippers and mates) are represented by the Officers' Federation[1] (Farmanna- og fiskimannasamband Íslands, or FFSÍ), and engineers have organized themselves in the Icelandic Engineer Officers' Association (Vélstjórafélag Íslands). Because they are divided between three different unions, members of crews of Icelandic fishing ships have tended not to speak with one voice. The three unions have expressed somewhat different attitudes towards the ITQ system. The FFSÍ has been the most pointed opponent to the ITQ policies, whereas the two other unions have argued that the system should be modified to secure the interests of their members.

Their employers are organized in the Federation of Icelandic Fishing Vessel Owners (Landssamband Íslenskra Útvegsmanna, or LÍÚ). The LÍÚ has been an active proponent of ITQs, and since quota reform has turned the federation into an exclusive club of quota holders, its relative influence has increased. Since 1985, owners of small boats have been organized in the National Association of Small Boat Owners (Landssamband smábátaeigenda, or LS). The LS was established to guard the interests of the owners of small vessels that were allowed to operate outside the quota system, a group that doubled in size in the late 1980s but was severely reduced in number during the 1990s. All these organizations were more or less represented on different task forces and committees appointed by Iceland's government to review the fisheries policy during the 1980s. Fish processors and marketing units are also organized and have in some cases been represented in government committees. Besides, since a large fraction of the fishing fleet is owned by vertically integrated companies, the processors are to a certain degree represented by LÍÚ as well. The stakeholders who have probably been least involved in decision making and policy design in the fisheries are the workers employed at the processing plants. They were, however, represented by their Workers Union of Iceland (Verkamannasamband Íslands, or VMSÍ) on two fisheries task forces during the 1980s (Pálmason 1992).

Another important organization is Iceland's Fisheries Association (Fiskifélag Íslands, or FÍ). Founded in 1911, the FÍ was originally an ideal organization for furthering fisheries development in general. Besides being a service organization and a semigovernmental office for keeping records of fisheries statistics, the FÍ has since 1942 constituted a forum for fisheries debates involving the different interests within the fisheries. The FÍ arranged the annual Fisheries Assembly (Fiskiþing), in which the different unions and organizations for the harvesting, processing, and marketing sectors were represented, along with the representatives from the regional units of the FÍ itself. These regional units were membership organizations, each entitled to elect representatives from their area to the annual Fisheries Assembly. Thus, the assembly was a mixture of branch and union representatives and representatives from the different geographic regions of the country. With its broad functional and regional representation, the Fisheries Assembly was an important forum for debates on fisheries issues and policy during the latter half of the twentieth century. It passed resolutions on topical issues, and in general, its recommendations strongly influenced the national fisheries policy.

During the 1990s, however, the influence of the Fisheries Association declined considerably. Its role as a semigovernmental body was discontinued in 1998, and at the same time, the regional units were abolished. Today, the annual assembly is a theme-based conference, with participants only from branch organizations and unions.

Fishing communities, or coastal municipalities, have been directly represented neither in fisheries task forces nor in the Fisheries Assembly, but up to the 1990s, most fisheries companies were closely linked to municipalities, some of them through ownership by the municipalities. In these cases, the companies were considered more or less representatives of the communities in which they were located.

The Introduction of Vessel Quotas

The decision to introduce vessel quotas in the demersal fisheries was made in the face of bleak prospects for the cod stock in 1983. During the previous years, there were prolonged debates over the issue within the stakeholder organizations, and these debates culminated at the Fisheries Assembly in 1983. A majority of boat owners and the regional representatives from Iceland's north, east, and south supported a vessel quota solution, with opposition concentrated among the regional representatives from the west and northwest (Westfjords) regions. At the 1983 assembly the opposition found itself in a minority position and finally acquiesced support of a recommendation to the government concerning vessel quotas. The assembly reached consensus on the issue by recommending a trial system of vessel quotas for one year. Vessel quotas were not totally new in Iceland, having been introduced in the herring and capelin fisheries a few years earlier.

On the basis of the recommendation from the assembly, the Icelandic Parliament passed the 1983 Fisheries Management Act in December of that year, almost without debate. The new law was very brief: it authorized the Ministry of Fisheries to work out the details of a vessel quota system for the demersal fisheries.

The ministry decided that the initial allocation of vessel quotas should be based on the catch history of each vessel for the three previous years (1981–83). Although regulation through catch quotas was introduced as the basis of the new management system, there was also another option: boat owners could also choose to fish within an "effort quota" system based on a limited number of days at sea. The effort quota was originally designed as a safety valve, an opportunity for owners of boats who had for some reasons been idle during the previous years (and therefore would have limited or no access to catch quota). But by 1985, it had become an attractive alternative for all those who felt discontent with their share of the TAC.

Small boats, those up to ten gross register tons (GRT), were not included in either of the systems but were initially allowed to fish practically without restrictions. From 1985 on, their catch was regulated by a sort of effort quota designed specially for this group. The idea behind this special treatment of small boats was to preserve the flexibility of Iceland's traditional small-scale fisheries. But within a few years, the small boats doubled in number, and the small-scale fishermen became a thorn in the side of quota holders.

Designing the new system while maintaining an atmosphere of consensus in the industry was a great challenge to the ministry. The solution was to involve stakeholders closely in the implementation of the system and to allow for a series of individual and group adjustments in the quota allocation. To deal with adjustment problems and to resolve disputes about unintended or unfair outcomes of the quota allocation, the ministry engaged the stakeholder organizations. It appointed the leader of the boat owners (LÍÚ) and a representative of the crew unions (alternating between the leadership of FFSÍ and SSÍ) to a Consultative Committee (Samrádsnefnd), along with a third committee member from the Ministry of Fisheries. As it turned out, this committee quickly became extremely busy, especially during the first half of 1984. From late January to the end of May of that year, it had no less than sixty-five meetings. It had to handle a great number of complaints from boat owners and took on the responsibility of correcting unfair outcomes and in some cases redistributing quota among vessel groups and individual vessels. By handing these problems over to the representatives of the major stakeholders in the harvesting sector, the government managed to maintain an atmosphere of consensus within the industry. The committee remained very active until 1990 but was discontinued in the first half of 1991, after the introduction of the 1990 Fisheries Management Act.

With some minor changes, the quota system that originated from the 1983 Fisheries Management Act was in force until 1990. Quota was transferable to a certain degree but could be transferred from one vessel to another only if the transferring vessel was permanently removed from the fisheries. Exchange and leasing of catch quota within a year was allowed and could freely take place within the same fishing community (municipality) or between vessels owned by the same company. Transfers involving vessels from different communities required an application to the ministry in each separate case and could not be carried out without the approval of the involved municipalities and local workers unions. This provision enabled municipalities to block such transfers in cases where they were perceived as threatening the local employment situation.

The quota system was changed several times between 1984 and 1990, but in essence it worked only in a limited sense as a system of ITQs. The regulations were relatively complex, and a number of loopholes allowed for catches beyond those permitted by the quota. Especially in 1986–1987, it seemed that the catch quota system would soon wither away, as a majority of boat owners opted for the effort quota alternative. At the same time small boats, which were subject to the more liberal regime described earlier, became extremely popular. Whereas about 1,000 small boats were registered in 1984, the number had increased to about 2,000 in 1990 (Pálmason 1992). As small boats became more numerous and more efficient their aggregate cod catches increased from 16.6 tons in 1984 to 47.7 tons in 1990, increasing their relative share of the total cod catch from 4 percent to 14 percent (Pálmason 1992).

Pálmason (1992) has studied the decision-making processes leading to changes in the system from 1984 to 1991, including stakeholder representation on five different preparatory committees (task forces) appointed by the government during this period. He finds that the committees that prepared the law revisions in 1984 and 1985 were small (seven persons), with representatives mainly from the Ministry of Fisheries and the harvesting sector. But by the end of the period (law revisions of 1988 and 1990) there was a tendency toward larger committees (up to twenty-four persons) with a broader range of representatives, including all political parties, research institutes, workers unions, Iceland's Federation of Employers (Vinnuveitendasamband Islands), processor organizations, and marketing units. He explains this development in terms of growing consciousness about the importance of the choices that were to be made in the fisheries policymaking among stakeholder groups outside the harvesting sector, as well as within the political parties. In other words, there was a tendency within the Icelandic fisheries administration, as time went on, to consider a broader range of stakeholder groups to be legitimate participants in the policymaking process.

The importance of participation in the decision-making process with regard to the quota system became more urgent for different stakeholder groups as the distributive effects of the system became more focused. It was gradually realized that fisheries management was no longer merely a question of temporary technical measures to protect fish stocks but one of a permanent solution to the problem of managing Iceland's fisheries. On the other hand, the tendency toward larger committees also indicates that the government was accepting a wider definition of who should be considered legitimate stakeholders in fisheries management. Fishing communities or municipalities, however, were still not included among stakeholders represented in these committees.

The consensus-based policies of the 1980s can be seen as result of the broad alliance between stakeholders in the fisheries created during the cod wars with Britain in the 1970s. Basing their statement on Gísli Pálsson's work, McCay and Acheson (1987, 33) characterized Icelandic fisheries management during this period as comanagement:

> The Icelandic management process is open and flexible, able to respond to and incorporate the interests of diverse actors and groups, in sharp contrast to the systems portrayed by Anderson [1987] and Pinkerton [1987] and those with which we are familiar in the United States.
>
> Co-management is a social reality in Iceland; it does not need a special label, nor need it be based on either homogeneity on the part of the users or total accord between users and managers. Accordingly, the state is trusted in Iceland. The boat quota system being considered in the 1980s will privatize rights to catch fish in a property mimicking way. It is controversial because of realistic worries about its effects on the structure of the industry. But if it is accepted, it will be

partly because the state is trusted to be impartial in the assignment of quotas, a trust that enables the continuation of the ideology of equal, or equitable, access and the effective management of a limited good.

McCay and Acheson describe quite well the situation in the management of Icelandic fisheries in the mid-1980s: the state could still build upon the trust established by the broad national consensus during the 1970s. The stakeholder organizations could also participate in the management discourse with more or less equal voices. The dividing line between quota owners and the others was not yet well defined. Jónsson (1990) has discussed the roles of the state and the Fisheries Assembly in decision making in Icelandic fisheries, focusing on the introduction of the quota system. He argues that facing the poor resource situation in the fisheries in 1983, the government took a rather passive role and demanded absolute consensus in the Fisheries Assembly before it would take any action. This put the stakeholders under great pressure to take on the responsibility of fisheries management.

Permanent Allocation and Transferability: The 1990 Fisheries Management Act

Developments during 1988–1989 seemed to indicate that the management system implemented after the 1983 Fisheries Management Act was not achieving to its goals. There were signs of economic crisis in the industry, and the loopholes in the system made it extremely difficult to enforce the established TAC for cod. Despite the government's restrictive policies, investments in the fishing fleet and consequently the total catch capacity continued to increase. The effort quota alternative was seen as particularly to blame for the continuing investments in increased catch capacity. The industry also complained that the system was too complex. In addition, the industry found long-term planning difficult, as the management system was subject to unforeseeable modifications on almost annual basis.

The 1990 Fisheries Management Act, which established an ITQ system in Iceland's fisheries management, came about after prolonged controversies in the Fisheries Assembly as well as in the Icelandic parliament. In comparison to the relative ease with which the debate in 1983 was able to conclude in an accord, it had become more difficult to reach consensus in the Fisheries Assembly, but the opponents of quota management still constituted a minority among the representatives. In the 1989 Fisheries Assembly, a draft of a new ITQ law proposal from the government was fiercely debated, but again, opponents of quota management finally opted for a consensus resolution. This resolution supported a law proposal that eliminated the requirement, in cases where a vessel with quota was to be sold out of a municipality, that such a transfer be approved by the municipality and the local trade union. The elimination of this requirement was on request of the LÍÚ, as it found quota transfers too difficult under the 1983 law. Many representatives at the assembly were concerned about the elimination of this requirement, and as a part of a consensus package, the assembly proposed another safety valve for the communities. According to this proposal, which eventually became a part of the law, municipalities could interfere with the sale of vessels that were about to be sold outside the community by buying them temporarily in order to find local buyers. If no local buyer would turn up, the municipality could keep the vessel. If a municipality intervened, if would have to meet the price the owner was offered by a potential buyer. This was supposed to replace the previous rule that had enabled municipalities to block quota transfers if they considered them as a threat to the local employment situation. In practice, the new rule turned out to be almost worthless, as it referred to vessels and not to the quota itself. With the new law, quota could be sold separately, and interfering in the sale of a vessel was of little use, if it had been stripped of its quota before the sale. It also turned out that with rising quota prices, lack of financial resources made it quite unrealistic for small municipalities to intervene in quota transactions.

A preparatory committee, with both stakeholder and political representatives, delivered a revised draft of the law to the government in January 1990, with several reservations on the part of individual committee members. These reservations referred to different parts of the draft, and there was little consensus among those who disagreed with the committee majority. Only five out of the twenty-four committee members were totally unwilling to support the draft, basically out of concern for fishing communities and workers in fish processing. Three of these were representatives of small political parties: the Women's Party (Kvennalisti), the Citizens' Party (Borgaraflokkur), and the Association for Equal Rights and Social Policy (Samtok um jafnrétti og félagshyggju). The two other dissenters were a union representative of workers employed in the fish processing industry (VMSÍ) and a representative from the Association of Salt Fish Processors (Samband fiskframleidenda). The union representatives from SSÍ and FFSÍ supported the draft with certain reservations about transferability of quota but noted that they were ready to contribute to a consensus solution to the problem. Small-boat owners also supported the draft with some reservations.

The overall impression that emerges from the debate in the 1989 Fisheries Assembly and the preparatory committee (1988–1990) is that the participants put a lot of effort into seeking consensus. Even though there were strong reservations about free transferability of quota, out of concern for the increased vulnerability of fishing communities under ITQs, the main idea of the new law appeared to have broad support among both stakeholder representatives and politicians. The issue of possible negative effects on fishing communities was only briefly commented upon in the draft presented by the committee. It said that such problems should be taken care of through other suitable measures, like government support to regions especially affected.

There may have been several reasons for the general unwillingness among the participants in the 1989 assembly and among the members of the preparatory committee to enter into an open conflict. There was a common understanding that the fishing fleet in Iceland was too big and that effective measures were needed to adapt the fleet to the nation's fish resources. There was also a common understanding that finding a solution to the problems of fisheries management was an urgent national task. There was a relatively broad consensus on the issue that this meant the continuation of a quota system in some form. Two parallel regimes, the catch quota and the effort quota alternative, had already been tried out, and the first one clearly seemed more promising for the purpose of reducing the fleet capacity. But as McCay and Acheson (1987) point out in the passage quoted earlier, it was probably quite important that at this point the state was trusted to be impartial in the assignment of quota. Stakeholders outside the harvesting sector, unions, and communities felt confident that their voice would count and continue to count in the future and that possible mistakes could be corrected at a later stage. People in fisheries communities also seemed relatively confident that the political leaders, in particular the fisheries minister, were concerned about their future. The role of Halldór Ásgrímsson, fisheries minister from 1983 to 1990, in this respect should not be underestimated. With a family background from a small fishing community and representing the Progress Party (Framsóknarflokkur), whose voters tend to live in rural areas and small towns, his credibility was hardly questioned. To sum up, it would probably not have made much difference in the outcome of the debates over the 1990 law if the fishing communities had somehow been more directly included in the decision-making process.

Proponents of the ITQ system received strong support from certain academic circles, fisheries economists in particular. By the late 1980s, theoretical models provided by fisheries economics were exerting an increased influence on the quota debate. The focus of interest shifted from resource protection to the question of economic efficiency in the fisheries. It was argued that in perpetuity allocation of quotas would provide conditions for long-term planning and sound investment behavior, and free transferability would provide flexibility and efficient use of capital. Inefficient vessels would be bought out, and efficient ones would be able to optimize their operations. Some economists also argued that the efficiency generated by ITQs could produce a basis for management by resource rentals. Resource rentals (annual payments from quota holders to the state in return for the privilege of harvesting the fish resources) could subsequently become an important source of revenue for society at large. The revenue from the rentals could be used to compensate communities and regions that might be negatively affected by the system. Thus the ITQ system was justified partly by practical reasoning on the part of the LÍÚ, such as the need for predictability and flexibility, and partly by theoretical reasoning by fisheries economists focused on efficiency and the potential benefits of the resource rent upon the national economy.

The 1990 Fisheries Management Act allocated TAC shares permanently to the boat owners by prolongation of previous allocations. According to section 1 of the act, Iceland's fish resources would remain the common property of all Icelanders,[2] and the rights allocated to quota holders could not be considered private property in the constitutional sense. The effort quota option was abolished; the only exception granted was to small boats, up to 6 GRT, which were still subject to an effort quota regime.

Another important change the 1990 law brought about was a broad liberalization of quota transfers. TAC shares became divisible and could in effect be transferred as a separate commodity, not just as a part of the market value of a fishing vessel. Quota transfers could, however, take place only between owners of Icelandic fishing vessels. Exchange and leasing of annual quota for any particular species was also liberalized and could in practice take place without consulting the Ministry of Fisheries or the involved communities and unions.

Capacity, Effort, and Changing Fleet Structure

Reviews by Icelandic fisheries economists (Árnason 1997; Runólfsson 1997, 1999; Runólfsson and Árnason 1999a, 1999b) indicate that the ITQ system has clearly improved the economic efficiency of Iceland's fisheries, as predicted by the proponents of the system. According to their analysis, there seems to be little doubt that the system works, meaning that fleet capacity and effort is reduced and that efficiency is subsequently increased. Increased efficiency can in this context be defined roughly as being able to get the same output (catch) with use of less input, also known as effort. These conclusions are not necessarily incorrect, but it is sometimes useful to ask for the exact meaning of these concepts, and the methods of calculating the numerical values they express. A frequently used shortcut for measuring fleet capacity is merely counting the number of fishing vessels. By this measure, the size of the fishing fleet has severely decreased since the introduction of ITQs. If one compares figures on GRTs and engine capacity, however, one finds that in these terms, fleet capacity actually increased gradually in the 1990s despite the ITQ system (table 6.1). From the introduction of ITQs in 1984 to the end of 1997, the capacity of the fleet expanded by 13 percent, or 14.1 thousand (14,100) GRT. The main increase occurred, however, in 1984–1990 (while the effort quota alternative was still available); since 1991, there has been only a slight increase. Desegregation of the data creates a more complicated impression. For the inshore fleet, defined as vessels from 12 to 200 GRT, there has been a great reduction, by tonnage: 7.9 thousand (7,900) GRT (27 percent).

Table 6.1
The Icelandic fishing fleet, 1984–1997 (thousands of GRT)

Vessel type	1984	1989	1995	1997	Change 1984–1997
Trawlers >500 GRT	16.0	24.1	39.7	43.5	+27.5
Trawlers <500 GRT	35.1	32.1	25.6	24.6	–10.5
Other >200 GRT	28.0	32.5	30.5	32.7	+4.7
Inshore 12–200 GRT	29.6	28.8	22.8	21.7	–7.9
Small boats <12 GRT	2.1	3.3	2.6	2.4	+0.3
Total	110.8	120.7	121.2	124.9	+14.1

Source: *Útvegur* 1984, 1989, 1995, 1997.

The trawler category has increased by 17 thousand (17,000) GRTs (33 percent). The increase here can be explained by the growth in the factory trawler fleet. Factory trawlers have space-consuming fish-processing facilities on board, and the relation between their tonnage and fishing capacity may not be comparable to that in other vessels. Considering the technological development in the fisheries, however, it is reasonable to believe that the fishing capacity of an average one thousand-GRT trawler in 1997 was somewhat greater than that of a similarly sized vessel in 1984. The size of the group of small coastal vessels, those below 12 GRT, increased in the late 1980s, but is now more or less back to the 1984 level.

Despite the incentives for concentration of quota on fewer vessels in the ITQ system, the restructuring of Iceland's fishing fleet would not have taken place without additional assistance. In January 1989, Iceland's fisheries minister proposed a fund to buy out excessive capacity, financed through annual fees from the industry and annual quota allocations. The idea was broadly rejected by the LÍU and did not get political support in Parliament. It was argued that such a program would be unnecessary; in an ITQ system, the market mechanism would take care of the problem of excess capacity.

Four years later, however, a similar fund, financed through fees from the industry, was established to buy out excessive capacity. When Fisheries Minister Þorsteinn Pálsson introduced the fund in September 1993, he argued that the aim was to strengthen the incentives for vessel owners to "reduce the number of vessels, concentrate quota on fewer boats and reduce the harvesting costs" (Anonymous 1993). During its time of operation (1994–1997), the fund bought out 459 vessels, with an aggregate tonnage of 7.8 thousand (7,829) GRT,[3] and twenty fish-processing plants. Most of the vessels (398) were from the small boat (<12 GRT) category. The main effect of the program was thus to reduce the number of small boats, which were still operating outside the quota system. The fund was also used to solve another problem: the existence of a number of vessels with little or no quota of their own, operating on a contract fishing basis. These vessels were generating a market demand for quota, creating an upward pressure on quota (leasing) prices (Eythórsson 1996a). The high leasing prices were considered unsound from the standpoint of resource management, especially since high quota prices caused a downward pressure on the income of fishing crew (see section on contract fishing). In terms of market prices, many of these vessels were practically worthless, but the buyout price was set at 45–80 percent of insurance value.

The buyout program has thus been an important contribution to the change in the structure of Iceland's fishing fleet in the 1990s. It is worth noting, however, that despite this substantial buyout of catch capacity, the aggregate tonnage of the fleet did not decrease during the fund's operation between 1993 and 1997.

Fishing effort is another parameter for measuring the effects of fisheries management through ITQs. Árnason (1997) and Runólfsson and Árnason (1999a, 1999b) have found a significant decrease in aggregate fishing effort in Iceland since the introduction of ITQs. The method they use to measure aggregate effort is simply multiplying capacity (aggregate GRT) by days at sea, adding up fishing days for all participating vessels (Árnason 1997). By this mode of analysis, it is obvious that if a large number of small vessels are replaced with a few supertrawlers, with aggregate gross tonnage held fairly constant, the result will inevitably show a reduction in aggregate effort, since a single vessel cannot operate for more days than there are in one year whereas numerous smaller vessels with the same gross tonnage in the aggregate can log as many as 365 times the number of vessels in terms of days fished. In the formula noted above, the smaller vessels will therefore weight the aggregate effort far more than a single vessel with an identical tonnage to the smaller vessels in the aggregate. Considering the change in the composition of the fleet between 1984 and 1997 (table 6.1), this result is therefore no surprise. Given this definition of the concept "fishing effort," the conclusion is correct, but the question is whether this definition of the concept makes it a suitable measure for the amount of resources used for the purpose of catching a given amount of fish.

Another way of measuring aggregate effort might be to check the input of capital and labor in the fisheries. The input of capital should be reflected in the investments in fishing vessels and the development of the aggregate value of the fishing fleet. According to Runólfsson (1999) and the Ministry of Fisheries (2001), the aggregate value of Iceland's fishing fleet decreased somewhat from 1991 to 1997. According to these figures, the value of the country's fishing hardware in 1997 had retreated approximately to the 1987 level. There are, however, some obvious difficulties involved in comparing the capital input in the fisheries before and after the introduction of ITQs. Before 1991, the market value of a vessel normally included the value of quota (or the value of days at sea allocated to the vessel), whereas after 1991, vessel and quota are separate items. With the current quota prices, a major part of the capital value of the harvesting sector is the market value of quota. Consequently, it is reasonable to include the aggregate value of quota as a part of the capital applied in the harvesting sector. Considering that there has been a steep increase in the aggregate market value of quota from 1991 to 1997, there is little doubt that the input of capital, represented by the aggregate market value of vessels *and* quota, has increased rather than decreased since 1991. Quota, along with vessels and equipment, represents capital investment that is necessary to catch fish. Those who received quota in the initial 1984 allocation thus received capital free of charge. But for new entrants and for companies that are buying quota in order to expand, investment in quota is quite similar to investment in fishing hardware. There is no available record of the exact figures for investments in quota, but records of quota transfers indicate that they are substantial. Another indicator of the importance of investments in quota is a sharp increase in the aggregate debt of the harvesting sector, by more than 70 percent from 1994 to 2000 (Ministry of Fisheries 2001, 24). It has also been pointed out by Valsson (1999) that only half of the aggregate debt increase in the fisheries sector from 1995 to 1998 can be accounted for by investments in fisheries hardware, indicating that the other half represents credits for the purpose of investment in quota.

The concept of efficiency is the usual measure of success in economic terms. As mentioned earlier, Árnason (1997) and Runólfsson (1999) find that the efficiency of the Icelandic fisheries has improved considerably under the ITQ system. As fishing effort is a component in the calculation of efficiency, any analysis of efficiency in Iceland's fisheries depends upon a reliable measuring of fishing effort. Árnason (1995) applies catch volume/effort and catch value/effort as measures of efficiency. If effort is defined as aggregate tonnage multiplied by days at sea, catching a certain volume of cod with a large (and capital-intensive) vessel will automatically appear as more efficient than catching the same amount with 50 (low-capital but labor-intensive) small boats. This applies even if the small boats are more profitable and bring a more valuable catch. Obviously, Árnason's methodology is flawed, and his conclusions really aren't very revealing.

In terms of average profitability, the harvesting sector in Iceland has done rather well in the late 1990s. There is no doubt that there is more flexibility and predictability in the harvesting sector under ITQs compared to under previous regimes. But there is reason to believe that the profit of new entrants, boat owners and companies that have to buy quota at the current market price, is rather low compared to those who got their quota free through the initial allocation. At present, there are two categories of actors in the fisheries operating under different conditions: a privileged group that received free quota during the initial allocation and a nonprivileged group that has to pay market price for the quota.

There has been a substantial concentration of quota shares within the larger, vertically integrated companies since the introduction of ITQs, especially since 1991. In 1990, the biggest ten fishing companies in Iceland held 21.9 percent of TAC in demersal species, but by 2000–2001 the ratio had gone up to 48.1 percent (Ministry of Fisheries 2001). Responding to the increasing concentration of quota ownership, Iceland's parliament set an upper limit on the number of TAC shares that can be held by a single owner in March 1998. A single owner can hold up to 10 percent of the total TAC shares for cod and haddock and up to 20 percent of the total TAC shares for each of the other demersal species, as well as for herring, capelin, and shrimp. The concentration process accelerated at the end of the 1990s, especially as a result of a more concentrated ownership structure and mergers within the industry (Guðjónsson 1999).

Along with a general liberalization of the economic policy in Iceland, there has been a trend toward an ideological shift within the industry, leaving behind the idea that fisheries and fish processing should be locally embedded in fisheries communities and replacing it with an attitude that the first priority of fisheries companies should be to maximize profits. Many fisheries companies have joined the Icelandic stock market, and ownership of these companies is in many cases not linked to any particular community. Investors without fisheries background are now well represented among the owners of quota-holding companies (Garðarsson 1999).

Resolution of Conflicts over Contract Fishing

During the early 1990s, new types of relations emerged in the fisheries as a consequence of quota leasing and contract fishing (Eythórsson 1996a). Quota leasing (renting) is an arrangement in which the right to catch a certain amount of a certain species within a given year is transferred from one vessel to another. This right can be paid for through (1) transfer of fishing rights of a corresponding value for another species, (2) direct payments according to a market price, or (3) different forms of contract fishing arrangements. Contract fishing, as it appeared in 1992–1998, was often referred to as "fishing for others." A typical fishing contract was between a vertically integrated company with large quota holdings and an inshore vessel with little or no quota of its own. The vessel was obliged to deliver its catch to the company in return for a fixed price (market price for raw fish minus the quota leasing price). In 1993, the average fixed price for raw cod in contract fishing was about half the market price in auction markets, the remaining half representing the payment for quota leasing (Eythórsson 1996a). This practice influenced the income of a vessel's crew in a negative direction, as they receive a fixed share of the fish price on delivery. The cost of quota decreases the landing price, since the owner of the quota is not the vessel owner, but the fish buyer, and the cost of renting the quota is taken out of the price paid for the fish. In the case of contract fishing, the fishermen are in a position similar to that of hired farmhands, bringing home the harvest from the farmer's land.

As contract fishing became more widespread, more crewmen experienced a drop in their income. According to the crewmens' unions, speculative leasing transactions (*kvótabrask*) were in some cases undertaken to reduce the wages paid to crew members. An example of a speculative transaction is if a vessel owner decides to lease his quota to a fish buyer in the beginning of the fishing season and then starts contract fishing on behalf of the fish buyer. The landing price he receives will be low, and consequently the crew will get a lower share. The group of fishermen that were directly affected by these practices was probably not very large, but the unions feared that the practice could spread throughout the industry. In these speculative arrangements, the leasing prices charged represented the rate of interest (or resource rent) that could be obtained by quota owners by leasing their quota to others. This rate would set the standard for the annual return from the quota capital in general (the resource rent collected by quota owners) for the whole fishing fleet, and this in turn would mean a higher share to the quota owner and a lower share to the crew.

The leasing prices in 1993–1995 represented approximately 20 percent annual return on quota capital, about twice the general rate of interest for bank credits. The fear that all fishermen would soon be affected was the background of the crewmens' strike in January 1994 and repeated strikes in 1995 and 1998. The strike in January 1994, and the events leading to it, in many ways marked the end of the consensus atmosphere in the fisheries. The unions of crewmen and officers joined forces and organized meetings in every fisheries region. Union leaders urged the members not to tolerate the new practices and to support the demand for a ban on all quota leasing and contract fishing.

The LÍÚ seemed unwilling to acknowledge the existence of the problem, and no progress was made in negotiations. During the strike, the situation became very heated. A prolonged strike in the fisheries represented a threat to the national economy, and as no progress had been made in the negotiations two weeks after the strike commenced, the government decided to intervene with a preliminary law making the strike illegal. The unions found this act rather provocative and accused the government of taking the LÍÚ's side. Considering the nature of the controversy, there was, however, little chance of successful negotiations. The roots of the problem were in the fisheries legislation itself, and there was apparently no alternative to government intervention. As a part of the preliminary law, the government appointed a committee of three high-level officials to look into the matter. Interestingly enough, no stakeholder representatives were appointed to the committee.

After two more strikes and many rounds of negotiation, a new institutional framework was set up in March 1998 to control prices, resolve disputes, and control leasing transactions. A new Share Price Office (Verðlagsstofa skiptaverðs) was established to control landing prices and thereby secure a fair renumeration of crew. A standing committee linked to the Share Price Office with representatives from organizations from both sides (Úrskurðarnefnd sjómanna og útvegsmanna) was set up to resolve price disputes between boat owners and crew. Finally, a stock market–like structure was set up to control leasing transactions: the Quota Exchange Market (QEM) (Kvótaþing). In essence, all quota leasing transactions, apart from exchange of species and transactions between vessels held by the same owner, now had to take place anonymously at the QEM,[4] which meant that in effect, contract fishing in the form described above was no longer allowed. Market prices at the QEM were extremely high, especially for cod quota (80–90 percent of landing price), a situation that indicated that quota leasing was viable only as a solution to adjustment problems of matching the composition of species in the catch to the quota holdings of a vessel.

According to the organizations of crewmen,[5] these institutional reforms eased the situation of crew members somewhat, as leasing transactions became less common. The secretary of SSÍ, however, concluded that the problem was not completely solved after the first year of experience (Jónsson 1999). The experience with the QEM was mixed, and as a part of a negotiated agreement following eight weeks of strike among crewmen in the spring of 2001, the government decided to close it down.

Besides bringing the tensions within the fisheries out in the open, the conflict in 1994 functioned as a learning process for stakeholders, politicians, and the public. New concepts like "quota profiteering" *(kvótabrask)*, "sealords" (*sægreifar*), and "serfs" (*leiguliðar*) became a part of the common vocabulary, linking quota transfers to dubious acts and a return to a "feudal system" in the fisheries. Understanding of the dynamics of quota leasing and contract fishing reinforced the general opposition against ITQs, even among those who had been supportive of the 1990 law. On November 27, 1993, an editorial (*Reykjavíkurbréf*) in Iceland's leading newspaper, *Morgunblaðið*, commented on the situation: "The public support for the ITQ system is losing hold. The earth is cracking under the feet of its proponents, where the leader of LÍÚ and the fisheries minister are in the forefront. During these weeks, the discontent with the system has been flaming up from every direction, especially within the fisheries themselves."

The system was condemned by a number of politicians whose parties had supported the law in 1990, and highly critical voices were heard from some representatives of the processing industry. Jón Ásbjörnsson, a senior fish exporter, was quoted on the issue in *Morgunblaðið* on March 27, 1994: "Icelandic politicians created the quota on request of the fishing capitalists and with the support of fisheries scientists. In this manner, huge valuables were created for the vessel owners, valuables they shamelessly buy and sell, even if the fishing grounds, according to the constitution, are the common property of the nation. Such an arrangement is accepted nowhere but in Iceland. All this is unbelievably strange." Magnús Jónsson, the head of a fisheries committee within the Social-Democratic Party (Alþýðuflokkurinn), wrote in the same paper on April 23, 1994: "In my articles on the quota system, I have brought forward strong arguments showing that the system is unusable because of waste, discrimination and socially destructive power—apart from the fact that its goals have not been attained. I am not the only one who despises this system more than anything else in our society, even if I have no interests to defend, other than my wish to continue to be an Icelander."

The conflict level remained high from 1994 to 1999, with new strikes, as noted earlier, in 1995 and 1998, but despite a strong critical focus on the system, no united opposition materialized. The political parties, both inside and outside the government, experienced internal disputes, but all the major parties had been favorably disposed to ITQs in 1990 and found it difficult to turn back the clock.

Fisheries-Dependent Communities

By the end of the 1990s, the effects of ITQs on the situation in certain fishing communities reappeared in the foreground of the debate. Some communities had become marginalized as a result of a loss of quota. The geographical and economic structure of Icelandic fisheries, with a number of remote fishing villages, organized as single-enterprise communities, means that alternative employment opportunities in those communities are very sparse. During the 1990s, the vulnerability of these communities became more visible as several fishing villages lost most of their quota as the owners moved or sold out. A comparison of different size categories of fishing communities gives a clear impression that small communities, those with less than 500 inhabitants, have on the average lost a much larger share of their quota than bigger communities and towns (Eythórsson 1996b; Garðarsson 1999).

Although the effects of ITQs have obviously contributed to the marginalization of a number of fishing communities, changes in technology and markets have contributed as well. Land-based frozen fish production is in decline, whereas processing at sea and export of fresh products have increased during the 1990s. Consequently, local freezing plants, most constructed during the 1960s and 1970s, are no longer a guarantee of employment and prosperity in fishing communities. But even if some communities would most likely have faced crisis also under a different management regime, losing the right to catch fish has a strong demoralizing effect on people living in fisheries-dependent communities. These communities are heavily dependent upon quota owners for their survival, and no one seems responsible for the victims of the system: the people living in communities abandoned by the quota owners.

Even if crises in fisheries communities also happened before the days of the ITQ system, these outcomes should not be treated as minor externalities to an otherwise smoothly working management system. The obvious unfairness they represent has been a major contribution to the controversies over ITQs in Iceland's fishing industry.

Most economists would probably agree that as far as these effects can be considered external costs produced by the ITQ system, they should somehow be internalized—that is, compensated from the revenues produced by the system. So far, little has been done to build a mechanism for internalization of these costs into the system. It has been argued that resource rentals could become such a mechanism, as the rental payments from the industry could be used to compensate for these losses. There has been little discussion, however, about what kind of compensation is needed and who should be entitled to compensation. As a result of the problems faced by certain communities, a small quantity of quota was withdrawn by the government in 1999 and allocated on a five-year basis to companies in two of the most severely affected communities. The decision was strongly opposed by the quota owners in LÍÚ, who fear that this kind of "politically motivated measure" will undermine the ITQ system.

Whose Property? The Litigation Process

The somewhat confusing legal status of quota as "semiprivatized" has evoked complicated debates over the issues of taxation, depreciation, and the use of quota shares as collateral for credits. In which sense is it possible to buy and sell something that is legally defined as public property? And would such entities be liable to taxation? Should banks accept public (or national) property as collateral for private loans? Before 1991, the value of quota shares was treated as a part of the value of the vessel to which it was attached, since vessels and quota shares could not be separated. In consequence, quotas could be depreciated at the same rate as the vessels and were treated as collateral for loans in the sense that they contributed to the market value of the vessel. After 1991, the situation became somewhat unclear. In some cases investment in quota shares was considered as an expenditure and quota holdings were not treated as real capital, which meant that they could not be used as collateral. But in 1993, the Icelandic Supreme Court[6] found that quota holdings should be taxed as private capital[7] and could be depreciated by the same rate as copyrights (20 percent annually). After a law revision in 1998, however, depreciation of quota shares has not been permitted.

The use of quota as collateral is still not formally accepted, but the collateral problem has been solved by mutual agreements between banks and boat owners to ensure that in cases in which quota shares have been used as collateral for loans, quota shares and vessels cannot be separated without consulting the bank. It has proved difficult to uphold the paradoxical status of quota shares as public or national property according to the law but private property for all practical purposes, as illustrated by a Supreme Court decision in December 1998. A suit was filed by a fisherman who had applied for a fishing license and a catch quota but had been turned down by the ministry based on the fact that the fisherman in question had not been an owner of a fishing vessel during the early 1980s, when fishing experience became converted into fishing rights.

Considering the Icelandic constitution, which claims equal employment rights for every citizen,[8] and the Fisheries Management Act of 1990, which defines the fish resources as public property, the majority of the Court (three out of five judges) found the denial unlawful and unconstitutional. In short, the Court found that by introducing the ITQ system, the government had given away, as perpetual rights (that is, without time limits), to a group of people who happened to be the owners of active fishing vessels at a certain point in time, exclusive rights to the publicly owned Icelandic fish resources. Such an act could not be justified by the need to preserve the resources or by the public interest:

> It is undeniable that this arrangement [perpetual allocation of quota based on catch history in 1984] is a discrimination between those who acquired their fishing rights from vessel ownership at a certain point of time and those who have not been, and have no opportunity to be, in such position. Even if it may have been justified to employ such arrangements as a temporary measure to prevent a collapse of fish stocks, an issue that is not judged in this court case, one cannot see a logical necessity of making permanent the inequality that follows from paragraph 5 in law no. 38/1990 about allocation of fishing rights, by law for times to come. The charged [the state] has not substantiated that there are no alternative ways of attaining the goal of protecting the fish resources in Icelandic waters. By this law paragraph, most of the people is prevented, other requirements fulfilled, from enjoying the same employment rights in the fisheries, or comparable benefits from the common property fish resources as the relatively few individuals who were owners of active fishing vessels when the system of catch restrictions was initiated.[9]

The ITQ system as such was not considered unconstitutional; the constitutional problem was linked to the perpetual allocation of quota and consequently the permanent closure of the fisheries in favor of a "guild" of quota owners. The Court's decision primarily addressed the question of fishing licenses (paragraph 5 in the law), however, and despite the clear condemnation of the method of quota allocation, cited earlier, it did not draw a clear conclusion on the issue of quotas and their distribution.

The Court's decision caused a renewed debate about the future of the quota system; some argued that the system had to be changed to gradually recover state control over quotas and that future allocations should be for a limited term. The result, however, was that the fifth paragraph in the fisheries management law was revised only to allow for the granting of fishing licenses to all new vessels, with or without quota,[10] leaving the ITQ system intact. A year later, another quota case,[11] in which a vessel had delivered catches without possessing corresponding quota, passed the Low Court. Referring to the Supreme Court decision just discussed, the court concluded that the charge against the vessel owner was based on a law that had been found unconstitutional and consequently that the offense was not punishable. But in April 2000 the decision was overturned by the Supreme Court.[12] The case was heard by seven judges, and this time the majority of the Court (four out of seven) found that perpetual allocation of quotas was not against the constitution. Public interests in resource protection were found sufficient to justify restrictions in equal employment rights, and as quota holdings were not formally defined as private property, they could be changed or made conditional by the legislator. The Court majority also argued that it was up to the Parliament to decide what kind of measures were most suitable for resource protection and for promoting economic efficiency. A minority of the Court, consisting of the three judges who constituted the majority in 1998, dissented. Two judges referred mainly to the 1998 decision, but one also referred to the long-established common property rights to the fish resources and the interests and employment rights of people in fisheries-dependent communities.[13]

In retrospect, it seems unlikely that permanent allocation of quota would have been supported by Parliament if politicians had realized the scope and magnitude of the implications in 1990. With a more cautious approach, such as through a temporary allocation of quota for five to ten years, the government might have avoided the problematic legal situation created by permanent allocation. With the allocation of perpetual rights, it can be argued that a "point of no return" has been passed. Even if the allocated rights could in principle be taken back without compensation, the economic implications for those who have invested in quota shares would somehow have to be taken into consideration.

Consolidation without Consensus

During the 1990s, the Icelandic fisheries were transformed from a strictly regulated industry with units of production embedded within local communities into a globally oriented free-market industry with highly mobile units of production. This process was certainly not generated by the ITQ system alone: a wide range of liberalization policies have, in sum, created a free-market environment in the fisheries. The transformation of fishing rights into capital, represented by quota value, has been an important contribution to the present economic strength of companies with large quota holdings. In terms of export value of fish products and profits made by leading fisheries companies, there is little doubt that ITQ management has been a success. Icelandic fishing companies are expanding overseas and demonstrating their competitiveness in terms of technology and know-how. Quota holders have consolidated their position through a series of court cases that have reinforced the status of quota shares as de facto private property.

Despite this apparent success, the controversies caused by the system divided the industry and the public probably more than any other issue in Icelandic politics in the 1990s. Repeated polls among the public have shown that a majority of Icelanders are either skeptical of or opposed to the system in its present form. Criticism of the system is in essence aimed at its distributional effects. The initial allocation of ITQs led to a gratis distribution of valuable rights to certain families, and in some cases these families have enjoyed great windfall gains from selling out their shares. The Supreme Court decision in 1998 calling the legality of this procedure into question reinforced this criticism.

The distribution of economic, political, and negotiating power within the fisheries, as well as in society at large, has also been influenced by the system. As an association of quota owners, the vessel owners' association (LÍÚ) is in a superior position compared to the crewmen's unions and other stakeholders' organizations. By control over the quota, they are able to decide the terms of negotiation with other stakeholders, they can chose how and where they use quota, without consulting other fisheries stakeholders. The position of quota owners vis-à-vis crewmen can be compared to the position of a land owners vis-à-vis the peasants. The practice of working out the fisheries management policy through broad debates and through seeking consensus in the Fisheries Assembly is no longer viable. The former practice of having new legislation prepared by task forces with broad representation from different stakeholder groups has also been abandoned. The role of the Fisheries Assembly in policymaking has been diminished; since 1998 the assembly has ceased to pass resolutions on policy issues and is now reduced to an annual fisheries conference. In 1998, two task forces were appointed to work on the question of revising the quota system: The Resource Committee (Auðlindanefnd) and a "consensus committee" for reviewing the fisheries legislation, whose mandate was the rather difficult task of resolving the controversies over the fisheries management system and trying to restore the atmosphere of consensus and trust. Given this mandate, it is remarkable that no stakeholder representatives were appointed to the committee: the appointed members were politicians, lawyers and economists. The consensus committee was however unable to reach consensus. In September 2001, a majority of four of the committee's seven members delivered its recommendation: a further liberalization of quota transfers, combined with a resource tax that mainly replaces other taxes. The minority of three opted for gradual recovery of quota by the state for resale or leasing and clearly defining the public property rights to quota. The minority did not want to abandon the ITQ system as such, but to reverse the privatization process and operate an ITQ system in which fishing rights are allocated on contract basis for a limited term (Ministry of Fisheries 2001).

The abandonment of the policy of broad stakeholder involvement in fisheries policymaking is an interesting outcome of the ITQ system. There is reason to believe that this change of policy has been an inevitable outcome of a process of power transfer to the boat owners (quota owners) generated by the ITQ system. Apart from the increased differences and mistrust between stakeholder groups and between stakeholders and the government, the change can be explained by the shift in the balance of power between the LÍÚ, which represents the quota owners, and other stakeholder groups. The control of quota is a formidable asset, not only in economic terms, but also in terms of power. With the increasing differences in power base between quota holders and other stakeholders, it has become increasingly irrelevant to enter into mutual deliberations with equal voices, let alone to vote with equal votes in a common arena of stakeholders, such as the Fisheries Assembly before 1998. Quota owners are not dependent upon support from other groups, and they can chose with whom to cooperate. They also hold a strong position vis-à-vis the government, as they represent the largest source of private capital in Iceland. For the same reason as when eighteenth-century landlords found it irrelevant to deliberate with equal voices with the peasants, quota owners see little point in participating in deliberations.

One of the reasons why the ITQ system has survived may be that crucial decisions made at the early stages of the system (from 1984 to 1991) have proved increasingly difficult (and expensive) to reverse as time passes. Another complicating factor is that all major political parties and many stakeholder organizations have at one point or another been involved in the design of the system and have to a certain degree been co-opted during the decision-making process. The opposition to the ITQ system has not been homogenous, and there has been little agreement about what the alternative should be. In a poll among the general public, published in *Ægir*, the journal of the Icelandic Fisheries Association (1999), only 7.1 percent of the respondents wanted to keep the present system unchanged. Only 17.3 percent, however, wanted to abolish the quota system altogether. One third (33.3 percent) of the respondents favored some kind of regional allocation or "community quota." Almost one third (29.2 percent) were favorably disposed to either resource rentals or quota auction, and 10.5 percent wanted a special tax on quota transactions.

Feasible Alternatives?

Despite their critical attitude toward the system, the basic principle of fisheries management by some sort of transferable quotas now seems widely accepted among the Icelandic public. According to the results of the poll cited in the last section, a consensus solution should take into account the insecure situation of fishing communities, it should safeguard the income of fishing crew, and it should include payments of resource rentals, taxes, or cost recovery by those who have benefited from the system.

The question of resource rentals has been debated since the introduction of ITQs. Although such rentals are favored by many economists, the quota owners disapprove of them, and until recently they have not been broadly supported by the public. Public support has gradually increased as people have realized what kind of valuables have been handed over to quota owners.

The industry tends to view resource rentals as another tax, reducing the competitiveness of Icelandic fisheries compared to foreign competitors. It is also argued that resource rentals, as a special tax on fisheries, would be an unfair burden upon fisheries-dependent communities and regions. Although they are simple in theory, the practical implementation of resource rentals might prove more complicated, in economic as well as in political terms. In New Zealand, a resource rental regime was discontinued and replaced by a cost recovery program in 1997, as resource rentals proved "politically unfeasible" and unacceptable for the industry (Gaffney 1997).

Cost recovery can be viewed as a compensation from the industry for the costs of fishery-related services provided by public agencies, such as stock assessment, monitoring, and control. In other words, it means that in principle the industry purchases necessary services. From an industry point of view, the cost recovery approach may have more of an appeal than resource rentals, which are supposed to capture the resource rent generated by the industry. Cost recovery may also fit with ideas about devolution of fisheries management authority and increased responsibility for rights holders in the fisheries. Ultimately, a cost recovery regime might develop toward a takeover of management responsibilities by the quota owners themselves. The LÍÚ could gradually take control of research and management institutions, indirectly through funding policies or directly by purchasing the research and management "products" they would find suitable for their needs from whoever might be able to deliver them. A regime of resource rentals, payments for unspecified purposes to the state, based on the market value of quota, would be less likely to produce such an outcome, since there would be no direct link between the rentals and the management institutions and the services provided by them. The resource tax proposed by the consensus committee discussed in the previous section is a combination of a cost recovery tax and a tax on aggregate profits.

From a certain point of view, such development might be labeled comanagement or self-management by "fishermen," "user groups," or even "stakeholders" and as such seen as desirable. This would imply a narrow definition of stakeholders that includes only vessel owners. Although such a solution would take the burden of fisheries management off the shoulders of government, it would also facilitate further transfer of power and control from public agencies to quota owners. In other words, it would mean not only a privatization of the rights to harvest the fish resources, but eventually a privatization of stock assessment, management, and control as well.

Another alternative that has been put forward several times in the Icelandic debate, most recently in the minority proposal from the consensus committee, is that the state could annually recover a certain percentage of quota shares from the quota holders without compensation, as an alternative to resource rentals. The "real owner" of quota shares, the public represented by the state, would then take back a certain portion of allocated quota on annual basis and thus gradually regain control over its rightful property. The recovered quota could then be resold at a market price for a limited term or redistributed by other criteria to communities or regions to reestablish access to the fish resources for marginalized fishing communities. A gradual recovery of quota shares to the state could allow the industry to adapt to the change. It could also be a way of clarifying the legal status of quota shares and of dealing with the most disturbing distributional effects of ITQs. Marginalized communities could be offered compensation in the form of quota or payments, and the basic principles of the system would remain intact. Although the industry is unlikely to actively approve of this solution, it might become a basis for a broader consensus on the quota issue.

Conclusion

Radical institutional reforms, especially reforms that involve a major redistribution of wealth, power, and income among the citizens of a democratic society, are bound to generate overt conflicts and fierce political debates when introduced. The implementation of fishery resource management through ITQs in Iceland is no exception. As Iceland was one of the first countries to adopt ITQs, it was difficult to predict all the effects of the system at the point when crucial decisions were made. The knowledge status of stakeholders, politicians, and the public however, improved gradually as experience was accumulated. As the reform entered new phases, new issues of conflict surfaced, and the sometimes exhausting public debate continued. Both conflict and consensus depend upon a certain degree of shared conceptualization of a situation; participants must share an understanding of what their agreement or disagreement is about—otherwise, communication is difficult. During the process in Iceland, the content and orientation of the debate changed as the conceptual framework for communication about the system developed.

The process of learning to understand the dynamics of the ITQ system and adapting to them involved not only fishermen, but also participants at different levels of society: communities, regions, trade unions, companies, political parties, experts and courts of law. In a sense, the 1990 Fisheries Management Act turned Iceland into a test site for a market-based fisheries management system. It provided a basis for a quota "stock market" that continuously redistributes fishing rights among vessel owners located in different communities and regions. Some of the effects of the system had been predicted by economists and policy-makers, but the ITQ system also produced some unexpected side-effects. One of these was the negative impact on crew income that followed from the practice of contract fishing.

The roots of this problem are in the design of the ITQ system: the transformation of fishing rights into capital. The rent on quota capital, the resource rent, has become a cost that affects the economy of the fishing operations. Unlike the Ricardian land rent, which was assumed to differ according to the productivity of the rented land,[14] the quota rent is also affected by demand created from idle fishing capacity, unemployment, and municipal interventions (Eythórsson 1996a). The rent is high if quota is in high demand. Excess capacity in the form of vessels with little or no quota of their own leads to demand for quota on leasing or contract terms, and as in any other market, high demand means high prices.

Proponents of ITQs often argue that they allow for smooth structural adjustments, as owners of inefficient vessels are compensated for leaving the fisheries when they sell their quotas. But vessel owners are not the only people who have invested in the fisheries. Those who have put their lifetime savings into building a home in a fisheries community and have paid municipal taxes to build common infrastructure also find their livelihood punctured when companies leave and take their fishing rights with them. In contrast to boat owners, fish workers, crew members, and other community residents hold no valuable rights. As Mikalsen and Jentoft (2001) put it, they have urgent and legitimate stakes in the fisheries, but their relative power is weak compared to that of the quota owners. Consequently, they get no compensation if the rights holders find that the community is no longer necessary for their operations. Neighbors who more or less considered themselves equals during the 1970s and 1980s, have since found themselves in totally different positions. Whereas the quota owner has valuable assets that allow for a comfortable retirement, his neighbor has lost both his livelihood and his lifetime savings that were placed in a house that is now impossible to sell.

The finding that a broad stakeholder involvement in policymaking has become very difficult under ITQs is also interesting. The system of broad stakeholder consultations in the Fisheries Assembly and in government committees, a system that McCay and Acheson (1987) characterized as comanagement, has now become history in Iceland. The reason, expressed within the conceptual framework of Mikalsen and Jentoft (2001), is that within the group of definitive stakeholders, some have become more definitive than others.

ITQs and similar measures that involve gross redistribution of property, income, and power in society cannot be considered merely technical arrangements for improving economic efficiency and resource conservation. The ethical and political values involved in ITQs were poorly addressed in the decision-making process in Iceland. The relatively broad support for the system at the early stages seems to reflect a lack of awareness among the stakeholder representatives, rather than a conscious choice among these values. Studying the Icelandic experience should offer a better opportunity to make informed choices concerning ITQs as a management tool in the fisheries and in management of other common-property resources.

In chapter 7, Alexander Farrell and M. Granger Morgan discuss the prospects of regional or global institutions for issuing tradable emissions allowances for greenhouse gases. Several aspects of the experiences with ITQ management in the fisheries in Iceland and New Zealand are also relevant to such institutions. The Icelandic experience demonstrates the need for carefully defining the legal status and durability of tradable allowances to avoid creating a privileged group of first owners who can make a fortune from perpetually allocated allowances by leasing them to new entrants.

Notes

1. The official name is the *Merchant, Navy and Fishing Vessels Officers' Guild.*

2. The first paragraph of the act reads as follows: "Nytjastofnar á Íslandsmiðum eru sameign íslensku þjóðarinnar. Markmið laga þessara er að stuðla að verndun og hagkvæmri nýtingu þeirra og tryggja með því trausta atvinnu og byggð í landinu" (The fish stocks on Iceland's fishing grounds are a common property of the Icelandic Nation. The aim of this law is to further protection and efficent utilization of these and thereby ensure secure employment and settlement in the country).

3. Approximately 6.4 percent of the total fishing tonnage in 1994.

4. Transactions were mediated by the QEM; buyers and sellers were not supposed to know about each other.

5. Interviews with H. Jónsson (secretary of SSÍ) and B. Valsson (secretary of FFSÍ) in September 1999.

6. Case no. 291/1993.

7. An interesting aspect of the taxation question is that quota shares that have been bought and paid for are liable to taxation, whereas quota shares that were allocated gratis in 1984 and have not yet changed hands are not liable to taxation.

8. This refers to paragraph 75 in the constitution. The right of every citizen to freely choose occupation was established in the first Icelandic constitution of 1874, which was modeled after the Danish constitution of 1849. This right was originally established by liberal governments in Europe to abolish the monopoly to certain occupations held by the closed guilds.

9. Decision no. 145/1998: *Valdimar Jóhannesson versus The State of Iceland.* The full text is found at <http://www.kvotinn.is/domur.htm>. The translation is by the author. The original text is as follows:

Er óhjákvæmilegt að líta svo á, að þessi tilhögun feli í sér mismunun milli þeirra, sem leiða rétt sinn til veiðiheimilda til eignarhalds á skipum á tilteknum tíma, og hinna, sem hafa ekki átt og eiga þess ekki kost að komast í slíka aðstöðu. Þótt tímabundnar aðgerðir af þessu tagi til varnar hruni fiskistofna kunni að hafa verið réttlætanlegar, en um það er ekki dæmt í málinu, verður ekki séð, að rökbundin nauðsyn hnígi til þess að lögbinda um ókomna tíð þá mismunun, sem leiðir af reglu 5. Gr. Laga nr. 38/1990 um úthlutun veiðiheimilda. Stefndi hefur ekki sýnt fram á að aðrar leiðir séu ekki færar til að ná því lögmæta markmiði að vernda fiskistofna við Ísland. Með þessu lagaákvæði er lögð fyrirfarandi tálmun við því, að drjúgur hluti landsmanna geti, að öðrum skilyrðum uppfylltum, notið sama atvinnuréttar í sjávarútvegi eða sambærilegrar hlutdeildar í þeirri sameign, sem nytjastofnar á Íslandsmiðum eru, og þeir tiltölulega fáu einstaklingar eða lögaðilar, sem höfðu yfir að ráða skipum við veiðar í upphafi umræddra takmarkana á fiskveiðum.

10. Previously, new vessels could receive a license only if the owner removed old tonnage corresponding to the tonnage of the new vessel from the fisheries.

11. The case, referred to as "Vatneyrarmálið", arose when the vessel Vatneyri BA delivered excess catch on purpose to try the issue of quota allocation in court.

12. Decision no. 12/2000: *The Prosecutor versus Björn Kristjánsson, Svavar Rúnar Guðnason and Hyrnó Ehf.* The entire text is found at <http://www.haestirettur.is>.

13. "Sératkvæði Hjartar Torfasonar" (The Vote of Hjörtur Torfason).

14. "Metaphors of Property: The Commoditisation of Fishing Rights" (Eythórsson 1998) offers a further discussion of the concepts of resource rent and land rent.

References

Anderson, E. N., Jr. 1987. "A Malaysian Tragedy of the Commons." In *The Question of The Commons: The Culture and Ecology of Communal Resources*, ed. Bonnie McCay and James M. Acheson, 327–343. Tucson: University of Arizona Press.

Anonymous. 1993. "Aðalfundur Samtaka fiskvinnslustöðva í Stykkishólmi. Ársverkum í fiskvinnslu fækkað um 2,500 á 6 árum. Sjávarútvegsráðherra kynnti frumvarpið um þróunarsjóðinn á fundinum" (Annual meeting of The Organisation of Fish Processors. A reduction of 2,500 jobs in fish processing in 6 years. The Fisheries Minister provided information about the Development Fund at the meeting). *Morgunblaðið* (news daily), Sept. 23. 1993.

Árnason, Ragnar. 1991. "Efficient Management of Ocean Fisheries." *European Economic Review* 35:408–417.

Árnason, Ragnar. 1995. *The Icelandic Fisheries: Evolution and Management of a Fishing Industry.* Oxford: Fishing News Books/Blackwell Science.

Árnason, Ragnar. 1997. "The Icelandic ITQ System and Its Consequences." In *Property Rights in the Fishing Industry*, ed. Gudrún Pétursdóttir, 71–101. Reykjavík: University of Iceland Press.

Copes, Parzival. 1994. "Individual Fishing Rights: Some Implications of Transferability." In *Proceedings of the Sixth International Conference of the International Institute of Fisheries Economics and Trade*, ed. M. Antona, J. Catazano, and J. G. Sutinen. Issy-les-Moulineaux, Framce: French Research Institute for Exploitation of the Sea (IFREMER).

Eythórsson, Einar. 1996a. "Theory and Practice of ITQs in Iceland; Privatization of common fishing rights." *Marine Policy* 20(3):269–281.

Eythórsson, Einar. 1996b. "Coastal Communities and ITQ Management. The Case of Icelandic Fisheries." *Sociological Ruralis* 36(2):212–223.

Eythórsson, Einar. 1998. "Metaphors of Property: The Commoditisation of Fishing Rights." In *Northern Waters: Management Issues and Practice*, ed. David Symes, 42–51. Oxford: Fishing News Books/Blackwell Science.

Gaffney, Kaitilin R. 1997. "Property Rights Based Fisheries Management: Lessons from New Zealand's Quota Management System. Master's thesis. Wellington, New Zealand: Victoria University.

Garðarsson, Halldór Jón. 1999. "Kvótakerfið. Áhrif þess á þróun lýðræðis" (The quota system. Its impact on the development of democracy). Bachelor's thesis. Reykjavíc: University of Iceland.

Guðjónsson, Stefán Broddi. 1999. "Hver verður stæstur? Uppstokkun í sjávarútvegi á fullri ferð (Who is going to be biggest? Restructuring the fisheries in a high speed). *Viðskiptablaðið* (Business Weekly), Sept. 8–Sept. 14, 4–9.

Hannesson, Rögnvaldur. 1991. "From Common Fish to Rights Based Fishing: Fisheries Management and the Evolution of Exclusive Rights to Fish." *European Economic Review* 35:397–407.

Helgason, Agnar. 1995. "The Lords of the Sea and the Morality of Exchange: The Social Context of ITQ Management in the Icelandic Fisheries." Master's thesis. Reykjavík: University of Iceland.

Jónsson, Halldór. 1990. "Ákvarðanataka í sjávarútvegi og stjórnun fiskveiða" (Decision making in the fisheries and fisheries management). *Samfélagstíðindi* 10:99–141.

Jónsson, Hólmgeir. 1999. "það stefnir í átök enn á ný (Another conflict is in the making)." *Fiskifréttir*, March 5.

McCay, Bonnie, and James M. Acheson. 1987. *The Question of the Commons: The Culture and Ecology of Common Resources*. Tucson: University of Arizona Press.

Mikalsen, Knut H., and Svein Jentoft. 2001. "From User-Groups to Stakeholders? The Public Interest in Fisheries Management." *Marine Policy* 25:281–292.

Ministry of Fisheries (Sjávarútvegsráduneytið). 2001. *Skýrsla nefndar um endurskoðun laga um stjórn fiskveiða* (A report from a committee for revising the law on fisheries management). Reykjavík, Iceland: Ministry of Fisheries.

Pálmason, Snorri Rúnar. 1992. "Allmenn tragedie på Island?" (A common tragedy in Iceland?) Cardidate thesis. Tromsø: Norwegian Fisheries College.

Pálsson, Gísli, and Agnar Helgason. 1996a. "Figuring Fish and Measuring Men: The Individual Transferable Quota System in the Icelandic Cod Fishery." *Ocean and Coastal Management* 28(1–3):117–146.

Pálsson, Gísli, and Agnar Helgason. 1996b. "Property Rights and Practical Knowledge: The Icelandic Quota System." In *Fisheries Management in Crisis*, ed. Kevin Crean and David Symes, 45–60. Oxford: Fishing News Books/Blackwell Science.

Pálsson, Gísli, and Agnar Helgason. 1999. "Kvótakerfið, kenning og veruleiki" (The quota system, theory and practice). *Skírnir 1999* (periodical). Reykjavík.

Pinkerton, E. 1987. "Intercepting the State: Dramatic Processes in the Assertion of Local Comanagement Rights." In *The Question of The Commons: The Culture and Ecology of Communal Resources*, ed. Bonnie McCay and James M. Acheson, 344–369. Tucson: University of Arizona Press.

Runólfsson, Birgir. 1997. "The Icelandic Fishing Industry: A Descriptive Account." Paper presented at Symposium on the Efficiency of North Atlantic Fisheries, Reykjavík, September 12–13.

Runólfsson, Birgir. 1999. *Sjávarútvegur Íslendinga. þróun, staða og horfur. Skýrsla til sjávarútvegsráðherra tekin saman í tilefni að ári hafsins 1998* (Icelandic fisheries. Development, status and prospects. A report for the Fisheries Minister 1998).

Runólfsson, Birgir, and Ragnar Árnason. 1999a. *Evolution and Performance of the Icelandic ITQ System*. Available at <http://www.hi.is/bthru/iceitq1>.

Runólfsson, Birgir, and Ragnar Árnason. 1999b. "Individual Transferable Quotas in Iceland." In *Fish or Cut Bait: The Case for Individual Transferable Quotas in the Salmon Fishery of British Columbia*, ed. Laura Jones and Michael Walker. Fraser Instutute, Books Online. Available at <http://www.fraiserinstitute.ca/publications/books/fish/fish2.html>.

Squires, D., J. Kirkley, and C. A. Tisdell. 1995. "Individual Transferable Quotas as a Fisheries Management Tool." *Reviews in Fisheries Science* 3(2):141–169.

Útvegur (a yearbook of fisheries statistics). 1984–1997. Reykjavik: Icelandic Fisheries Association.

Valsson, Benedikt. 1999. "Skuldir sjávarútvegs" (The debts of the fisheries). *Morgunblaðið*, September 22.

7

Multilateral Emission Trading: Heterogeneity in Domestic and International Common-Pool Resource Management

Alexander E. Farrell and M. Granger Morgan

Introduction

Marketable emissions allowance systems[1] have been proposed as efficient means of managing emissions of carbon dioxide (CO_2) and other greenhouse gases (GHGs) to mitigate global climate change. The U.S. Acid Rain Program to control sulfur dioxide (SO_2) emissions from power plants is often held up as an example for international CO_2 control efforts (Solomon 1995, 1999; Stavins 1997). There are serious limitations to this example, however, including the fact that the Acid Rain Program was developed within a national framework of domestic environmental laws, whereas any agreement on GHG control will need to be agreed to by several (perhaps most) nations of the world, which vary enormously in terms of wealth, political culture, and many other characteristics (Fort and Faur 1997; Victor 1991; Grubb, Vrolijk, and Brack 1999, 210–213). In this chapter, we examine two examples of efforts to establish air pollution emission trading programs among several states in the eastern United States that present different and perhaps less limited insights for international GHG emission trading than those available from the Acid Rain Program since they were not developed through a system of central government. In particular, this chapter examines heterogeneity among political jurisdictions attempting to come to agreement on a joint approach to environmental protection by looking at how variation among the states involved affected these two efforts.

The atmosphere and the rest of the climate system can be characterized as a common-pool resource (CPR), which implies that their management may be problematic, since many different actors must negotiate and agree on any management scheme. A particularly difficult issue in the development of a management scheme is the heterogeneity of the actors. The literature on CPR dilemmas is fairly large (see Ostrom et al. 1999 for a review), but only recently has the issue of heterogeneity among the actors in CPR disputes been discussed in any detail (Connolly 1999; Hackett 1992; Mitchell 1999; Schlager and Blomquist 1998).

Several relevant hypotheses have been generated from this prior research. First, the position of each actor with respect to the resource itself and on other dimensions can vary significantly, so that some actors are advantaged and others disadvantaged. Schlager and Blomquist (1998) give hypothetical examples of institutionally differentiated actors and deduce the outcomes but present no examples. Second, Mitchell (1999) shows that for a variety of reasons CPR dilemmas are likely to be more common and more difficult to manage in the international domain than in the domestic. Third, Connolly (1999) shows that an important feature of negotiations about common resources is how the perception of self-interest can change depending on whether an actor favors developing CPR use or not. These authors provide some evidence for their hypotheses but none for cases of multilateral emission trading.

Illustrative examples of how such multilateral marketable emissions permit schemes may or may not emerge in the heterogeneous international setting come from the interstate markets for the control of nitrogen oxides (NO_x) in the eastern United States. These programs, designed to combat smog in the eastern part of the country include one successful example, the Ozone Transport Commission (OTC) NO_x Budget, and one highly troubled example, the NO_x State Implementation Plan (SIP) Call (Farrell, Carter, and Raufer 1999; Farrell 2000; Environmental Protection Agency 1998c). Comparisons between these examples and potential international emission trading are of course limited by the fact that they were developed within a federalist political structure, but the examples do have the key desirable feature that makes them somewhat parallel to the international case: they could be implemented only through the voluntary efforts on the parts of states (i.e., the federal government could not impose a NO_x emission trading program). As is argued below, both of these examples had relatively favorable conditions for the creation of a multilateral emission trading system, yet only one of them succeeded. Explaining the difference in outcomes here is thus important if we are to have any hope of understanding the potential for international GHG emission trading.

Finally, we note that using the conceptual framework laid out by Schlager and Ostrom (1992), this chapter examines what they identify as the constitutional level of action, in which the methods to devise collective-choice rules are decided upon, but not the rules themselves. In particular, we are interested in how different jurisdictions can agree to create and govern a multilateral emission trading system, but not what sort of jurisdictions should be participants in that system.

Emission Trading

Several types of marketable emission trading systems exist.[2] We will focus on the "cap-and-trade" variety in this chapter because the most successful examples use this type of system and because Article 17 of the Kyoto Protocol essentially envisions such a system (Farrell 2000; Klier, Mattoon, and Prager 1997; Stavins 1997).[3] In cap-and-trade programs, regulated firms are allocated a fixed number of allowances and are required to redeem one allowance for every ton of pollution emitted. Each firm's allocations cover a level of emissions that is smaller than its historical level of emissions, so regulated firms have four basic options: (1) control emissions to match their allocation, (2) "undercontrol" and buy allowances to meet the redemption requirement, (3) "overcontrol" and bank allowances for use in future years (when even fewer allowances may be allocated), or (4) overcontrol and then sell their excess. Cap-and-trade systems have gained support over traditional command-and-control regulations from various actors because they greatly improve the likelihood of meeting emission reduction goals and at the same time are more flexible and lower in cost than traditional approaches. Several practical considerations must be accounted for, however. It must be possible to:

1. *Define and accurately measure the pollutant(s)* of concern, their sources (both natural and anthropogenic), and their atmospheric fate and transport (i.e., understand the science).
2. *Agree on the quantity of emissions* that will be allowed (the cap), a value that typically declines over time.
3. *Account for differences* (if any) in environmental impact for all pollutants being traded to normalize damages across emission locations and times and across pollutant types.
4. *Create emission allowances, a distribution mechanism, and an enforceable redemption requirement* for regulated sources such that they

can obtain, but then must surrender to the government (i.e., redeem), one allowance for each unit of emission they release.
5. *Operate a market* with enforceable contracts and rules that assures competitive behavior.

It is important to recognize that there are roles for both private and public actors, and for both technical and political factors, in meeting the requirements set out above. For instance, measuring emissions is essentially a technical issue, whereas deciding on allocation allowances is essentially a political one, and accounting for differences is both. Similarly, government takes most of the steps listed above, but both government and private industry are crucial to establishing an effective market. Actors negotiating a common-property regime may specify other requirements beyond those listed above. For example, to obtain agreement among the actors, it may prove necessary to demonstrate some minimum level of "burden sharing," that is, to show that all emitters are doing something to reduce their own pollution, not just buying permits from others, even if this raises overall costs somewhat.

Several outcomes of existing cap-and-trade programs are worth noting. First, all the U.S. examples have a highly coercive character: government, not private industry, determines (sometimes through the legislative and rule-making processes at the federal level, occasionally by similar processes at the state level, and sometimes through interstate negotiations) the rules for the emission trading programs and specifies what counts as a regulated source. Of course, the views and concerns of private industry are taken into account by government, but in the end the government has the final say. Indeed, cap-and-trade programs can be construed as simply the most flexible form of command-and-control regulation, and many of the cost savings observed in the SO_2 example are due to this new flexibility alone (Burtraw 1996).

Second, the property of ensuring absolute emissions limits, given adequate monitoring and enforcement provisions, is usually ascribed to cap-and-trade programs. One of the earliest applications of a cap-and-trade framework, the phaseout of lead from gasoline in the United States provides an example of the types of problems that could arise along these lines (Loeb 1990; Nichols 1997).[4] The main problem in this program was overreliance on self-reported data to ensure compliance with the program requirements. As a result subsequent cap-and-trade programs in the United States have had very strict monitoring requirements and have been very successful in reducing emissions. An important deviation on this point is worth noting: because of the potential for very high costs in any CO_2 cap-and-trade program, price caps have been suggested (Kopp et al. 1999; Victor 2001). In such an approach, sometimes referred to as a "safety valve" approach, the government would print new allowances and sell them at a fixed price that would escalate over time. This would invalidate the cap unless the sale price rose beyond the cost of emissions control but would retain many of the desirable flexibilities of emission trading programs.

Third, in all of the cap-and-trade programs implemented so far in the United States, the allowances themselves have been distributed free of charge to existing sources (a practice called grandfathering), generally based on previous emissions. The advantage of this approach is that politicians can literally use allowances as bargaining chips to help arrange the necessary support to pass legislation enacting such systems (Joskow and Schmalensee 1998).

Potential for an Emission Trading Program for Greenhouse Gases

The nations of the world are currently in the process of negotiating what may become an international regime for the control of GHG emissions, through the Kyoto Protocol to the United Nations Framework Convention on Climate Change (UNFCCC). The Kyoto Protocol contains several "flexibility mechanisms" that can be interpreted as authorizing various sorts of international emission trading. The approach that is most relevant to this discussion would follow from Article 17 and primarily involve industrialized countries (Grubb, Vrolijk, and Brack 1999, 89–96, 194–217). This approach involves only the industrialized nations of the world, which are those that are identified in Annex 1 of the (still unratified) protocol as having obligations to reduce GHG emissions.

Emissions Trading under the Kyoto Protocol The United States is one of 160 signatories to the UNFCCC, which was ratified by the Senate on October 7, 1992. Although the agreement commits the nations of the world to work to stabilize atmospheric concentrations of GHGs at a level that will "prevent dangerous anthropogenic interference with the climate system," it specifies no quantities, timetables, or strategies. These were worked out in a series of conferences of the parties (COP) meetings, the earliest of which are analyzed in detail by Grubb, Vrolijk, and Brack (1999). Various environmental organizations monitor COP meetings on an ongoing basis (see, e.g., <http://www.iisd.ca/linkages>, which contains detailed reports on each COP session). The negotiations in the COP sessions led, in December 1997, to the Kyoto Protocol, which calls for the developed (or Annex 1) nations to collectively reduce their GHG emissions in the 2008–2012 period to a level about 5 percent below 1990 emission levels, although each nation is given a different specific target. For instance, the U.S. target was 7 percent, whereas the European Union accepted an 8 percent cut, which was subsequently divided up (very unevenly) by member nations during the June 1998 Environment Council. For example, under this "Burden Sharing Agreement," France, Germany, and Greece accepted emission allocations representing 0 percent, –21 percent, and +25 percent changes, respectively, in GHG emissions from business as usual. Although the Kyoto Protocol sets targets for six GHGs, for the most part efforts to develop emission control programs focus on CO_2 emissions (Reilly et al. 1999).

Several provisions were included in the Kyoto Protocol to provide flexibility in how the targets are met. The first of these is a provision for bilateral or multilateral emissions trading for Annex 1 countries that essentially follows the cap-and-trade approach. The second is a pair of programs that basically follows the emission reduction credits design (see note 2), called joint implementation (JI) and the clean development mechanism (CDM). This pair of programs is designed to involve developing (i.e., non–Annex 1) countries in the emission reduction effort on a project-by-project basis. As we discuss below, however, inclusions through these programs create fundamental difficulties for the design and implementation of trading systems.

Since the Kyoto Protocol was signed, the signatory nations have held several further COP meetings to hammer out the details of how the agreement will be implemented. The signatory nations (except for the United States, which abandoned the process in the spring of 2001) essentially resolved all the major implementation issues by the end of COP-7 in Marrakech in November 2001. One significant change during the

Kyoto-to-Marrakech period was that the emission reduction commitments of the Kyoto Protocol have become much less burdensome for the nations that have remained part of the agreement. Two factors account for this. First, the departure of the United States, which was expected to be the largest (by far) net buyer of CO_2 emission allowances or credits, has decreased the demand for Annex 1 allowances or JI/CDM credits. Second, changes in how terrestrial carbon sinks (e.g., forests) are accounted for has dramatically raised the number of Annex 1 emission allowances that several participating nations can now sell (e.g., Russia, Canada, and Australia). These two effects have lowered the expected price of a CO_2 emission allowance from about $100 per ton of carbon (with the United States in) to about $10 per ton of carbon (with the United States out) and have also made it less likely that JI and CDM will be used very much. Thus possible net beneficiaries under the original agreement (e.g., Russia and large South American and African nations) will have less near-term interest in the Kyoto Protocol.

Early Action in the United States Even though the United States signed the Kyoto Protocol, it was never likely to ratify it, given the Senate's 95–0 passage of the Byrd-Hagel resolution, which essentially rejected it even though President Clinton did not submit it for a ratification vote.

Early in the Clinton administration a voluntary program was started through the Department of Energy to attempt to meet the Kyoto targets (Clinton and Gore 1993). This plan, however, had little effect: U.S. GHG emissions through 2000 continued to increase, and projections show a similar trend (Energy Information Administration 2001; Environmental Protection Agency 1998a). Following its abandonment of the Kyoto Protocol, the Bush administration has suggested that the United States may attempt other approaches to reducing GHG emissions, but none have been forthcoming through 2001 (Bush 2001). The administration's energy policy devotes less than one page to climate change and stresses scientific and technological research, with no mention of limiting emissions (National Energy Policy Development Group 2001). The subsequent "Clear Skies" initiative proposes a voluntary approach to current trends (Bush 2002). Nonetheless, there have been some limited actions in the United States toward GHG control policies.

Several bills have been introduced in Congress to grant credit in any future regulatory system to firms that undertake control actions today (Nordhaus and Fotis 1998). Although in principle such "credit for early action" sounds like a good idea that might get the country moving while Congress slowly builds the political confidence to act, a look at the details leaves one far less confident. Most current proposals would create complex auditing and accounting systems that in some cases would treat different industrial sectors differently. In the interest of giving credit for actions taken now, they would impose substantial constraints on the freedom of action available in the future design of a national regulatory program. In some proposals, out-of-country activities similar to the JI and CDM actions described in the Kyoto Protocol would be allowed to create emission credits. The more complexities and differentiated sectoral treatment in a domestic program (whether imposed through legislation directly or by subsequent regulatory actions), the greater the difficulties of integrating that program into any international emission trading system that is developed. In addition, depending upon how regulatory arrangements develop subsequently, such credits could constitute a very large wealth transfer to those who earn early credits.

There has also been activity on GHG controls in state capitals, some of which have moved before the federal government on local and regional pollution issues in the past. Bills that would require firms to monitor or control CO_2 emissions have been introduced in a few state legislatures, although these are widely acknowledged as symbolic. (See also the discussion later in the chapter on the Conference of New England Governors' international involvement.)

Early and Voluntary Efforts Internationally To prevent global climate change, worldwide emissions of GHGs will have to be controlled, to perhaps one tenth of the levels they would otherwise reach, but policies to control GHGs do not have to *start* on a global basis. They do not even have to start on an international level. It may be far better to build them from the bottom up, from national or subnational efforts (Morgan 2000). Variations on this theme are already becoming evident as GHG control polices have begun to emerge without an international agreement. Several nations have already implemented internal taxes based on CO_2 emissions, and one subset of Annex 1 countries (the European

Union) has proposed what can best be called a system for linking national cap-and-trade CO_2 programs to allow for international trading. (Many research- and information-oriented climate policies have been developed as well, but these are ignored here.)

Starting in 1990, Denmark, Finland, the Netherlands, Norway, and Sweden began introducing taxes based on carbon content of fuel consumption or based directly on CO_2 emissions. This type of approach tends to be rather heterogeneous from country to country, and such taxes tend to be introduced as other national taxes (which themselves vary among nations) are reduced. For instance, Sweden introduced an energy tax and a CO_2 tax in legislation in 2000, somewhat offsetting reductions in labor taxes. Norway's CO_2 tax, which started in 1991, covers more than 60 percent of national emissions and partly substituted for an oil depletion tax. (Norway is a large oil exporter.) In 1992, the Netherlands changed the basis of an existing environment tax from production to carbon emissions and energy consumption, and has introduced additional relevant taxes since them. All of these taxes have provisions for significant exemptions, however.

National-level CO_2 emission trading policies have also emerged. Denmark was first, beginning a CO_2 cap-and-trade program for its electricity sector in 2001. The program is designed to reduce CO_2 emissions from that sector by 21 percent from 1990 levels. It replaces a very strong command-and-control regulatory regime in which the government could more or less order electricity companies to use specific technologies and fuels.

The United Kingdom started up a CO_2 emission trading system in the spring of 2002. This is a voluntary program, but preexisting commitments and financial incentives are also used to encourage participation. The system will run in 2002–2006 and use 1998–2000 emissions as the baseline. The preexisting commitments are "climate change agreements" that the U.K. government has entered into with some companies in the past. The U.K. CO_2 system gives these companies a more flexible way of complying with these commitments by establishing a rather typical emission reduction credit system. The financial incentives consist of a fund of £215 million (a little over $300 million), which will be used to pay companies voluntarily reduce emissions. That is, in this part of its system, the U.K. government has created an emission reduction credit program for CO_2 emissions in which it is the purchaser. Companies covered by a climate change agreement or those that produce electricity for consumption off-site are ineligible for these funds.

The European Union recently proposed what can best be called a system for linking national cap-and-trade CO_2 programs to allow for trading among EU member countries (European Commission 2001). This is an important step: previously the European Union had opposed the use of emission trading. The proposed EU system applies to CO_2 emissions from specified large sources: electricity generation, petroleum refining, and the manufacture of iron, steel, cement clinker, ceramics, glass, and pulp and paper. Oil and gas production, solid-waste incineration, and chemical manufacturing are not included, nor are transportation or residential energy consumption. The proposed system would allow for trading among companies, to be tracked by national governments, which would establish registries for CO_2 emission allowances. A trade across a national boundary would require offsetting entries (one addition and one subtraction) in the two national registries involved.

The EU proposal is also important because it sets important precedents for international CO_2 emission trading: it would create the largest CO_2 emission market in the world, and other countries that wanted access to that market would probably have to follow the procedures set down by the European Union. A key provision is that emission credits created by JI and CDM activities could not be used in the EU cap-and-trade system. This feature could complicate efforts to include developing (non–Annex 1) countries or developed countries that allow the use of emission reduction credits, and it obviously creates a compatibility problem for the EU and U.K. systems. Further compatibility problems may arise from the differences in preexisting energy or CO_2 taxes within the EU member nations.

In North America, numerous experiments with voluntary and mandatory GHG control policies are being undertaken at the subnational level. Over thirty cities and counties in the United States have set GHG control targets and implemented local action plans. Local actions typically consist of monitoring and reporting emissions, changes to building codes, and funding for efficiency and renewable energy projects. A few states (e.g., Oregon and Massachusetts) have passed or may soon pass legislation requiring emission reductions from selected source categories (typically electric power plants) that can meet their requirements through various emission trading mechanisms while providing incentives or standards for other sectors. Further, the Conference of New England Governors and Eastern Canadian Premiers, which has a successful track record of international cooperation to promote acid rain policy, issued a resolution in July 2000 that defined global warming as a regional concern and as mostly associated with the combustion of fossil fuels (Conference of New England Governors and Eastern Canadian Premiers 2000). It subsequently issued a wide-ranging Climate Change Action Plan that adopted a short-term goal of returning GHG emissions to 1990 levels by 2010, with subsequent reductions to follow (Conference of New England Governors and Eastern Canadian Premiers 2001). This plan recommended creating a standard GHG emission inventory, conducting several studies and public education efforts, promoting renewable energy and efficiency, and developing a common set of rules for a multilateral GHG emission trading program. Although many of these activities are based on nonbinding "goals," some are mandatory and have significant costs. These first steps that indicate that at the subnational level things are beginning to happen with respect to GHG emission reduction in North America, and they demonstrate a trend toward using multi-lateral emissions trading.

In addition, a few industrial firms have begun to experiment with emission trading. Most notably, BP-Amoco has started an effort to control CO_2 emissions and has decided to employ an internal (business unit–to–business unit) emission trading program to do so. Some financial-services companies (generally those already involved in the U.S. pollutant emission allowance markets) have begun to facilitate emission trades of various sorts, usually bilateral deals between a U.S.-based or transnational firm and an organization (often associated with a national government) in a less developed nation. Other efforts have also begun. For instance, one Canadian and four U.S. electricity companies recently formed Energy for a Clean Air Future group, which is proposing multi-pollutant legislation in the United States that would reward voluntary CO_2 emission reductions.

Types of Heterogeneity The applicability of a cap-and-trade system within this context is summarized in table 7.1, based on the five criteria outlined at the beginning of this section. Two types of difficulties in applying such systems are apparent, both instances of heterogeneity, a factor that can complicate the management of CPRs and emission trading systems generally (Ben-David et al. 1999; Hackett 1992; Schlager and Blomquist 1998).

The first type of difficulty arises from heterogeneity among the actors, which can be characterized as differences in capabilities (assets), in preferences, in information and beliefs, and in decision making (Keohane and Ostrom 1994). In this case, heterogeneity springing from differences in capabilities is possibly the most important. The nations of the world vary enormously with respect to historical GHG emissions, population, level of development, economic output, and other parameters. Since GHG emission allowances (or requirements to reduce emissions) are thought to be very valuable (or costly, in the case of reduction requirements), arguments about the appropriateness of allocation arrangements based on any of these parameters are highly contentious (Baer et al. 2000). Other sources of heterogeneity among the actors are also important, however, such as differences in how important various jurisdictions judge climate change is and even how reliable they feel climate change–related information is. Because international emission trading may bring significant savings, it is likely to be sought, highlighting the importance of heterogeneity due to differences in rules and decision making, since property rights and trade rules are not uniform globally. For instance, the EU proposal discussed earlier prohibits the use of JI or CDM credits, conflicting with the U.K. program and the interests of developing countries and possibly with a domestic U.S. program (should one be developed).

The second type of difficulty arises from heterogeneity of the components of the physical system in which the CRR is embedded, as well as current scientific understanding of those components. For instance, in the case of climate, many of these difficulties stem from complexities and uncertainties regarding GHGs themselves. The GHGs vary extremely widely in atmospheric lifetime and in warming potential, so comparing them is difficult and including them in the same emission trading program and accounting for their differences within that program is

Table 7.1
Criteria for a cap-and-trade system and two applications

Criterion	Applicability to OTC NO_X budget	Applicability to GHGs
1. Define and accurately measure the pollutants, their sources, and their fate and transport.	Sources are easily identifiable and most already had continuous emissions monitors for NO_X as part of the SO_2 Acid Rain Program monitoring requirements. Fate and transport are reasonably well understood, although they vary from source to source and with weather conditions somewhat.	Source identification and emissions measurement are relatively straightforward for CO_2 emissions due to fossil fuel combustion but much more difficult for other gases, such as agriculturally produced methane. Fundamentals are well understood, but sinks are still a source of considerable uncertainty for many GHGs, especially CO_2.
2. Define the quantity of emissions that will be permitted and thus available for trading (the cap).	Specified by OTC NO_X memorandum of understanding, which was a product of multistate negotiations with significant input from photochemical models and engineering-economic estimates.	Internationally, this would require bilateral or multilateral international agreements that could be implemented within a framework of national laws.
3. Account for differences in environmental impact across emission locations and times and across pollutant types.	Area of greatest disagreement. Spatial variation was shown to be unimportant to achieving air quality goals, given the configuration of sources and deep emissions reductions. Temporal differences were largely ignored in the analysis, although their effects are uncertain.	Spatial and locational differences have no effect, but GHGs vary significantly in warming functions and atmospheric lifetimes, implying important intergenerational judgments for multigas trading.

Table 7.1
(continued)

Criterion	Applicability to OTC NO_X budget	Applicability to GHGs
4. Create emission allowances, a distribution mechanism, and an enforceable redemption requirement.	Accomplished by state law, although allowances created by one state are recognized by all. Allocation methods vary significantly. Enforcement mechanisms are simple, strong, transparent, and certain.	Same as 2 but complicated by the existence of multinational firms that operate in a number of national jurisdictions.
5. Operate a market with enforceable contracts and rules to ensure competitive behavior.	Well-defined law and practice for interstate trade based on state contract law, federal commerce clause, and other business law.	Could be easily established within any nation. Existing international trade law could support (or be developed to support) GHG emissions trading; new bilateral or multilateral agreements could further facilitate this development.

fundamentally an arbitrary choice (Reilly et al. 1999). This problem expands if other components of the climate system are included, such as CO_2 sinks (e.g., growing forests) or land use.

Interstate NO_x Trading

This section of the chapter examines two efforts to control NO_x emissions from large, stationary sources (mostly coal-fired power plants) in the United States. One has been quite successful, and the other mostly a failure, at least in terms of achieving a negotiated solution among different political jurisdictions. Emissions of NO_x are controlled for several reasons, but for these two cases, the rationale is to ease the problem of photochemical smog (usually measured in terms of ground-level ozone concentrations) in the eastern United States. The law, economics, and atmospheric chemistry of this issue have evolved over the last half century, and at least a minimal understanding of these factors is necessary to understand the successes and failures of interstate NO_x emission trading programs in the United States. Among other things, they describe and explain the types of heterogeneity among the actors in these cases.

Both the OTC NO_x Budget and NO_x SIP Call are designed to help regions meet clean-air quality standards for tropospheric ozone (or photochemical smog). Ozone is present naturally in the lower atmosphere in low background levels (15–45 parts per billion, or ppb), but in polluted atmospheres, these levels can rise to several times background values (80–150 ppb are not uncommon in many areas of the country) and have significant negative health impacts. In the troposphere, ozone is conveniently thought of as a secondary pollutant formed through reactions of two classes of primary pollutants, volatile organic compounds (VOCs) and NO_x, in the presence of sunlight. This simple description, however, belies the complex chemistry that is actually involved (Seinfeld and Pandis 1998, 234–336). Ozone is actually formed via the photolysis of nitrogen dioxide (NO_2). Instead of playing a direct role in the formation of ozone, the presence of VOCs affects only the speed with which NO_2 forms ozone. The photolysis reactions compete for NO_2 with other processes, including the formation of nitric acid, peroxyacetylnitrate, and other organic nitrates, and with washout by rain, which eventually removes nitrogen from the ozone formation cycle. Ozone is also eventually removed (or cleansed) from the troposphere by further photolysis, reactions with NO_x or VOCs, washout by rain or surface deposition.

The ozone problem is typically described in the forgoing language, referring to emissions and concentrations, and has never (to our knowledge) been framed as a CPR problem. Indeed, pollution problems in general are considered distinct from resource problems; however, in order to apply CPR theory to the ozone problem, we must identify the resource at issue. An important exception is Barb Connolly (1999, 131), who applies CPR theory to acid rain and notes that the key issue is the finite ability of the environment to absorb emissions without causing damage (which can be called "assimilative capacity"). For tropospheric ozone process, the factors that limit the ability of the environment to absorb emissions without causing damage are the removal (or cleansing) processes, which have essentially fixed chemical reaction rates.[5] By definition, therefore, when emissions of NO_x and VOCs rise the rates at which the resultant ozone begins to harm health or the environment, the assimilative capacity for ozone has been depleted. More generally, pollution can be defined as the point at which emission rates rise higher than assimilative capacity for long enough to cause undesirable effects. The most applicable term from CPR theory for this effect is probably "congestion."

Since it is effectively impossible to control the removal processes of the ozone formation cycle, management of this CPR is usually viewed as an emission control problem. That is, sources are required to limit their emissions, and thereby limit the demands they place on the assimilative capacity of the environment. The primary pollutants contributing to ozone formation, NO_x and VOCs, are emitted from a variety of different anthropogenic and biogenic sources. Anthropogenic NO_x emissions are almost entirely due to combustion processes in which high temperatures oxidize the nitrogen in the ambient air to form NO_x. Once emitted into the atmosphere, emissions are mixed, advected, dispersed by winds, and eventually are removed by the processes described above. All these processes (those associated with both ozone formation and pollutant removal), however, may take several days to complete, during which time pollutants travel with the wind; thus photochemical smog in any given location is the result of emissions from local as well as distant sources. This phenomenon can be called "ozone regionality" or "ozone transport." However, the effect of distant and local emissions will vary significantly due to the complex chemistry of smog and the variations in atmospheric mixing that distant and local emissions will be subject to. This can be referred to as spatial heterogeneity and is in the second type of heterogeneity discussed in the introduction—heterogeneity associated with the components of the physical system in which the CPR is embedded. The net result of ozone transport and the spatial heterogeneity of the relationship between emissions and ozone concentrations means that it is possible for one jurisdiction to consume significant portions of the assimilative capacity in other, though different, proportions.

Ozone transport was discovered, however, only after the basic structure of U.S. air quality policy was developed, and air quality policy has been slow to adapt to the phenomenon (Farrell and Keating 2002). Under U.S. law, ozone is regulated pursuant to Title I of the Clean Air Act, for which Congress created a governance structure called "conjoint federalism." In this system, the federal government (specifically the Environmental Protection Agency, or EPA) is responsible for setting air quality standards, creating and enforcing some emissions standards for new sources, and the state environmental agencies are responsible for controlling emissions from existing sources and for operational controls such as automobile inspections.[6] To carry these activities out, states are required to develop SIPs, which detail the steps they will take (in addition to federal control measures) to attain the ambient standard. The EPA has oversight authority over the states and must approve their SIPs as adequately demonstrating (through a series of modeling steps) that the state will attain the relevant air quality standard, and the EPA has strong enforcement capabilities if they do not.

Over time, the preparation of SIPs and their enforcement has become the focus of state air pollution regulatory agencies, and states with little need for emissions controls (because of their good air quality) have not developed the same capabilities in this area that their more polluted counterparts have had to. This heterogeneity has had important effects. Perhaps most important, states with clean air did not monitor air pollution in much detail in the 1970s and 1980s, and they have not developed the ability to estimate reliably emissions of pollutants for in-state sources or the ability to use complex atmospheric models created for air pollution policy analysis during the 1980s. These differences have been key sources of heterogeneity in the technical capacity to study air quality and in beleifs about ozone regionality among the states, because both air quality monitoring and the development of accurate photochemical models were crucial to the discovery and understanding of ozone transport. Thus, by the mid-1990s, heterogeneity in air quality had led to heterogeneity in capabilities and information.

A crucial feature of Title I is that the EPA does *not* have the authority to regulate existing sources of emissions directly.[7] Instead the states control existing sources through a system of air quality permits. When new evidence warrants, the EPA *can* announce a "SIP call," which presents new requirements of states. In particular, an SIP call can define total emission reductions a state must make, but it cannot create specific requirements for any of the source categories over which the states have authority. In contrast, the EPA is given explicit authority in Title IV to create a national SO_2 trading program, and the states have had little to do with the implementation of the requirements of this program.

Changes to the Clean Air Act in 1977 and 1990 added provisions for states to pursue legal means to force other states to control sources from which they believe pollution is entering their airshed. These provisions, called Section 126 petitions, were added with sulfur dioxide pollution in mind but they have been consistently rejected by the courts when applied to this purpose (Grumet 1998). In 1997, however, some northeastern states used an untested part of the 1990 revisions to apply Section 126 petitions to the ozone transport problem, and the EPA subsequently issued a rule based on them (Environmental Protection Agency 1998b, 2000). This time, the appeals court upheld the Section 126 petitions, a decision that the U.S. Supreme Court declined to review (DC Circuit Court 2001).

The states are thus put into a very odd position in which they are required individually to meet an externally imposed environmental standard for a pollutant over which they (in many cases) have only partial control.[8] This has helped create a sharp division among the states that can roughly be characterized as separating them into two groups—"upwind" versus "downwind" states—depending on whether they tend to contribute to ozone pollution in other states or tend to receive it.[9] (This distinction will be discussed further later in the chapter.) Adding to this division is the variation in ozone levels among the states and the variation in the ways that the Clean Air Act treats them. Although ozone transport is an important phenomenon in tropospheric ozone, it has a strong local characteristic as well. Urbanized areas, especially those along the mid-Atlantic coast from Washington to Boston, tend to have higher pollution levels than other areas, largely because of car and truck exhaust. These areas are subject to more stringent federal requirements than rural upwind areas, so power plants in rural areas (of which there are many) tend to have little or no state-level NO_x emission control requirements. Thus, the combination of an ozone policy that essentially recognizes only local characteristics and the reality of ozone transport creates a counterintuitive situation in which NO_x sources (particularly sources with tall smokestacks) in relatively clean rural areas contribute to photochemical smog in relatively dirty urban areas. The political question, framed as a CPR issue, is, Which sources should have access to the limited assimilative capacity for ozone? This question is especially difficult because the upwind and downwind states are highly heterogeneous. More prosaically, how should the burden of cleaning up the dirty areas be shared between upwind and downwind sources?[10]

States do have some common interests in NO_x control. For one thing, they would all like to attain the ozone standard, both because of the federal enforcement mechanisms and because of internal pressure from voters (although this tends to be a stronger force in states that face a pollution problem, i.e., downwind states). In addition, they would all like to minimize the apparent costs to voters and the real costs to firms within their borders. Emission trading systems can accomplish this since their principal virtue is greater efficiency than command-and-control regulations. Further, the more sources there are in an emission trading program, the greater the available efficiency gains, so states considering such an approach have an incentive to join a multistate program rather than rely only on in-state ("domestic") emission trading.

Comparing this arrangement to the politics of climate change, one can see that the case of NO_x control is more like potential CO_2 control efforts will be than the SO_2 case was, but it is still an imperfect comparison. The biggest similarity between the two cases is that states are largely independent when it comes to establishing regulations for existing sources, and all the more so if they can support the claim that their emissions do not affect downwind states, much as nations will claim the right to control domestic CO_2 emissions in any way they want in an international agreement. There are dissimilarities as well. One is that all the states in the OTC have relatively strong incentives to control NO_x emissions. An even larger difference is that U.S. states operate within an authoritative legal system and a single economic system that permits virtually unfettered capital and trade flows among them. Thus, although there is less heterogeneity among the U.S. states than among the nations of the world, even these relatively weak differences go far in explaining the formation or lack of formation of interstate emission trading, as we will see.

The OTC NO_x Budget

The first example of an NO_x control effort that we will look at is a successful one, the OTC NO_x Budget, that essentially applies to electrical generating units rated at 25 megawatts or larger and similar-sized industrial facilities (such as process boilers and refineries). About 90 percent of the NO_x emissions covered by the program come from electric power plants. The NO_x Budget covers emissions from May through September in eight northeastern states (of the eleven in the OTC). There are over 470 individual sources in the program, owned by 112 distinct organizations (mostly private firms). The program has three phases. The first was essentially a relabeling of a federal program that the states were required to implement anyway, the NO_x Reasonably Available Control Technology (RACT) program. The second and third phases use a cap-and-trade emission allowance program to reduce total emissions by 55–65 percent (compared to uncontrolled sources) for 1999–2002 and by 75–85 percent starting in 2003. For electric power plants, these restrictions are most often discussed by referring to the equivalent emissions rate limit corresponding to the final, most stringent emission reduction requirement, measured in terms of heat input to the plant's boiler. For the OTC NO_x Budget, this value is 0.15 pounds of NO_x per million British Thermal Units (i.e., 0.15/mmBtu).

As explained earlier in the chapter, the federal government could not impose the OTC NO_x Budget directly, since Title I of the Clean Air Act gives it no authority to regulate existing NO_x sources directly. Instead, it emerged in 1994 from cooperative action by several northeastern states that had been working together on air quality issues (especially acid rain) for some time. This longstanding, informal relationship was institutionalized in the 1990 revisions to the Clean Air Act, and an attempt was made to expand the concept of regional cooperation on air quality to other parts of the country. To do this, Congress added Section 176, which defined "Transport Commissions" and encouraged their formation, and Section 184, which defined the "Ozone Transport Region" as twelve northeastern states and the District of Columbia and the OTC as the Transport Commission for that region.[11] The OTC was charged in the 1990 amendments, with "developing recommendations for additional control measures to be applied within all or part of such transport region if the commission determines such measures are necessary."

The EPA supported the development of an OTC emission trading program and used several approaches to stimulate cooperative action by the OTC states, such as funding several studies of multistate emission trading and supporting several multistate organizations dedicated to regional air quality management (Environmental Protection Agency 1992).[12] Further, the EPA offered to operate systems to track NO_x allowances and monitor NO_x emissions for any emission trading program the OTC developed and implemented.

After the OTC was created, it still took over five years for the states to develop the NO_x Budget, a process that occurred in two important steps. First, the states (with one exception, discussed below) signed a Memorandum Of Understanding (MOU) on September 27, 1994, that committed them to emission reductions as stated above. The states were not ready to agree to emission trading yet, however, so the MOU presumed command-and-control regulation but also provided for the development of a "region-wide trading mechanism." The intent was for the states to negotiate the specifics of an emission trading regime and come to a mutually agreeable solution.

Although the OTC members (membership consisted of the chiefs of each state's environmental protection agencies, who are typically political appointees, with significant support by the air quality group chief, typically a career civil servant, and associated professional staff) supported emission trading in general, they found the development of regulations to implement such a program much more difficult than was expected at first. A number of issues account for these difficulties, most of which are more complex than the OTC members originally thought, and about which they had a poor understanding when the MOU was signed in 1994. In many cases these issues were places in which heterogeneity among the OTC members could be found.

One of the most basic issues was that some regulators felt uncomfortable abandoning the traditional command-and-control plus enforcement approach. They felt that regulated companies would abuse any flexibility given to them, based on long experience of misleading rhetoric, extensive and drawn-out lawsuits and duplicitous behavior: for example, falsified reporting in the emission trading part of the lead phaseout program (Nichols 1997) and the abuse of various provisions of the existing regulations (Perez-Pena 2000).

Another important issue was the ability of the states to retain as much control as possible in the program. In particular, the states demanded that they individually be allowed to allocate allowances to sources as they chose (rather than using a uniform formula). Each state was also concerned about a first-mover *dis*advantage; what if they went ahead and implemented a tough emission reduction program only to find that some of the other states had backed out of, or weakened, their commitment? Until the OTC states agreed to allow each state to allocate emission allowances as it saw fit, it was not clear that any allocation scheme would gain sufficient support.

Probably the most difficult issue, however, was the potential impact of cross-border trades in emission allowances. Because of the directionality of the ozone problem, downwind regulators were concerned that firms in their states would overcontrol (and pass the cost on to in-state consumers), only to sell their excess allowances to upwind facilities. The upwind facilities thus would not have needed to control their emissions, which would then be transported into the downwind state. To the downwind-state regulators, this seemed like the worst of all possible outcomes, since costs to their states would be higher whereas the upwind sources would not be "doing their share" to reduce emissions. With this phrase, downwind regulators seemed to imply that upwind sources must reduce their emissions at least somewhat. This rather misses the point of emission trading: that firms with high control costs can "do their share" by buying emissions from firms with low control costs, who cut emissions more than they would have otherwise had to. However, the directionality of the photochemical smog problem in the northeastern United States does provide some rationale for concerns about cross-border trades. (That is, in the language of environmental economics, NO_x, VOC, and ozone are not "uniformly mixed.") Regardless of the scientific basis (or the lack thereof) for concerns about cross-border emission trading in any particular instance, the view that burden sharing is an important goal in itself, even if it comes at a cost, is widely held.

The question of burden sharing also arises in discussions about controlling GHG emissions. Nations are just as likely as U.S. states to want all parties in an emission trading program to share the burden of cutting emissions, and they may also prefer that burden sharing take the form of similar emission reductions among all participants rather than similar costs of control, even if this drives up the total cost. To some degree, concerns along these lines are expressed within the framework of the Kyoto Protocol in discussions about "supplementarity" (Grubb, Vrolijk, and Brack 1999, xxxvii, and 217–24).

The OTC states and the EPA took several steps to solve the question of burden sharing. First, the EPA funded studies of emission trading programs that showed no major geographic effects (ICF Resources 1995). Importantly, the states participated in the design of these studies, so they knew that their questions had been addressed, and they had reasonable confidence in the accuracy of the research (Farrell 2001). Second, the OTC states solved the image problem of emission trading by emphasizing the regional emission reduction, not the effects on in-state facilities specifically. Third, the states cooperated to develop a model emission trading rule that all could adopt but that was flexible enough to match the peculiarities of each state's legal framework and gave each state control over how to allocate emissions (Carlson 1996). The actual amount of emissions available for allocation was fixed ahead of time in the agreement to control emissions in the first place (in the MOU) and was not part of the negotiations on the emission trading rules.

It took several years of work by the OTC to resolve all the issues associated with the implementation of a NO_x emission trading program, and when they were resolved, the result was a not quite uniform program. Figure 7.1 shows which OTC states have joined the NO_x Budget (solid shade with diagonal lines) and which have not (solid shade). The pattern is interesting; the states at the extreme upwind and downwind have tended not to participate. Vermont and Maine (two of the most downwind states) decided to operate traditional permit-based programs, because the small number of sources involved (less than three in each state) and their regulatory status did not justify the administrative burden of developing an emission trading program. At the upwind end, Virginia did not join the NO_x Budget, but it has not taken any other action to regulate the sources that would have been part of the program. In fact, Virginia has been an uncooperative participant in the OTC negotiations all along: It is the only state that did not sign the original MOU in 1994 and has obstructed or ignored many other OTC activities.

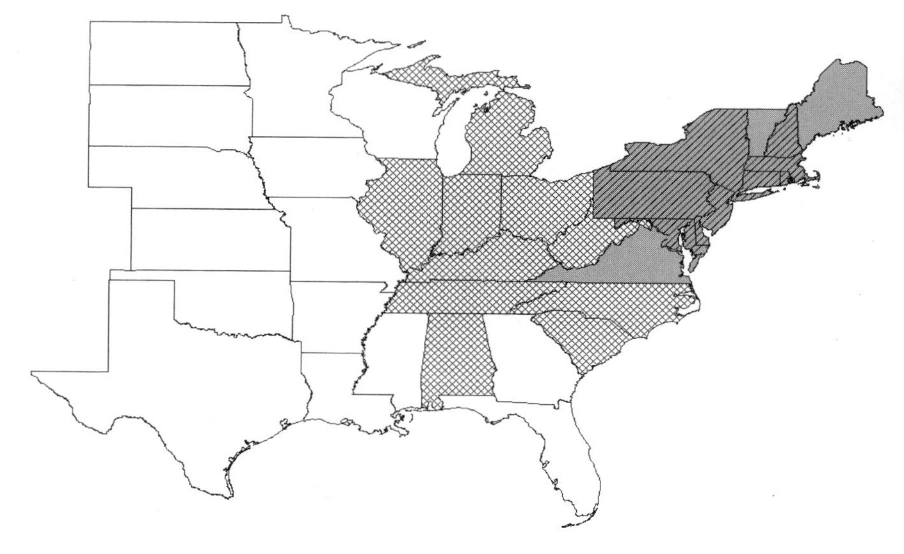

Figure 7.1
Ozone Transport Assessment Group states and NO_x control program status. ▩ OTC states *not* in NO_X Budget; ME and VI have similar command-and control programs. ▨ OTC states in the OTC NO_X Budget; all OTC states are affected by the NO_X SIP Call, except NH, ME, and VT. ▦ States subject to the initial NO_X SIP Call but *not* in OTC. WI will not be affected by the *final* NO_X SIP Call outcome; all, part, or none of AL, GA, MI, and MO may be affected by the *final* NO_X SIP Call outcome. ▭ States in not in OTC or affected by the *original* NO_X SIP Call.
Note: All states shown participated in the Ozone Transport Assessment Group.

In general, this pattern of participation in the NO_x Budget matches the pattern of interests of the states. Those that participate fully both have cities on the eastern seaboard with severe ozone pollution problems and are both upwind and downwind of other states in the OTC. The states that do not participate lack one of these two characteristics.

OTAG and the NO_x SIP Call

Although the OTC NO_x Budget is an important step forward in the management of a large and complex CPR (the assimilative capacity of North America for photochemical smog and its precursors), by no means has it solved (or even truly addressed) the upwind/downwind problem described earlier. In this section, the two most important efforts to deal with that issue are presented, the Ozone Transport Assessment Group (OTAG) and the NO_x SIP Call. They followed, and to some degree were based upon, the successful process that resulted in the OTC NO_x Budget program but they had very different results.

The OTAG process preceded the NO_x SIP Call and came about through an unusual set of circumstances (Farrell and Keating 2002). It was created in response to a crisis in air quality management that occurred in November 1994, just after the OTC MOU on NO_x control had been signed. This crisis consisted of a combination of a failure of states to submit new SIPs, which were due that month under a provision of the 1990 amendments to the Clean Air Act, and the election of a new Republican congressional majority with an overt antiregulatory, anti–federal government agenda (Gillespie and Schellhas 1994; Pagano and Bowman 1995). The downwind states failed to submit SIPs because their own air quality modeling had shown that even if they implemented expensive, unpopular emissions control programs mandated in the 1990 amendments (e.g., automobile emission inspections), they would not necessarily achieve federal clean-air standards because of incoming ozone (and precursors) from the upwind states. Of course, upwind states felt they had no serious air quality problems and had little reason to develop expensive SIPs, much less implement NO_x controls. This created a stalemate.

The EPA, upwind states and environmental groups did not pursue traditional methods of addressing this type of problem, such as new regulations or lawsuits, because they feared reprisals by the newly elected Congress. The belief of state environmental agencies that they, not the federal government, were the appropriate group to address this problem was also important in this regard. Thus, in the spring of 1995, the EPA quietly agreed with a few key states and nongovernmental organizations (NGOs) to hold off on issuing orders or lawsuits and instead engage in a "consultative process" to be completed by the end of 1996. The purpose of the OTAG process would be "to reach consensus on the additional regional, local and national emission reductions that are needed for . . . the attainment" of the ozone standard (Nichols 1995, 1).

The director of the Illinois EPA and vice-chair of the newly formed Environmental Council of the States (ECOS), Mary Gade, became the chair of OTAG, which quickly gathered momentum and took on a character and direction of its own. By August 1995, there were over 300 participants and eventually, there would be approximately 1,000 people involved at some level. This effort involved thirty-seven states, all of which are shown in figure 7.1. The participants disagreed, however, about why they were there. To downwind states facing statutory deadlines, OTAG was a mechanism for delaying expensive and unpopular emission control programs and for obtaining long-sought emission reductions in upwind states. To the upwind states, however, which were not subject to statutory deadlines, OTAG was a mechanism being used by downwind states that *were* subject to these deadlines to *unnecessarily* extend these same expensive and unpopular emissions control programs to them. Their goal was to prevent any such thing from emerging out of what quickly became viewed as the highest-profile regional ozone study ever conducted. The essential point is that despite lengthy cooperative studies and negotiations in OTAG, the upwind states never came to believe that it was in their interest to cut NO_x emissions as the downwind (OTC) states had decided to do.

Instead, OTAG's recommendations were very vague; it recommended a level of control from the status quo (i.e., nothing beyond what was already in the Clean Air Act) through the tight limits of the NO_x Budget, which represent approximately an 85 percent reduction in NO_x emissions. In addition, the recommendations supported emission trading generally, but OTAG was not able to develop any specific proposals for developing or implementing such a trading program. This allowed a wide range of states that essentially did not agree to nonetheless "come to consensus" on these conclusions. In this way the OTAG recommendations look very much like an initial agreement in international environmental negotiations: a relatively soft statement that reaffirms the status quo (Keohane, Haas, and Levy 1994; Victor, Raustiala, and Skolnikoff 1998).

The end of the OTAG process was followed quickly (too quickly, in the opinion of some state leaders) by the NO_x SIP Call, in which the EPA attempted to create a cap-and-trade program much like the OTC NO_x Budget (including an 85 percent emission reduction), but through the SIP process described earlier in the chapter. It was formally proposed in an EPA announcement in the *Federal Register* on October 10, 1997 (finalized on October 27, 1998, *Federal Register* 57356–57538).

Whereas the OTC NO_x Budget was created by a group of states to help them meet federal clean-air requirements with which they were all challenged, the NO_x SIP Call is designed primarily to reduce NO_x emissions from upwind states to help the downwind states (who are the OTC states, of course) meet those same requirements. In its announcement, the EPA identified twenty-two states, including all the OTC states (less Maine, New Hampshire, and Vermont) plus the states shown in cross-hatch in figure 7.1, that would be required to reduce NO_x emissions because of their "significant contribution" to ozone pollution in down-wind states and called on them to revise and resubmit their SIPs to accomplish these reductions. Not all twenty-two states will be affected by the NO_x SIP Call and associated final outcome of regulation and litigation. Thus, the NO_x SIP Call was proposed after the OTC NO_x Budget had begun to take shape, but before emission reductions or emission trading had started.

In practice, the NO_x SIP Call would extend the 0.15 lb./mmBtu requirement embodied in the NO_x Budget to all twenty-two states it covers. The EPA is not allowed to specify such a requirement, of course, since it has no authority to regulate existing sources directly, so instead it developed a "budget" for each state and required that the states develop SIPs that would meet this budget. In calculating these budgets, the EPA estimated the total emissions from each state assuming existing control programs would remain in place and that additional cost-effective emissions controls would be used on all sources. Among the additional controls that EPA identified as cost-effective were those that would bring coal-fired electric power plants down to 0.15 lb./mmBtu, plus a few others. Most importantly, the EPA applied *uniform* controls across all twenty-two states, ignoring the directionality and cross-border issues discussed earlier.

To encourage the formation of an emission trading program, the EPA included in its *Federal Register* announcement a provision that it would automatically approve SIPs that contained emission trading provisions. It also volunteered to take on many of the administrative and monitoring tasks, just as it had for the OTC NO_x Budget, and sponsored studies of emission trading over the larger geographic area as well (Dorris et al. 1999; ICF Kaiser 1996; Environmental Protection Agency 1998b).

It also appears that the NO_x SIP Call more or less conforms to the emission reductions that the EPA had internally decided would be needed even before OTAG started, based on a previous set of studies (Milford et al. 1994; Possiel and Cox 1993; Possiel, Milich, and Goodrich 1991; Roselle and Schere 1995). It is important to note that the analysis conducted subsequently under OTAG did not contradict these previous findings; rather, it tended to increase the number of people (especially those outside the EPA) who were familiar with the results (Keating and Farrell 1999). It is also worth noting that the EPA at the time was predisposed to support emission trading, which is most easily implemented with a uniform requirement for emissions reduction (Nichols 1999).

The NO_x SIP Call was spectacularly unpopular and generated a large number of lawsuits by the upwind states, who claimed that the EPA did not have the authority to issue the NO_x SIP Call and that the analysis underlying it was flawed in any case (Midwest Ozone Group 1996; Flannery 1997; Flannery and Spatafore 1998). Although most non-OTC states subject to the SIP Call planned to control power plant NO_x emissions to help maintain air quality in their own states, and seemed willing to impose emission trading requirements equivalent to 0.25–0.20 lb./mmBtu, they refused to go further (Arrandale 2000).

Many of the downwind states filed Section 126 petitions at about the same time the NO_x SIP Call was announced, further adding to the dispute. Some time later, the EPA decided to grant several of these petitions, which asked for actions similar to the requirements of the NO_x SIP Call (Wald 1999).

After the NO_x SIP Call and Section 126 petitions were filed, the upwind states and power companies operating there filed still more lawsuits.[13] The Courts eventually rejected the arguments presented in these lawsuits, so the emission reductions envisioned in the NO_x SIP Call will indeed be made (Kelley 2000). The upwind states have never really agreed to this approach, however: they were literally forced to follow it by an authoritative federal government. Implementation plans currently being developed by the states appear to be headed towards a complex outcome; many states will participate in multi-state emission trading, some may introduce in-state emission trading, and others may stick to command-and-control approaches.

The NO_x SIP Call, Section 126 petitions, and associated litigation and SIPs may all be made obsolete by legislative action that imposes a federal emission trading program. Several Democratic and Republican proposals along these lines have been made, both in the form of legislation introduced into the 107th Congress and the "Clear Skies" proposal by the Bush White House (Bush 2002). None of these outcomes would offer a very informative example of how a multilateral cap-and-trade system for atmospheric emissions could be developed, since they would all be authoritative actions of a domestic government.

Interestingly, the Canadian province of Ontario recently proposed an NO_x cap-and-trade system that it would like to partially integrate with at least some of the U.S. systems. However, the Ontario proposal would allow the use of emission reduction credits, which the U.S. programs forbid, so it is not clear how such an integration would work. Further, environmentalists in both countries have criticized the Ontario proposal. The interesting aspect of this situation is that Ontario is now in a position similar to one in which non-EU countries (and possibly even the United Kingdom) might soon find themselves: desiring to use an efficient international emission trading mechanism for CO_2 emissions control but finding that the ground rules (e.g., sources covered, monitoring and enforcement, levels of control) have already been set by others.

Discussion

The political and economic conditions for the creation of a multilateral cap-and-trade system for NO_x among the states of the United States are more favorable than they are likely to be in most international settings. In large part, this is due to the much more limited heterogeneity among the states compared with the larger differences that exist among the nations of the world. Yet even these relatively small differences played a large role in both programs for emission control discussed in the previous section. This section discusses the main issues for multilateral emission trading brought out in the comparison of the OTC NO_x Budget and the NO_x SIP Call, many of which spring from heterogeneous features among the actors.

Coordination

Coordination among different political jurisdictions is central to any multilateral emission trading program, and the OTC NO_x Budget shows that such coordination is possible, a useful finding in itself. The coordination of regulatory development in that program proceeded from a prior history of cooperative technical assessment and interaction among regulators. Even more importantly, the effort needed to develop a coordinated multistate cap-and-trade program was undertaken only *after* a political agreement to control emissions was made. The original MOU committed the signatory states only to NO_x controls; it explicitly left the option of emission trading open, but it was not at all clear in 1994 when the MOU was signed that efforts to develop an effective emission trading program would be successful.

In the case of the NO_x SIP Call, a larger number of states with very divergent interests were able to coordinate a technical assessment, but they were not able to agree on a firm pollution control strategy. The lack of specific recommendations by OTAG on an emission trading program may be related to a lack of any feeling of necessity on the part of the participants. Subsequently, as the final outcome of the NO_x SIP Call continues to emerge, it appears that emission trading will proceed on a piecemeal, state-by-state basis, if at all, most likely raising the costs of compliance.

Existing Regulations

Several authors have claimed that the absence of prior regulations was an important contributor to the success of the U.S. SO_2 control program (Stavins 1997). This is not true for the NO_x Budget, nor is it true for the SO_2 program either! Power plant SO_2 had been controlled for human health reasons beginning in the 1970s, at the federal level for new sources and the state level for existing ones (Ackerman and Hassler 1981). Moreover, the emission trading provisions of the 1990 Clean Air Act amendments clearly state that emission trading cannot result in any violations of the Title I health-based standards for SO_2. Of course, it is true that there were no federal SO_2 controls on existing power plants and that controls designed to address acidification were new. In any case, the NO_x Budget is clearly an addition to preexisting regulations at both the state and federal levels and is designed to achieve long-standing human health goals.[14] Thus it appears that cap-and-trade programs can be used to replace existing command-and-control regulations, or simply added on top of them. What is undesirable is to combine the two sorts of regulations (Foster and Hahn 1995) so that the flexible market-based system is burdened with regulatory complexities.

Symbolic Power

Emissions trading programs are sometimes thought to have less symbolic power than command-and-control programs; politicians supposedly cannot earn the same level of admiration and support for enacting an emission trading program as they can for "getting tough with polluters" through command-and-control approaches. Although there may be some truth to this hypothesis, it appears that in the United States at least a reasonable portion of the public understands that emission trading programs do in fact have teeth. This can be seen in the positive media stories about the NO_x Budget and the vociferous language used by states attempting to avoid the NO_x SIP Call.

It may be difficult to generalize this observation to the international community, however, since the United States is quite singular in its use of emission trading. Other nations (and especially public opinion in other nations) may continue to misperceive emission trading as an ineffective means of emissions control, particularly since most countries are less willing than the United States to rely on the market for things like health care or labor supply. In addition, buying emission allowances looks to many like avoiding any responsibility for the problem (despite paying for cleanup elsewhere), and this may present fundamental political impediments to the successful implementation of any emission trading program in these countries.

Simplicity and Flexibility

Two of the standard prescriptions for emission trading are simplicity of design and flexibility for the participants, however, the interstate NO_x cases show that it is all too easy to be *too* simple and *too* flexible. The uniform standard embodied in the NO_x SIP Call is a case of an overly simple design. Had the EPA responded to the concerns of the upwind (mostly Midwestern) states about spatial heterogeneity some form of emission trading might already be in place. In fairness, if it is in fact true that 85 percent reductions in NO_x emissions across all twenty-two states covered in the NO_x SIP Call would have been necessary to achieve the ozone standard in the cities of the eastern seaboard, then the EPA's proposal had considerable merit. (To some degree, of course, a debate about what is *scientifically necessary* is misleading and misses the point. Many combinations of local and regional emissions control plans would attain the standard; the real question is which one is most easily accomplished *politically*.)

Further, the flexibility that firms have in timing their emissions within the five-month season of the NO_x SIP Call may prove to be excessive (Farrell 2000; Farrell, Carter, and Raufer 1999). Although there is less of a temporal and spatial pattern in the global warming system than in the tropospheric ozone system, the basic lesson still holds: there must be a close match between the regulations to control an environmental problem and the physical and social phenomena that create it in the first place.

Expertise and Leadership

A common view of the way an international emission trading program could be developed is for the United States to "lead" by establishing a domestic emission trading program that other countries could copy (Solomon 1995). However, the evidence provided by the NO_x Budget and the NO_x SIP Call strongly counterindicates this claim: the program that was centrally sponsored (the SIP Call) failed, whereas the program that was cooperatively developed (the Budget) has turned into a success. Further, during the OTAG discussions on emission trading, the disparities in the positions of the OTC states (and the EPA) on one side, and the remainder of the states on the other, were very large (Keating and Farrell 1999, 70–80). The OTC states were comfortable with the idea of emission trading and understood the policy implications of various options, whereas the upwind states were much less familiar with the concepts and were concerned about being tricked into an agreement that was disadvantageous because they were less well informed. In addition, the problem of uniform controls emerged here as well. The OTC states assumed (or asserted) that uniform controls were necessary for the creation of a successful emission trading program, whereas the upwind states insisted on finding a way to enable emission trading among regions with differentiated control requirements to reflect the spatial heterogeneity of the relationship between distant and local emissions with ozone concentrations in the downwind states.

The lesson is clear: expertise and experience in emission trading does not automatically translate into a leadership position; rather, it is far more likely that advocacy of an emission trading approach by one jurisdiction will look like an effort to pressure other jurisdictions into an arrangement that they may not understand as well and that may be disadvantageous to them. This is especially true of how the upwind states viewed the NO_x SIP Call. In the NO_x Budget case, the key to a successful program seems to have been cooperation in analysis and policy development, not advocacy disguised as "leadership." With the U.S. reputation as a dominating participant in international negotiations in addition to its position as the leading advocate of emission trading, this problem seems particularly relevant to attempts to develop an international CO_2 trading regime.

Reading Market Signals

If creating cap-and-trade programs is a relatively new and uncertain endeavor for government, so, too, does it present industry with new roles and new challenges. These programs are vastly different from traditional environmental regulations, typically leading to the creation of new markets for emission allowances that bear some resemblance to financial markets. The newness of these markets, plus the changes in the regulation of the electric power sector, have produced a number of questionable interpretations of experiences with current emission trading programs. First, it has been asserted that emission trading markets are "thin" (i.e., lightly traded) and thus inefficient or subject to high transaction costs. The fact that these markets are quite concentrated (a small number of firms own many allowances and many power plants), however, means that many firms can achieve considerable savings simply by reallocating allowances internally, something that would not be picked up in market transaction data (Burtraw 1996). Some observers seem to think that a large volume of traded emissions is a necessary condition for success of an emission trading program, partly because many early models of such programs forecast this outcome. Second, periods of low allowance prices during an early part of the SO_2 market have been widely misinterpreted, and the advantageous conditions that produced those bargains may not occur in other markets (Schmalensee et al. 1998; Smith, Platt, and Ellerman 1998). Third, some have suggested that transaction costs in these markets are high, yet none of the participants have made such complaints.

Several consistent and convincing signals from existing emission trading programs have been observed, though: (1) consistent low allowance prices (volatile periods aside), (2) an increasing reliance on the market and an increasing sophistication in how it is used, and (3) significant emission reductions (Ellerman 1998; Ellerman and Montero 1998; Farrell 2000; Klier, Mattoon, and Prager 1997; Mueller 1995). The combination of these three factors suggests that despite the oddities of these markets and short-term glitches, the major cap-and-trade programs operate more or less as advertised to reduce total emissions at relatively low costs. Even more encouraging are the signs that these programs have stimulated technological change that will help bring down costs even more in the future (Conrad and Kohn 1996; Farrell 2000).

Participation and Interests

An important finding in the study of CPRs is that the position of an actor relative to resources can affect its willingness to participate in solving CPR problems (Schlager and Blomquist 1998, 102; Connolly 1999). Both of the programs examined in this chapter strongly support this contention. Further, this study suggests that heterogenity among the actors and of the components of the physical system in which the CPR is embedded has an important influence on their interests.

In the case of the NO_x Budget, Virginia, through the fact that it has not joined the program, provides a good example of an actor that perceives itself as immune to the adverse effects of others (to use Shlager and Blomquist's phrase). But Maine and Vermont have chosen not to join for a different reason: their level of industrialization is so low that joining seems to them to involve too much of an administrative burden to be worthwhile. The former is an example of heterogeneity in the physical system, the latter, among the actors.

Even stronger examples exist in the NO_x SIP Call case. Here, not only did state environmental agencies vary in their support for emission trading according to their location, but so did firms. The usual monolithic "no" to more regulation shattered during the OTAG negotiations, as firms located in downwind states came to recognize that a larger emission trading program would be to their advantage, and they became stronger supporters of such a program (Keating and Farrell 1999, 93–94).

Trust

Participants in the NO_x Budget negotiations and OTAG assessment all highlight the importance of developing "trust" among the participants. In general, this came from working closely together on problems that were to some degree shared. A closer examination, however, shows that in practice, "trust" meant different things in each case. More importantly, states came to trust that the *process* they were participating in could not be manipulated against their interests. In this sense, they trusted other states only insofar as they could verify their actions.

While they were developing the NO_x Budget, the OTC members came to trust that emission trading would work to solve their problems and to trust each other to accurately represent their own situations during group meetings. This trust developed over the course of several years as a result of repeated face-to-face interactions and the ability to examine the data and analysis used by other members. This ability is often referred to as "transparency." The OTC members came to realize that duplicitous behavior could be detected relatively easily through these steps. Third parties (such as the EPA) aided in these steps. Thus, when they finally agreed to control emissions, OTC members took steps to verify that emissions were actually being controlled. This included submission of texts of new regulations, emissions inventories, and reports of progress on control technology deployments. A key verification feature, however, was the emissions-monitoring regime that was well established by Title IV of the Clean Air Act by the time the OTC NO_x Budget was being negotiated. This clearly defined transparent protocol imposed by federal law was considered reliable by all parties.

Similar mechanisms were at work in the OTAG process, with similar outcomes, despite the greater distrust between the states beforehand and the shorter time available to overcome that distrust (Keating and Farrell 1999, 138–139, 144–46). The fact that no permanent follow-up was created after the OTAG process ended and thus no mechanisms were available for verifying compliance helps explain why the NO_x SIP Call has generated so much opposition.

State Control

It is hardly surprising that states want to retain as much freedom as possible to implement a multilateral emission trading program. In particular, in the NO_x Budget case, states insisted on retaining control at the individual-state level over how the emission allowances were allocated within each state. As it turns out, the states have adopted very different processes for allocation (for instance, some held public meetings, others did not), but these variations have had no observable effect on the performance of the system. This suggests that there are limits as to how much central control may be feasible in any multilateral emission trading program, especially in one with the potentially large economic consequences of controlling CO_2 emissions.

Conclusions and Recommendations

Despite our pessimistic view on the prospects for a global emission trading system based on the UNFCCC, we feel that it is certainly feasible for emission trading to become part of the international response to climate change. The lessons from the study of CPRs and from the cases presented in this chapter of efforts to develop interstate emission trading within the U.S. federal system clearly show reason for optimism. As we have stressed, understanding and managing heterogeneity among the actors is a crucial part of the process of developing an emission trading program, although there are other issues as well. We now offer some closing remarks.

This chapter presents evidence that successful multilateral emission trading is possible. The key conditions for success are a common belief among the different jurisdictions participating in the program that emissions control is needed and a formal structure in which coordinated analysis and policy can be developed. Nonetheless, crafting a multilateral emission trading program, even in a best-case scenario, is exceedingly difficult; it requires solving the institutional, resource allocation, and coordination problems described in the chapter. The prospects for further progress on multilateral emission trading in the United States may be limited, since it is quite rare for a large number of states to have similar enough interests to meet the first necessary condition. Even if the EPA is eventually successful in forcing a large-scale NO_x control program through the SIP process, it is not at all clear that this process will meet the second necessary condition.

Surprisingly, the prospects may be brighter internationally, as many different countries have come to see regional and global environmental issues as a common threat, and many routes for international cooperation on environmental science and policy now exist. The history of international agreements, in areas as divergent as environmental protection and trade, shows quite clearly that effective regimes start slowly. International trade agreements, represented by the General Agreement on Tariffs and Trade and now the World Trade Organization, arguably the most successful international regime to date, grew and evolved over time, adding new countries and new goods slowly, carefully resolving conflicts among national interests (Jacoby, Prinn, and Schmalensee 1998). Developing countries often received special phase-in arrangements and even perpetual opt-outs of the most demanding requirements. The World Trade Organization embodies a powerful agreement, but reaching that agreement took fifty years of hard work.

Once an agreement to reduce CO_2 emissions is in place among two or more states, we expect emission trading systems would arise among similar nations, where the most relevant dimension on which to measure similarity is national capability to implement emission trading.[15] The EU will probably develop the first international system for CO_2 emission trading, but such a system could be created outside of the Kyoto framework as well, possibly as a simple bilateral program at first. Nations that share (at least partly) energy system infrastructures (i.e., electrical generation capacity or petroleum product supply chains) and strong economic ties may be the most likely to undertake an international emission trading system. Other relevant dimensions of similarity among participating nations may be the presence of countries fossil resources, the structure and size of their energy taxes, the ability of their domestic economies to produce innovation and allow labor adaptations, and the role of environmental issues in their national politics.

We would *not* necessarily expect, however, that countries that created a GHG emission trading program would have similar CO_2 control costs (certainly the NAFTA countries would not). Indeed, emission trading saves the most money when control costs vary most among participants, so nations that see an opportunity to meet emissions control goals more by inducing reductions in another country than at home might well join, as well as countries that see an opportunity to improve their balance of trade and possibly stimulate energy system investments by over-controlling their emissions and selling the excess. Nonetheless, there are surely limits to the amount of money nations would be willing to see leave the country to provide a global public good such as climate stabilization, perhaps one or two multiples of current foreign aid budgets. For these reasons (and for others stated earlier), we would expect that either the prices or the volumes of internationally traded CO_2 emissions allowances would be relatively small. This observation suggests that balance-of-trade concerns might be an additional factor (besides those traditionally mentioned by economists) that could reduce the efficiency of international emission trading systems.

One particularly intriguing possibility is that as the dominating partner in the North American Free Trade Agreement (NAFTA), the United States could convince the other two member countries (Canada and Mexico) to an implement some sort of GHG control program. Conveniently, there is already an environmental side agreement to NAFTA and a small secretariat, the North American Commission for Environmental Cooperation. A NAFTA-affiliated GHG control program might have particular advantages for the United States, as it would have much greater control over NAFTA negotiations than it does over the COP process. Canada's position in such a system would have to be carefully considered, as it is likely to be complying also with the Kyoto Protocol. It might find itself with two GHG emission markets in which it needs or wants to participate, markets that might be incompatible or even mutually exclusive. Mexico would also be in an unusual position since it is not required to reduce emissions under the Kyoto Protocol. This might create some opportunities for GHG emission trades for sale to the United States or Canada, but, overall, the NAFTA members probably have too much heterogeneity among them to expect an emission trading system among all three nations. A bilateral agreement between the United States and Canada might be more likely.

Our most important conclusion, however, is that there is absolutely no need to assume that comprehensive (i.e., global) top-down international emission trading programs that involve significant binding commitments are the only way to develop an effective, efficient global GHG reduction strategy. Indeed, for at least the next decade, such approaches will almost certainly fail, since key countries such as the United States, China, and India will not agree to participate. Even the top-down approach embodied in the final agreements reached at the Marrakech COP represent a significant weakening of the emission reductions originally envisioned by the drafters of the Kyoto Protocol. The political leadership and diplomats of the Annex 1 countries (save the United States) have done a reasonably good a job in the COP process of creating a viable framework for international GHG trading. Getting such a program started will be very valuable (even if its very limited) in terms of establishing CO_2 emission trading as a standard business practice. It is unlikely, however, that all the world's major states will simultaneously be prepared to sign up for a serious program of CO_2 emissions control.

Skolnikoff (1999, 8) has argued that the United States, especially Congress, will be slow to become an active participant in any

> issue in which the UN and the international community must play a central role. There is a climate of xenophobia in the Congress, reflected to some degree in the electorate that is challenging the role of the nation in world affairs and particularly in the work of the UN and its associated bodies. . . . [T]he current mood, often reflected in Congressional statements and votes, sees a vocal portion of the public turning away from foreign involvements . . . and rejecting policies that are perceived as in any way infringing American sovereignty. In this context, an agreement negotiated under the auspices of the UN that if carried out would certainly have an impact on the American economy is immediately suspect.

U.S. domestic political concerns are not the only problem. As Jacoby, Prinn, and Schmalensee (1998, 61) have noted, developing international institutions that will facilitate policies to minimize the cost of reducing GHG emissions

> requires solving the monitoring and enforcement problems necessary to implement efficient international trading of rights to emit [GHGs. It also] . . . requires an institutional structure that can exploit the cheapest abatement opportunities, wherever they may be found. . . . This is a tall order. The international trade regime developed under the General Agreement on Tariffs and Trade, now the World Trade Organization, hints at the difficulties involved. This regime grew and evolved over time, adding countries and goods along the way, peacefully resolving conflicts between national interests. . . . By the standards of

international affairs, the WTO has been a stunning success, but it took 50 years of hard work.

Jacoby and his coauthors, and many others have argued that, because GHGs are global pollutants, they cannot be managed without an overarching international accord. Fortunately, as both the World Trade Organization example just noted and many examples in the literature on the management of CPRs suggest (Ostrom et al. 1999), a top-down international framework may not be the only route to a successful global regime for managing GHGs. The success of the OTC NO_x Budget shows it is possible for independent jurisdictions to agree on how to implement an emission trading system, but the limits in this example, and the outright failure of the OTAG process to establish a multistate agreement to control NO_x emissions, warn us that it is not easy.

Indeed, a top-down approach may not even be the best route. As detailed in this chapter, several countries have begun to take unilateral action toward GHG emission control. Although some observers dismiss these actions as limited and self-serving, they reflect the genuine political commitment of the citizens in these countries to solving the emissions problem. The history of international environmental protection shows quite clearly that effective regimes start slowly. The diplomatic community needs to figure out how to encourage the growth of local and regional regimes and how to encourage their coordination so that ultimately they can coalesce into a set of global arrangements that encompass all major states (Morgan 2000).

An evolutionary bottom-up strategy for emissions control has several benefits. Concerned states and regions can start today. As different early adopters try different strategies, the world will get an opportunity to evaluate alternative approaches and learn from mistakes. Early adopters can provide the inspiration and proof of concept to inspire or shame citizens in other countries to take action. Some will argue that a bottom-up approach can never work, because nobody will go first for fear of free riders. National environmental policies, however, are often not primarily driven by economic considerations. Growing numbers of people believe that the world must act and are willing to assume some extra burden and provide an example for others.

If a bottom-up strategy is going to work, the diplomatic community needs to take concrete steps to support and encourage subglobal carbon management efforts. For example, early adopters may want to impose a domestic carbon emissions tax on power plants, on process industries, and on the production or use of transportation and heating fuels. These states might be willing to have their industries face a modest competitive disadvantage in world markets. They will certainly not want to disadvantage domestic industries significantly, however. Thus they will want to impose nondiscriminatory border adjustment tariffs on the GHG releases that are implicit in imports. This might be accomplished through a set of default values for the GHG content of traded goods that importers can replace, at their option, with real values verified by some impartial international auditing entity. Such a system would have to be made compatible with World Trade Organization rules, which today might disallow adjustment tariffs on the grounds that they are discriminatory or inappropriately consider process. But trade rules are always in flux, and multilateral agreements are treated more favorably than unilateral initiatives. With some effort, several nations might be allowed border adjustment externality tariffs on global pollutants, even if not on local pollutants.

The diplomatic community could also help by opening a forum for discussions among states that want to act now. As more states begin to develop control strategies, there will be growing needs to coordinate, to reconcile tax-based approaches with cap-and-trade approaches, to figure out how to treat multinational firms, to determine how to promote the basic technology research needed to create the intellectual capital that the market will need to develop future clean-energy systems, and ultimately, to coalesce the voluntary network of controls into a more binding international system that includes all major industrialized and industrializing states.

Research into CPRs has shown that societies of all sorts have managed to develop sustainable means for managing vital resources but that many have failed to do so and perished. We need to act now to encourage initiatives by individual states and regions so that we can learn how best to move the world's economies toward a lower-impact, more sustainable future. Fortunately, it may be possible to develop such strategies from the ground up.

Acknowledgments

The research reflected in this chapter was made possible through support from a number of organizations, including the Center for Integrated Study of the Human Dimensions of Global Change, which was created through a cooperative agreement between the National Science Foundation (SBR-9521914) and Carnegie Mellon University and has been supported by additional grants. The research presented here was also supported by the Global Environmental Assessment Project at Harvard University, which is funded by the National Science Foundation (BCS-9521910), with supplemental support by other federal agencies. The authors thank A. Denny Ellerman, Terry J. Keating and David G. Victor for comments and suggestions. All remaining errors remain the fault of the authors. A previous version of this chapter was presented as a paper at the eighth biennial conference of the International Association for the Study of Common Property in June 2000 at Indiana University in Bloomington.

Notes

1. The phrase "emission trading" will be used in this chapter as a shorthand for marketable emissions allowance systems because of its frequency in the literature and among practitioners.

2. Other types of emission trading are referred to as emission reduction credits (Foster and Hahn 1995; Solomon and Gorman 1998) and open-market trading (Ayres 1994; Goffman and Dudek 1995). For a review of experience with market-based instruments in general, see Stavins 2000.

3. The discussion in the chapter assumes that the principal participants are firms that are accountable to a government body. For a discussion of how this sort of system might be implemented internationally for GHG, see Grubb, Vrolijk, and Brack 1999, 194–196, 206–213.

4. In this program, the amount of tetraethyl lead that petroleum refiners were allowed to add to motor gasoline was reduced and finally eliminated.

5. Reaction rates do change with variations in temperature, insolation, and pollutant concentration, but only the last of these is controllable by human action by limiting pollutant emissions. And in practice, primary-pollutant concentrations are managed not to control reaction rates but to control concentrations of pollutants, so these effects can be ignored.

6. For some pollutants, such as toxics, the EPA regulates existing sources, but these come under Title III rather than Title I.

7. On the other hand, *new* stationary sources are regulated by the federal New Source Performance Standard. Thus, the discussion presented here pertains only to sources that existed prior to the implementation of this standard. By the mid-1990s, a considerable portion of total NO_x emissions came from these sources.

8. This situation is very similar to the one Connolly (1999, 131) describes as a "differentiated position with respect to the resource."

9. In the eastern United States, air tends to move east and north, and the states of the eastern seaboard are generally considered downwind of nearby Midwestern and southeastern states.

10. It is more convenient to discuss this problem using the traditional language of pollution control policy rather than in terms of CPRs, so this framework will be adopted for much of the rest of the chapter.

11. It is worth noting that congressional representatives from the downwind states most supportive of the OTC introduced this concept into the law and worked for its passage.

12. For the OTC itself, see <http://www.sso.org/otc/>. The other organizations are the Northeast States for Coordinated Air Use Management and the Mid-Atlantic Regional Air Management Association.

13. Simultaneously, many of the same interests were simultaneously engaged in a separate legal battle over the health-based standard.

14. The same can be said for California's Regional Clean Air Incentives Market (RECLAIM) program as well (Lents and Leyden 1996).

15. We can assume such nations also have a similar (high) national interest in CO_2 control, else they would not have joined such an agreement to begin with.

References

Ackerman, B. A., and W. T. Hassler. 1981. *Clean Coal/Dirty Air, or, How the Clean Air Act Became a Multibillion-Dollar Bail-Out for High-Sulfur Coal Producers and What Should Be Done about It.* New Haven, CT: Yale University Press.

Arrandale, T. 2000. "Balking on Air." *Governing* (January):26–29.

Ayres, R. 1994. "Developing a Market in Emission Credits Incrementally: An 'Open Market' Paradigm for Market-Based Pollution Control." *BNA Environment Reporter* 25(31):1526.

Baer, P., J. Harte, B. Haya, A. V. Herzog, J. Holdren, N. E. Hultman, D. M. Kammen, R. B. Norgaard, and L. Raymond. 2000. "Equity and Greenhouse Gas Responsibility." *Science* 289 (September 29):2287.

Ben-David, S., D. S. Brookshire, S. Burness, M. McKee, and C. Schmidt. 1999. "Heterogeneity, Irreversible Production Choices, and Efficiency in Emission Permit Markets." *Journal of Environmental Economics and Management* 38(2):176–194.

Burtraw, D. 1996. "The SO_2 Emission Trading Program: Cost Savings without Allowance Trades." *Contemporary Economic Policy* 14 (April):79–94.

Bush, G. W. 2001. "Letter to Members of the Senate on the Kyoto Protocol on Climate Change." March 13. Washington, DC: The White House. Available at <http://www.access.gpo.gov/nara/nara003.html>.

Bush, George W. 2002. *Clear Skies Initiative*. Washington, DC: The White House. February 14. Available at <http://www.whitehouse.gov/news/releases/2002/02/clearskies.html>.

Carlson, L. 1996. *NESCAUM/MARAMA NO_x Budget Model Rule*. Boston: Northeast States for Coordinated Air Use Management.

Clinton, W. J., and J. Albert Gore. 1993. *The Climate Change Action Plan*. Washington, DC: The White House.

Conference of New England Governors and Eastern Canadian Premiers. 2000. "Resolution Concerning Global Warming and Its Impacts on the Environment." Boston: Conference of New England Governors. 2 pages. Available at <http://www.cmp.ca/res-25-9-en.htm>.

Conference of New England Governors and Eastern Canadian Premiers. 2001. "Climate Change Action Plan." Boston: Author. Available at <http://www.cmp.ca/CCAPe.pdf>.

Connolly, B. 1999. "Asymmetrical Rivalry in Common Pool Resources and European Responses to Acid Rain." In *Anarchy and the Environment*, ed. J. S. Barkin, and G. E. Shambaugh, 122–149. Albany: State University of New York Press.

Conrad, K., and R. E. Kohn. 1996. "The U.S. Market for SO_2 Permits: Policy Implications of the Low Price and Trading Volume." *Energy Policy* 24(12): 1051–1059.

DC Circuit Court. 2001. *Appalachian Power Co. v. EPA*. 251 F.3d 1026. June 8.

Dorris, G., J. Agras, S. Burrows, M. McNair, and S. Reynolds. 1999. *Development and Evaluation of a Targeted Emission Reduction Scenario for NO_x Point Sources in the Eastern United States: An Application of the Regional Economic Model for Air Quality (REMAQ)*. Washington, DC: U.S. Environmental Protection Agency.

Ellerman, A. D. 1998. "Electric Utility Response to Allowances: From Autarkic to Market-Based Compliance." Working paper. Cambridge: Center for Energy and Environmental Policy Research, Massachusetts Institute of Technology.

Ellerman, A. D., and J. Montero. 1998. "The Declining Trend in Sulfur Dioxide Emissions: Implications for Allowance Prices." *Journal of Environmental Economics and Management* 36(1):26–45.

Energy Information Administration. 1999. *Annual Energy Review 1998*. Washington, DC: U.S. Government Printing Office.

Energy Information Administration. 2001. "Emissions of Greenhouse Gases in the United States 2000." Washington, DC: U.S. Government Printing Office.

Environmental Protection Agency. 1992. *The United States Experience with Economic Incentives to Control Environmental Pollution.* Washington, DC: Environmental Protection Agency, Office of Policy, Planning, and Evaluation.

Environmental Protection Agency. 1998a. *Inventory of U.S. Greenhouse Gas Emissions and Sinks: 1990–1996.* Washington, DC: Environmental Protection Agency, Office of Policy, Planning and Evaluation.

Environmental Protection Agency. 1998b. *Regulatory Impact Analysis for the NO_x SIP Call, FIP, and Section 126 Petitions.* Washington, DC: Environmental Protection Agency.

Environmental Protection Agency. 1998c. "Findings of Significant Contribution and Rulemaking for Certain States in the Ozone Transport Assessment Group Region for Purposes of Reducing Regional Transport of Ozone." *Federal Register* 6 (October 27):57356–57536.

Environmental Protection Agency. 2000. "Findings of Significant Contribution and Rulemaking on Section 126 Petitions for Purposes of Reducing Interstate Ozone Transport." *Federal Register* 62 (January 18): 2674–2767.

European Commission. 2001. "Proposal for a Framework Directive for Greenhouse Gas Emission Trading within the European Community." COM (2001)581. Brussels: European Commission.

Farrell, A. 2000. "The NO_x Budget: A Look at the First Year." *Electricity Journal* 13(2) (March):83–92.

Farrell, A. 2001. "Multi-lateral Emission Trading: Lessons from Inter-state NO_x Control in the United States." *Energy Policy* 29(13):1061–1072.

Farrell, A., R. Carter, and R. K. Raufer. 1999. "The NO_x Budget: Costs, Emissions, and Implementation Issues." *Resource & Energy Economics* 21(2):103–124.

Farrell, A., and T. J. Keating. 2002. "Transboundary Environmental Assessment: Ozone Regionality in the United States." *Environmental Science and Technology* 36(12).

Flannery, D. M. 1997. "Midwest Chides Northeast for Hypocrisy." Charleston, WV: Midwest Ozone Group.

Flannery, D. M., and M. A. Spatafore. 1998. *Comments to the "Findings of Significant Contribution and Rulemaking on Section 126 Petitions for Purposes of Reducing Interstate Ozone Transport."* Charleston, WV: Jackson & Kelly.

Fort, J., and C. Faur. 1997. "Can Emission Trading Work beyond a National Program? Some Practical Observations on the Available Tools." *University of Pennsylvania Journal of International Economic Law* 18(2):463–475.

Foster, V., and R. W. Hahn. 1995. "Designing More Efficient Markets: Lessons From Los Angeles Smog Control." *Journal of Law and Economics* 38(1):19–48.

Gillespie, E., and B. Schellhas. 1994. *Contract with America.* New York: Times Books.

Goffman, J., and D. Dudek. 1995. "Comments on the Proposed Open Market Trading Rule." Washington, DC: Environmental Defense Fund.

Grubb, M., C. Vrolijk, and D. Brack. 1999. *The Kyoto Protocol: A Guide and Assessment*. London: Earthscan.

Grumet, J. 1998. "Old West Justice: Federalism and Clean Air Regulation 1970–1998." *Tulane Environmental Law Review* 11(2):375–413.

Hackett, S. 1992. "Heterogeneity and the Provision of Governance for Common-Pool Resources." *Journal of Theoretical Politics* 4:325–342.

ICF Kaiser. 1996. "OTAG Trading Analysis with EPA/IPM." Policy case paper. Washington, DC: U.S. Environmental Protection Agency.

ICF Resources. 1995. *Estimated Effects of Alternative NO_x Cap and Trading Schemes in the Northeast Ozone Transport Region*. Arlington, VA: ICF Kaiser.

Jacoby, H. D., R. G. Prinn, and R. Schmalensee. 1998. "Kyoto's Unfinished Business." *Foreign Affairs* 77(4):54–66.

Joskow, P., and R. Schmalensee. 1998. "The Political Economy of Market-Based Environmental Policy: The U.S. Acid Rain Program." *Journal of Law and Economics* 41(1):37–83.

Keating, T. J., and A. Farrell. 1999. *Transboundary Environmental Assessment: Lessons from the Ozone Transport Assessment Group*. Knoxville, TN: National Center for Environmental Decision-Making Research. Available at <http://www.ncedr.org>.

Kelley, T. 2000. "Appeals Court Upholds EPA Rules to Reduce Smog in the Northeast." *New York Times*, March 4, p. A1.

Keohane, R. O., P. M. Haas, and M. Levy. 1994. "The Effectiveness of International Environmental Institutions." In *Institutions for the Earth: Sources of Effective International Environmental Protection*, ed. P. M. Haas, R. O. Keohane, and M. A. Levy, 3–24. Cambridge: MIT Press.

Keohane, R. O., and E. Ostrom. 1994. "Introduction." In *Local Commons and Global Interdependence*, ed. R. O. Keohane, and M. A. Levy, 1–26. Thousand Oaks, CA: Sage Publications.

Klier, T. H., R. H. Mattoon, and M. A. Prager. 1997. "A Mixed Bag: Assessment of Market Performance and Firm Trading Behaviour in the NO_x RECLAIM Programme." *Journal of Environmental Planning and Management* 40(6): 751–774.

Kopp, R., R. Morgenstern, W. Pizer, and M. Toman. 1999. *A Proposal for Credible Early Action in U.S. Climate Policy*. Washington, DC: Resources for the Future. Available at <http://www.weathervane.rff.org/negtable/Dom_Early_Action.html>.

Lents, J. M., and P. Leyden. 1996. "RECLAIM: Los Angeles' New Market-Based Smog Cleanup Program." *Journal of the Air & Waste Management Association* 46:195–206.

Loeb, A. 1990. "Three Misconceptions about Emission Trading." Paper presented at 83rd Annual Meeting & Exhibition of the Air & Waste Management Association, Pittsburgh, PA.

Midwest Ozone Group. 1996. *Ozone Attainment, Proceeding in the Right Direction: Will Sound Science and Objectivity Prevail?* Charleston, WV: Author.

Milford, J. B., D. F. Gao, A. Zafirakou, and T. E. Pierce. 1994. "Ozone Precursor Levels and Responses to Emissions Reductions—Analysis of Regional Oxidant Model Results." *Atmospheric Environment* 28(12):2093–2104.

Mitchell, R. B. 1999. "International Environmental Common Pool Resources: More Common than Domestic but More Difficult to Manage." In *Anarchy and the Environment*, ed. J. S. Barkin, and G. E. Shambaugh, 26–50. Albany: State University of New York Press.

Morgan, M. G. 2000. "Managing Carbon from the Bottom Up." *Science* 289 (September 29):2285.

Mueller, P. 1995. "An Analysis of the First Year of the RECLAIM Program." Paper presented at 88th Annual Meeting of Air & Waste Management Association, San Antonio, TX.

National Energy Policy Development Group. 2001. *National Energy Policy*. Washington, DC: U.S. Government Printing Office. Available at <http://www.whitehouse.gov/energy/> or <http://bookstore.gpo.gov/>.

Nichols, M. 1995. "Memorandum: Ozone Attainment Demonstrations." Washington, DC: U.S. Environmental Protection Agency, Office of Air and Radiation.

Nichols, A. 1997. "Lead in Gasoline." In *Economic Analyses at EPA: Assessing Regulatory Impact*, ed. R. D. Morgenstern, 98–114. Washington, DC: Resources For the Future.

Nichols, M. 1999. Telephone interview with the author. March 11.

Nordhaus, R., and S. C. Fotis. 1998. *Early Action & Global Climate Change: An Analysis of Early Action Crediting Proposals*. Arlington, VA: Pew Center on Global Climate Change.

Ostrom, E., J. Burger, C. Field, R. Norgaard, and D. Policansky. 1999. "Revisiting the Commons: Local Lessons, Global Challenges." *Science* 284 (April 9): 278–282.

Pagano, M. A., and A. Bowman. 1995. "The State of American Federalism 1994–1995." *Publius: The Journal of Federalism* 25(3):1–21.

Perez-Pena, R. 2000. "Power Plants to Cut Emissions Faulted in Northeast Smog." *New York Times*, November 16, p. A1.

Possiel, N. C., and W. M. Cox. 1993. "The Relative Effectiveness of NO_x and VOC Strategies in Reducing Northeast U.S. Ozone Concentrations." *Water, Air, and Soil Pollution* 67:161–179.

Possiel, N. C., L. B. Milich, and B. R. Goodrich. 1991. *Regional Ozone Modeling for Northeast Transport (ROMNET)*. Research Triangle Park, NC: U.S. Environmental Protection Agency, Office of Air Quality Planning and Standards.

Reilly, J., R. G. Prinn, J. Harnisch, J. Fitzmaurice, H. D. Jacoby, D. Kicklighter, P. H. Stone, A. Sokolov, and C. Wang. 1999. "Multi-gas Assessment of the Kyoto Protocol." Working paper. Cambridge: MIT Joint Program on the Science and Policy of Climate Change.

Roselle, S. J., and K. L. Schere. 1995. "Modeled Response of Photochemical Oxidants to Systematic Reductions in Anthropogenic Volatile Organic Compound and NO_x Emissions." *Journal of Geophysical Research—Atmospheres* 100(D11):22929–22941.

Schlager, E., and W. Blomquist. 1998. "Heterogeneity and Common Pool Resource Management." In *Designing Institutions for Environmental and Resource Management*, ed. E. T. Loehman, and D. M. Kilgour, 101–112. Northhampton, MA: Edward Elgar.

Schlager, E., and E. Ostrom. 1992. "Property-Rights Regimes and Natural Resources: A Conceptual Analysis." *Land Economics* 68(3):249–262.

Schmalensee, R., P. Joskow, A. D. Ellerman, J. P. Montero, and E. M. Bailey. 1998. "An Interim Evaluation of Sulfur Dioxide Emission Trading." *Journal of Economic Perspectives* 12(3):53–68.

Seinfeld, J. H., and S. N. Pandis. 1998. *Atmospheric Chemistry and Physics: From Air Pollution to Climate Change*. New York: John Wiley.

Skolnikoff, E. B. 1999. "From Science to Policy: The Science-Related Politics of Climate Change in the U.S." Working paper. Cambridge, MIT Joint Program on the Science and Policy of Climate Change.

Smith, A. E., J. Platt, and A. D. Ellerman. 1998. "The Costs of Reducing Utility SO_2 Emissions—Not as Low as You Might Think." *Public Utilities Fortnightly* 136(10) (May 15):22–29.

Solomon, B. D. 1995. "Global CO_2 Emissions Trading—Early Lessons from the U.S. Acid-Rain Program." *Climatic Change* 30(1):75–96.

Solomon, B. D. 1999. "New Directions in Emission Trading: The Potential Contribution of New Institutional Economics." *Ecological Economics* 30: 371–387.

Solomon, B. D., and H. S. Gorman. 1998. "State-Level Air Emission Trading: The Michigan and Illinois Models." *Journal of the Air & Waste Management Association* 48(12):1156–1165.

Stavins, R. 1997. "What Can We Learn from the Grand Policy Experiment? Lessons from SO_2 Allowance Trading." *Journal of Economic Perspectives* 12(3):69–88.

Stavins, R. 2000. "Experience with Market-Based Environmental Policy Instruments." In *The Handbook of Environmental Economics*, ed. K. Goren-Maler, and J. Vincent, 453–474. Amsterdam: North-Holland/Elsevier Science.

Victor, D. G. 1991. "Limits of Market-Based Strategies for Slowing Global Warming: The Case of Tradable Permits." *Policy Science* 24(2):199–222.

Victor, D. G. 2001. *The Collapse of the Kyoto Protocol and the Struggle to Slow Global Warming*. Princeton: Princeton University Press.

Victor, D. G., K. Raustiala, and E. B. Skolnikoff, eds. 1998. *The Implementation and Effectiveness of International Environmental Commitments*. Cambridge: MIT Press.

Wald, M. L. 1999. "EPA Is Ordering 392 Plants to Cut Pollution in Half." *New York Times*, December 18, p. A1.

IV

Financial, Social, and Political Capital: Managing Common-Pool Resources and Shaping the Macropolitical and Macroeconomic Environment

8

Shaping Local Forest Tenure in National Politics

Rita Lindayati

Introduction

This chapter explores how the legal framework of local forest tenure is shaped by macro politics, with specific reference to Indonesia. Its objectives are twofold. The first is to demonstrate that formal forest management rights and access patterns in a country are a function of the broader national political process; accordingly, structural changes in the country's political system will directly affect forest institutional arrangements. The second is to investigate policymaking—where ideas and interests over local forest access rights are contested—and the role of ideas in the policymaking process.[1] Policymakers' ideational conceptions are, arguably, key to understanding policy adoption. Earlier studies have described the continuing influence of colonial forestry ideology—and its conception of local forestry systems—on developing countries' contemporary forest management and tenurial policy. Why this is so, and how it is manifested, are seldom addressed.

Most research on micro-macro resource management linkages focuses on understanding local perceptions, experiences, and strategies in responding to national policy enforcement. More often than not, the process of policy implementation has disrupted long-established local-based resource management practices.[2] This creates what Bromley (in 1985) called "institutional dissonance," in which formal state institutional resource arrangements are incongruent with local informal rules and resource allocation customs (cited in Cramb and Wills 1990). This chapter is not another case study of the impacts of policies on local forest access and rights. Rather, it goes a step further by investigating the *process* of how a particular forest management policy, one that shapes legal forest tenure, is created. Many earlier studies have analyzed macro variables from the perspective of the periphery; in those studies, state policy and actions have often been assessed in terms of their differential impacts on various social groups at different times and locations. The policy *process* (which created the policy outcome in the first place) has often been treated as a black box, with many questions critical to understanding local forest management systems remaining unanswered. How is local tenure defined by national politics? What conceptions of local forest access rights influence policymakers? How do these conceptions develop into policy proposals that are contended in policy politics? What are the factors that drive a particular policy proposal at some times to be adopted, whereas at other times they cause it to be rebuffed?

This chapter attempts to address these questions by investigating how local forestry practices are continuously perceived by policymakers (particularly the state), how this perception is debated in policy politics, and how the ensuing political struggle results in policy outcomes that eventually determine the legal framework of forest tenure. Building upon such previous work as Peluso 1992 and Guha 1990, the chapter's basic assumption is that state forest management ideas are important in shaping forestry policy, that is, the legal framework that shapes de jure local tenurial systems. The way these ideas are interpreted and framed into policy arguments, however, and the influence they have on particular policies varies across time and political environments. The introduction and institutionalisation of state-based scientific forestry in Indonesia, beginning in the late nineteenth century, served primarily the expanding political and economic interests of colonial rulers. They encountered strong opposition from those who were in favor of traditional, locally based resource control, although the level of influence of this opposition at any given time was contingent upon the strategic power of those in the opposition at that time.

Therefore, it is argued here that political economic interests at a particular time affect the influence of ideas on policy outcomes. A synergy between the ideas and interests of key decision makers increases the likelihood of ideas' eventually being translated into policy (Hansen and King 2001). Furthermore, politics involves power, with its distributional patterns shaped by political structures. Thus, policymaking and outcomes are arguably shaped as much by the interactions of policy ideas and interests as by the political structure through which the interactions occur. Political structural changes have an impact on the "terms of reference" of these interactions, and most likely, on the forestry policy outcomes. Forestry lawmaking, from colonial times to the present, provides rich empirical grounds for investigating how multiple interests and forest management ideas are defined and contended within structured power relations.[3]

Policy Origin: The Importance of Ideas

Forestry policymaking is a means of understanding how local forestry systems are shaped by national politics. Explanations of the origin, content, and pursuit of public policy (forestry and other sectors) have so far been dominated by accounts of material-driven interests (e.g., economic, political). Human activities are viewed as an inherent struggle to satisfy human wants within a context of limited resources. Such an approach overlooks the possibility of policy reform's taking place as a result of such forces as policymakers' learning processes, leadership, and values.

Without undermining the importance of material interests, some analysts have argued for the importance of ideas and belief systems in shaping policy processes and outcomes (see, e.g., Odell 1982; Goldstein 1988; Hall 1989, 1993; Sikkink 1991; Goldstein and Keohane 1993; Sabatier and Jenkins-Smith 1993; Howlett 1994; Hansen and King 2001). Their basic argument is that ideas are not epiphenomenal to interests, although the formulation and enactment of ideas are usually not separate from interests and power. Ideas may initially be adopted because they serve the interests of the powerful, but their effects may persist even after the interests that espoused them in the first place have faded. Ideas, according to Goldstein and Keohane (1993), have the potential to influence policy when they (1) act as "road maps" that provide causal links between and normative principles governing goals and the means of reaching these goals, (2) serve as "focal points" or "coalitional glue" to facilitate group cohesion, and (3) become embedded within powerful political institutions.

In parallel, the literature on developing-country resource management has overwhelmingly discussed the persistent influences of colonial forestry ideology on many postcolonial forestry policies. These influences have been especially facilitated by the institutionalization of scientific forestry into state forest doctrine and organizational procedures, including forestry training and education. Colonial forestry ideology—centered on uniform, state-centralized, scientific forestry—replaced site-specific customary management systems. It offered a clear policy prescription: the state (i.e., the forestry agency) is the sole legitimate (and capable) forest manager, state-based forestry serves the greatest good for people (through conservation as well as sustainable forest production for economic growth), and scientific forestry is an efficient and rational form of resource use (Peluso 1992). Customary local forestry practices alien to the scientific forestry doctrine were considered problematic and thus to be overcome. Embedded within state forestry institutions, this ideology became both a "road map" and a shared belief for collective action.

This chapter benefits from earlier works on professional foresters' systematic forest management ideas and their correlation with current forest management policies (Peluso 1992; Guha 1990). This is not to suggest, however, that ideas always shape policy and politics (i.e., ideological determinism). Describing the connections between ideas and policy change does not explain the process under which ideas are enacted (Goldstein and Keohane 1993). Equally important are the underlying conditions under which these causal connection exist. As Krasner (1993) rightly put it, ideas are more likely to be politically efficacious when they are in conjunction with changes in interests, power relations, or both. Legal recognition of customary forest tenure and management rights, inconceivable during Indonesia's New Order, was passed into law by the succeeding government. This was possible because of the new government's policy orientation toward more equally distributed forest-based economic benefits, as well as because of the increased political power of local-forestry proponents.

Explanations of macro political influences on local forest management systems generally follow similar arguments: enactment of particular local forest management ideas is greatly affected by the relationship of those ideas to interests and power. Specifically, it is argued that political-economy interests—both institutional (i.e., the forest department's political economy of forestry) and individual—influence the impact of ideas on policy outcomes. A synergy between the ideas and interests of key decision makers increases the likelihood of those ideas' being translated into policy (Hansen and King 2001). Decision making also involves power whose configuration and relational patterns are conditioned by the nature of a given political structure. Therefore, it is suggested that policymaking and outcomes are shaped as much by the interactions of policy ideas and interests as by the political structure within which the interactions occur. Accordingly, political structural changes have an impact on the terms of reference of these interactions, and thus, on the ideas' impacts on forestry policy outcomes. In more open or democratic systems, competing ideas are more likely to have a space in policymaking, leading to more possible policy change. But in more closed and authoritarian governments, nonstate views may be suppressed, and the "old" may continue to influence policy for a long time.

The discussion presented in this chapter consists of four sections, organized chronologically: those on Dutch Colonial (late nineteenth century–1942), Transition (1942–1967), New Order (1967–1998), and post–New Order (1998–) governments. The colonial origins of contemporary state forestry ideology in Indonesia are discussed in the first section, with particular attention to its introduction in the Outer Islands.[4] In each historical period, the struggle over different local forestry ideas during different time periods and political systems is investigated with respect to forestry law formulation. The influence of state-sanctioned and non-state-sanctioned forestry ideas on local forestry policy is discussed within the context of their confluence with political economic interests and power relations.

The Colonial Period (Late Nineteenth Century–1942)

Philosophical Origins of the State Forestry System

The introduction, in the mid-nineteenth century, of state-based scientific forestry ideas in Indonesia is inextricably linked with the colonial government's ambition to expand its political and territorial control, along with its revenues. The *domein* doctrine of scientific forestry gave the state authority to seize and control large territories of "unused" land. Prior to colonialization, the territory that is now called Indonesia was divided into self-governed principalities that favored local customary, or *adat*, law-based resource rights and usage. *Adat*s are unique—different from place to place, community to community, and time to time—and have traditionally been the principal indigenous sociopolitical institutions that have shaped local human ecology. Feudal kings and nobilities did claim the forests, but not in the Western ownership sense. With a relatively low population, combined with an inaccessible landscape and limited technology, local rulers were less concerned with controlling resource-based territory than with controlling people and production surpluses (Peluso 1992). Accordingly, resource access rights and usage were locally governed. Customary property rights, called *hak ulayat*, prevailed all over the archipelago. The law that governed the relationship between people and resources was called *hukum adat* or *adat* law.

The Dutch colonial intervention altered *adat*-based property rights systems in ways never before experienced. Western-based ownership concepts replaced complex, site-specific, and usually flexible customary tenurial arrangements over forest land and products. Yet institutionalization of state forestry did not take place at the same time and pace all over the colony; colonial forest control rapidly penetrated Java, whereas most parts of the Outer Islands, at least until the late nineteenth century, largely remained out of Dutch control. The following sections examine this institutionalization process, focusing on the Outer Islands. Java will be discussed only briefly—for it has been examined extensively elsewhere (see Peluso 1992)—primarily to highlight the main principles of state forestry ideas. Outer Island forests have been particularly important since independence, as they constitute 97 percent of Indonesia's forests and shape the nature of Indonesia's forest management politics.

The Institutionalization of State Forestry in Java The bankruptcy of the Dutch East India Company (or VOC: Vereenigde Ost-Indische Compagnie) in the late eighteenth century ended nearly two centuries of monopoly over Java's teak, with the concurrent establishment of a colonial government a turning point in Java's forestry history.[5] Unlike the VOC, whose primary concern was to secure access to Java's teak and labor, while forest land remained under local rulers' control and community forest access and rights were unrestrained, the colonial government went further by instituting direct control over Java's forest land and management (Peluso 1992).[6] To entrench its power base, the colonial government employed the same political economic strategies as at home, including the internal territorialization of land and resource control. Large tracts of "unused" land were unilaterally designated as forest reserves in which the colonial state proclaimed exclusive management rights. The underlying ideology was the concept of scientific forestry, centered on the state as resource developer and custodian and designed to promote long-term commercial timber production as a source of state revenue.

The development and institutionalization of scientific forestry was facilitated by the establishment of forest bureaucracies (i.e., forest departments) that were affirmed as the sole legitimate managers of Java's forests. This occurrence marked the beginning of centralized state forest management (which is still continuing), engendering a radical reorientation of long-established local resource arrangements. Centralized state forest control undermined local and traditional forest access and management rights and has often resulted in grassroots protests—overtly and covertly—in many parts of Java.

The 1865 colonial Basic Forestry Law and the 1870 Basic Agrarian Law, which asserted the colonial *domeinverklaring* doctrine (in which all "waste" and "unused" land would be declared state-owned) laid the foundation for the scientific forestry that is still practiced in Java today (Peluso 1992, 50). Guided by its management ideals, the Dutch colonial government managed Java's forests according to scientific silviculture principles, demarcating forest zones according to designated utilizations, and prosecuted those who disobeyed the management ideals. After nearly five decades of trial and error, which culminated in the enactment of the 1930s forestry laws, most of the basic principles of Java's forest management were in place. Barber (1989, 120–121) summarized these principles as encompassing the following themes:

1. The state owns and controls, and has the right to restrict (including with the use of force) public access to, forest lands and resources.
2. Forests are managed by a civil service bureaucracy whose primary role is to sustain timber production (with scientific silviculture techniques) for state profit.
3. Forest protection is a secondary goal for forest management.
4. Forests are managed according to laws and regulations written primarily by the forest service itself.
5. Java forest area cannot be decreased.

6. Forest management should provide benefits for the adjacent communities.

These principles, as demonstrated later, remain important in shaping postcolonial governments' political economy and policy orientation with respect to forestry.

The Assertion of State Forestry in the Outer Islands

Political Economy Trends Efforts to institutionalize the state forestry system in the Outer Islands grew in tandem with increasing colonial capitalist interests in expanding territorial control (particularly to acquire land for agricultural crop exports). Before the turn of the twentieth century, the Dutch, preoccupied with Java's invaluable teak, paid meager attention to the forests of the Outer Islands. Government attempts to invest in large-scale logging operations in these regions often ended with disappointing results. Lack of labor, deadly disease (e.g., malaria, dysentery), poor infrastructure and transportation, difficult geographical terrain that required massive capital and sophisticated technology, and—most importantly—potential conflicts with local rulers were among the factors that undermined these operations.[7] Commercial timber extraction was undertaken mainly through private initiatives—from small-scale businesses run by middlemen (who hired local timber cutters and then sold the timber to big merchants in the coastal markets), to middle-class businesses like the *panglongs*,[8] to a few foreign companies as well[9]—with their activities taxed by the government and local rulers.

In addition to a lack of economic incentives, Dutch forestry intervention was further hampered by the Outer Islands' unique customary land tenure system and the types of treaties the colonial government had concluded with local rulers. The Dutch could not unilaterally alter the provisions (e.g., about forest and land control) of any treaty involving a self-governed area without the local princes' agreement.[10] Even in directly governed areas, many local communities retained considerable autonomy to manage their *adat* territories (Soepardi 1974a, 40). Hence, the *domein* doctrine was mainly in effect in directly governed lands (e.g., Banjarmasin Sulatanate), whereas the many self-governing local rulers, although recognizing Dutch authority, maintained their traditional control over forest use, extraction, and disposal (Potter 1988).

By the late nineteenth century, the government was paying increasingly serious attention to Outer Islands forests. Aside from its ambition to expand state sovereignty through territorial and political-administrative control, this increased attention also reflected the course of political events in Holland and Europe. Liberal governments came into power, leading to criticism over colonial policy and abolishment in 1870s of Java's Cultivation System (i.e., forced labor and cash crop production), the system that had supported four decades of a government-monopolized economy. The adoption of a more open, market-oriented economy quickly attracted foreign (mostly European) private capital,[11] and the sources of this capital were permitted to lease land from the government on a long-term basis. The Outer Islands, unlike Java, were perceived as having ample cultivable and undeveloped areas. Demand for land for investment in export crops, mining, and timber soared, reinforcing the government's ambition to expand control of the Outer Islands forests. At the same time, the Dutch attempted to reduce commercial dependence on British-controlled Singapore (the center of Outer Islands trade after the fall of Malacca), which was flourishing with the intensifying world trade induced by the 1869 opening of the Suez Canal (Dick 1990). The end of the Acehnese war in 1890 also meant that the government had more resources to bring to bear in exerting its political and administrative control outside of Java (Dick 1990).

The 1870 Agrarian Law marked the beginning of liberal economy policy in the Indies. The law, through the *erpacht* right, allowed foreign investors to lease large tracts of agricultural land both from the government and from locals for up to seventy-five years. The fact that colonial laws were effective mainly in Java and Madura, however, and the *domein* doctrine in directly governed areas (both Java and the Outer Islands) hampered government territorial expansion. Some parts of the Outer Islands had different forms of colonial land and forestry regulations—dealing with logging procedures, forest protection, shifting cultivation, or general agrarian affairs—but these regulations were generally weak and in consistent and did not affect all directly governed colonies (Departemen Kehutanan RI 1986a, 84). The confusing and sometimes contradictory rules (i.e., different sets of central and local government regulations, local nobilities' rules, customary *adat* practices) that governed the Outer Islands' forest uses and access rights were perceived to hinder government efforts to "develop" and "protect" Outer Islands forests. Great concern was also voiced over local farming practices, particularly shifting cultivation, which, because of its use of vast amounts of land and its land-clearing methods, was believed to be ecologically destructive and a potential obstacle for *erpacht* issuance. This farming method was conceived to have high ecological and societal costs (e.g., removal of trees to make way for farm land, grasslands creation from ex-swiddens, loss of business opportunities through ineffective agrarian law) (Departemen Kehutanan RI 1986a, 175), and its practitioners were viewed as being lazy, as opposed to hard-working wet-rice agriculturalists (Masthoff cited in Potter 1988). Some foresters asserted that uniform forestry laws that enforced centralized state forest administration and management control, as in Java, were indispensable.

In the 1920s, some elements in the colonial government insisted that *domein* doctrine be enforced in the Outer Islands. The government's political economic orientation (i.e., expansion of political, economy, administrative, and territorial control) was the primary drive behind this insistence. The first step necessary, promoters of *domein* enforcement argued, was passing a uniform forestry law as the legal basis for this enforcement.[12] Policymakers, however, were split between those who advocated centralized state management ideas and those who favored local *adat*-based control. As the following section will demonstrate, whether or not the *domein* doctrine and scientific forestry principles could formally replace *adat* forestry systems was a function of political interaction, through which these state mainstream policy ideas and interests were asserted over others.

Policymaking Processes Efforts to establish uniform forest management rules for the Outer Islands proved troublesome as disagreements arose over the potential benefits (or lack thereof) of state forestry.[13] This reflected conflicts between local and colonial interests over the use of forest lands and its products as well as broader issues of forest control and sovereignty. State forestry advocates believed that the *domein* authority should not be questioned and that forests could be extracted wisely and maintained ecologically only by enforcing consistent regulations with direct guidance from the central forest service. Opponents (from both inside and outside government) asserted that local people should be allowed to maintain their rights to traditional forest land and products, since forest-based activities (e.g., shifting cultivation, timber cutting, gathering of nontimber forest products) remained the main means of local livelihood.

Most official foresters remained adamant in their belief in the value of a uniform forest management legal framework. The first draft of a forest management law for the Outer Islands, debated in 1923–1924, was conceptually similar to Java's 1927 forest ordinance, which divided Java (and Madura) into highly managed "teak" forests and "wild" forests, the latter consisting mainly of hydrological reserves and inaccessible mountains.[14] The majority of Outer Islands forests fell in the "wild" category; a few of these were intended for reserves, whereas most were left to local authorities. The government hoped to generate revenue from "unclassified" forests (i.e., "wild" forests outside the reserves), and thus these zones were retained for commercial exploitation (preferably with long-term concession contracts) even though local subsistence activities were still permitted. The overall management purpose was to produce profit for the state, maintain ecological conservation, and still serve local peoples' subsistence interests (Potter 1988, 138).

The law's first draft was rejected partly for technical reasons, partly for its failure to incorporate the Outer Islands' administrative governing system. In 1927, the Agricultural Department (to which the Forest Service belonged) submitted another proposal, which was strongly challenged by those (opponents of *domein*) who believed that uniform state land laws would undermine diverse local *adat*-based livelihood strategies. Chief Inspector of Forests J. Gonggrijp insisted that the state was the sole forest "sovereign" and thus had the right to levy taxes on forest-related activities, even though a portion of such taxes could be conferred to local *adat* communities. Local governments, however, refused the *domein* claim and denounced such taxes as illegitimate. To deal with this unresolved issue, the government sought advice from the Agrarian Commission (established in 1928), which (three years later) recommended that the government should respect local customary forest tenure rights or *hak ulayat* (based on the *adat* law) and that existing state forestry regulations did not conform to local agrarian systems.

Unsurprisingly, foresters who were more concerned with commercial exploitation and state forest sovereignty were unhappy with these recommendations and asserted *domein* legitimacy: the state's right to "unused" land was believed to be self-evident. The 1932 Dutch Forester Association's congress passed a resolution declaring that centralized legal mechanisms were urgently needed in Indonesia to implement state forest territorial claims and management plans[15] and that the Agrarian Commission's recommended forest regulatory scheme was applicable neither to Java nor to the Outer Islands (Departemen Kehutanan RI 1986a, 85). Among the foresters themselves disagreements existed, as some argued that the government (i.e., the Forest Service) did not have sufficient resources to directly manage such a huge territory (Haga cited in Poffenberger 1990).

Attempts to reformulate the proposed ordinance resumed in 1933, with the previous fierce debates causing the government to adopt a cautious stance. To avoid a decision deadlock, the government issued an order to avoid further debate over *domeinverklaring* and to respect, though not explicitly promote, *adat* law. In 1934, a new draft of the ordinance was submitted to the Indonesian-controlled People's Consultative Assembly (Volksraad), with the *domeinverklaring* debate reappearing, and Sumatra's representatives, Soangkupon and Mochtar, strongly challenged the premise that *domeinverklaring* would be the basis for future Outer Islands agrarian systems (Departemen Kehutanan RI 1986a). After nearly two years of legislature discussions, the People's Consultative Assembly finally approved an amended bill that obliged the government to confer all forest exploitation levies on *adat* communities living in the designated logging zones; the governor general would determine the portion the communities should remit for forest management costs. The government, however, objected to these amendments for their potential to undermine state *domein* authority. This situation remained undecided until war erupted in 1942. Despite a lack of official regulations, the government began to establish a new forestry administration, including the commencement of surveys and mapping for forest planning and reserve demarcation.

The relatively strong position of *domein* opponents to counteract the Forest Service's policy crusade may have been influenced by Holland's political climate at the time. Aside from laissez-faire economic policies, increasingly powerful liberals in Holland also insisted that colonial policy should be based more on humanitarian considerations (the

Cultivation System was especially blamed for having deteriorated Javanese peasants' living standards). By the turn of the twentieth century, the Ethical Policy was adopted in an attempt to raise the colony's general welfare and also to "protect" natives from the harmful consequence of Western economic penetration (Wertheim et al. 1985). This policy also led to the creation of representative bodies (*Volksraad*), with land rights a hotly debated issue from 1900 through 1930 (Boomgard 1989). van Vollenhoven, a leading *adat* scholar, and his followers were particularly critical of the government's *domein* crusade. In postcolonial forestry lawmaking, state- versus *adat*-based land control controversies remained, with the legitimacy of the various arguments shifting according to prevailing societal values and political economy trends.

The Transition and Old Order Periods (1942–1966)

Political Economy Trends

During Japanese occupation (1942–1945), the government's energy was devoted to war efforts (which involved massive tree felling for shipbuilding and construction), to the near exclusion of forestry development plans and other legal frameworks.[16] After Indonesia declared its independence from Holland in 1945, attempts were made (in 1947) to sort out the Dutch regulatory legacy and formulate new statutes (Soepardi 1974b, 65). This process was interrupted, however, by the Dutch invasion to reinstate Holland's power: the so-called Police Actions I (1947) and II (1948).

Soon after the 1949 Dutch handover to the newly created Indonesian government (the Old Order regime), the previously aborted attempts to formulate Outer Islands forest regulations were resumed. The process was clouded with great uncertainty, as the government changed frequently. During these early postindependence years, the central Forest Service was also preoccupied with internal consolidation and confronted with logistical problems, human resources scarcities, and internal conflicts (those who had cooperated with the Dutch during the struggle for control were resented by those who chose to fight [Soepardi 1974b, 83]).[17]

Debates over forest control remained at the center of national forest politics. Local forest access rights promoters received a boost when many began to believe that forest-derived profits were unjustly and primarily accruing to big (especially Western) investors and the government. Foresters at this time were apparently struggling to reconcile the "old, established" ways of state control with the nation's new revolutionary spirit. This could be seen in the Forest Service's conflicting policy orientations. At first, pursuant to the newly independent country's revolutionary ideology, which attempted to replace Dutch and Japanese large-scale, export-oriented tree cutting policies with a policy of distribution, at a fair price, of wood to people, the Forest Service gave top priority to distributing, at a fair price, wood to people. It soon became apparent, however, that this innovative strategy was not easily translated into practice (Peluso 1992). The early 1950s saw forest management emphasis shift from the early postwar "wood for people" rhetoric into industrial policy. The government was convinced that transforming agrarian forestry into industrial-based forestry was an essential step toward achieving economic development and national prosperity (Departemen Kehutanan RI 1986b, 48). With the assistance of an FAO staff member, J. A. von Monroy, the Industrial Forest Design Working Group (or PPHI: Panitia Perancang Hutan Industri) produced its industrial forest master plan for Indonesia, which became the foundation for the Forest Service's Five-Year Development Plan Guidelines (1956–1960). For the Outer Islands, the plan aimed to transform the natural forests (whose mixed vegetation was perceived to contain mostly "worthless trees") into economically valuable industrial plantations; under the plan, some twenty-eight species would be promoted (Departemen Kehutanan RI 1986b, 50–51).

During the Old Order, changing state forest political economic orientation did not abruptly lead to significant changes in the Outer Islands' de jure and de facto local forest use and access rights. The Dutch intent to control and profit from Outer Islands timber was continued by the Old Order, but political economic instability obstructed implementation efforts. It was during the New Order, which reinforced its predecessor's industrial policy orientation, that these efforts became far-reaching.

The Forestry Policy Process

The Old Order maintained the colonial state's political and administrative structure, including the Dutch-created forestry bureaucracy. Forestry affairs were administered centrally by the central Forest Service (in Bogor), which had separate divisions for Java-Madura and the Outer Islands. In 1957, Government Regulation no. 64/1957 attempted to divide authority over the forests, including the granting of concession rights, between central government and provincial forest services. The liberal democracy era of the 1950s created what was perhaps a more favorable political climate for this form of decentralized resource management. Besides, the newly created central government was not strong enough to exert absolute control over semiautonomous regional rulers. The regulation did not last long, as it was soon revoked by the succeeding regime.

At the same time, forestry officials did not abandon their ambitions of having a uniform forestry law for Java and the Outer Islands. As economic growth and industrial forestry regained policy prominence (in the 1950s), the central government found implementation difficult within the differing existing forest management frameworks for Java and the Outer Islands. In addition, the newly independent nation was eager to emancipate itself from the Dutch legacy, including its system of legal pluralism. A 1964 working group, drawn from the provincial and national forestry bureaucracies, mass organizations, and forestry experts, was established to prepare a uniform nationwide forestry law; in 1967, the House of Representatives passed the resulting Basic Forestry Law (BFL) no. 5/1967. Ironically, the *domein* doctrine, strongly challenged during the 1920s–1940s for its potential to undermine local communities' customary forest access, was readily adopted as the foundation of the 1967 BFL.

The 1967 BFL, which shaped three decades of Outer Island forestry, embraced the colonial forestry ideology of state-based forest control, and this was communicated according to political language at the time. The law was formulated during the last years of the Old Order, at a point during which President Soekarno, under his "guided democracy," was increasingly tightening and centralizing his political grip. The state (and its parastatal corporations) was the only legitimate forest developer (i.e., timber extractor) and protector, and local *adat* forest practices, barely mentioned in relevant forest protection articles (article 17), were interpreted as dealing primarily with the gathering of forest products and permitted as not contrary to the law's "purposes."

The social and political economic turbulence of the time, which culminated in the Communist Party's (PKI's) abortive coup in 1965 and the subsequent emergence of Soeharto's New Order government, likely had something to do with the relatively "smooth" law process that led to the formulation of the BFL. The years prior to the coup saw a worsening Indonesian economy and fierce ideological polarization between communist and anticommunist camps.[18] Foresters (especially in Java) increasingly adopted polarized ideological orientations, with some advocating state forest control and others supporting peasants' forest land distribution rights.[19] The bloody revolt in 1965, triggered by the killings of seven army generals, changed the national political configuration. Thousands of peasant activists and other PKI sympathizers were jailed or executed without trial, and for years to come, communism remained taboo and was effectively manipulated by the state to get rid of dissidents. Thus, the 1967 BFL was formulated when anticommunist feeling was ascending, when peasants' interests could be equated with communism, and when economic growth was perceived to be the only way out of national bankruptcy. The 1967 BFL was in tune with these trends; it emphasized forest economic and ecological roles according to the "old" philosophy of state-controlled forest production and conservation.

The New Order Period (1967–1998)

Political Economy of Forestry

The New Order maintained its colonial predecessor's conception of modern state building—including its political, economic, and territorial power bases—yet with more coercive force. Like many other postcolonial states, the New Order regime was determined to emulate developed countries' economic development through natural-resource extraction and transformation of indigenous-resource management institutions. The government nurtured its predecessors' view of forests as a source of state revenue that should be exploited efficiently and rationally to fuel national economic growth and modernization. The means was through large-scale commercial exploitation, controlled by the state, with villagers' customary forest practices considered inefficient and illegitimate. President Soeharto, unlike his predecessor Soekarno, welcomed foreign capital. A set of friendly investment policies was passed in the late 1960s to attract foreign capital and technology for the job of extracting timber from otherwise unaccessible Outer Islands forests. Forest administrative and political control was centralized to expedite implementation of the government's economic development policies, and Government Regulation no. 64/1957 dealing with resource management decentralization was officially revoked.

Many aspects of Dutch forestry planning and administration were maintained and strengthened by the New Order government's Forest Service. Through the 1967 BFL, large tracts of the Outer Islands' forests, most of them controlled by *adat* communities, were nationalized and converted into state property. Before World War II, the Dutch had declared approximately ten million hectares of Outer Islands forests as state territory. Under the New Order, the size of this territory increased to 114 million hectares, or about 75 percent of the country's total land mass, all under the direct control of the Ministry of Forestry (MoF). The 1967 BFL provided the state (i.e., the MoF) with the legal authority to plan and regulate all forest tenure and use arrangements within its jurisdiction. The law recognized only two types of forest tenure: those under private ownership and those with no formal ownership claim (article 2). The latter included most *adat* lands, since *adat*-based ownership was typically not officially registered, and thus the lands involved became subject to direct government control.[20]

Based on the 1920s Dutch map of the area, whose inaccuracies are evident, the Outer Islands' forest boundaries were delineated, divided, and granted to concessionaries (Potter 1988). The number of timber concessions skyrocketed from only 25 in the late 1960s to 574 units in the 1990s, involving a total area of more than 58 million hectares. During the same period, the country's forest products–based foreign-exchange earnings jumped from US$2 million in the 1960s to some US$3 billion in the 1990s, ranking second only to those from oil and natural gas. In the 1990s the forest industry accounted for 20 percent of Indonesia's nonoil exports and 7 percent of its national gross domestic product. Aside from timber industries, forests provided land for other development activities such as urbanization, mining, transmigration, plantation estates, and various forms of physical infrastructure (e.g., roads, dams). Forests also served the regime's political purposes; much of the forest-generated capital was channeled to small circles of elites, with the purpose of procuring civilians', bureaucratic, and military loyalty. Concession rights were allocated with secrecy, and access to such rights depended on one's proximity to the power center (especially the president).

As the political and economic importance of forests grew, so did the MoF's organizational size, personnel, budget, authority, and territorial control. Official and professional foresters had complete belief in the state as the superior forest manager and guardian, with large-scale scientific mechanization the most rational forest exploitation method. Local forest practices and property systems were largely perceived as a threat to state economic development and political interests. At first, under the HPHH (Hak Pengusahaan Hasil Hutan, or Forest Products' Collection Rights) system, some 20–30 percent of a total of 64 million hectares of production forest was allocated for small-scale logging and local customary use under provincial government supervision (Departemen Kehutanan RI 1986c). In the 1980s, however, the government completely revoked HPHH because it was neither economically or ecologically feasible nor easily controlled.[21] Despite official government policy to nationalize forest resources, many communities retained de facto claims over the forests surrounding them and continued their customary forestry practices. This often created conflicts between the state (and its backed business) and villagers' interests over forest use and tenurial rights. When these conflicts turned violent, the government used its military might to "resolve" them.[22] Repressive policies toward local *adat* forest practices were not always effective and partly depended on the in situ presence of elites' economic interests (i.e., stringent enforcement usually occurred where economic stakes were high).[23] Without secure user tenure, no incentives existed to utilize resources wisely, and this often resulted in open-access situations. Resource depletion ensued as the commons became a free-for-all in which each tried to harvest as much as possible before others did (Bromley 1992). Estimated deforestation rates during this period range from 600,000 to 1.3 million hectares annually, depending on whose numbers are used.[24] The government's view was that resource degradation was caused by villagers' destructive forest use and farming methods, particularly shifting cultivation. Accordingly, forestry officials tried hard to keep villagers away from the forests by some combination of outlawing access, resettlement out of the forests, or "education" on the virtues of sedentary farming over shifting cultivation.

The 1980s, in response to national and international events, saw the government paying increased attention to environmental issues. As the international community became increasingly concerned with the adverse ecological consequences of current development practices, many donor agencies incorporated strong environmental criteria into their development project portfolios. Donors, led by the Ford Foundation, also began promoting social forestry as an alternative rural development program that embraced environmental values and local participation. The catastrophic 1982 forest fires in Indonesia devastated the country economically and ecologically, provoking public debate over deforestation as a cause, with blame again directed at shifting cultivators. At the same time, the spread of democratization and globalization in many southern countries, including Indonesia, allowed many environmental (and later human rights) NGOs to flourish. These groups began to criticize the massive state-sponsored forest exploitation and unequal forest distribution benefits in Indonesia, including the government's harsh policies toward forest communities. They were also the loudest advocates of granting indigenous forest management and property rights. WALHI (Indonesian Forum for the Environment) and SKEPHI (NGO Network for Forest Conservation) were among the pioneers that supported a local-based forestry system, and their advocacy work (as well as that of other NGOs) was probably one of the major forces that later pushed the community-based forest management issue onto the formal policy agenda in Indonesia. The MoF, to a limited extent, tried to accommodate the demands of these organizations by coopting them into its policy framework but categorically refused to recognize local and customary forest tenure and management rights.

Although NGOs were hardly a threat to the regime (those who became too critical were often simply repressed) the government, for various reasons, could not totally turn a deaf ear to their demands. First, the grassroots services provided by NGOs helped the resource-poor government (e.g., government projects sometimes were executed by NGOs); second, these groups' strong international links meant that heavy-handed policies could invite international criticism, which in turn might lead to withdrawal of some development assistance (Eldridge 1995, 29).

Although, generally speaking, the New Order ruled with an iron fist, relations between NGOs and the government were not clear cut: at times there were conflicts, at times collaboration. Which occurred in any particular instance largely depended on the ideological predisposition of the NGOs concerned (i.e., whether they were more collaborative or confrontative with the government), the type of issues encountered, personal relations through strong alumni networks, and individual viewpoints.[25]

In addition to external demands for policy change, the MoF also faced internal pressures: some bureaucrats urged the department to modify its policy direction to keep up with existing and anticipated future socioeconomic and forest conditions. The late 1980s saw the beginning of incremental policy changes toward local forestry practices. Several ministerial decrees launched different social forestry programs that granted local participants limited forest user rights.[26] Social forestry programs derived from ministerial decrees do not enjoy a very strong legal basis, however, as the same ministers who originated them or any succeeding ministers can revoke them at any time. More importantly, ministerial decrees cannot provide access and tenurial rights beyond those sanctioned by the 1967 BFL. It is this law, which grants only limited local user rights (i.e., collection of nontimber forest products), that has the most profound and permanent impacts on formal and informal forest institutional arrangements in Indonesia. The next section will describe how process of changing this law, despite some internal pressures toward such a change, was a long and arduous one. The conception of local forest access and tenurial rights drawn from state forestry doctrine remained a powerful force that shaped most actors' policy preferences. Several reform stages were involved, each involving different actors with different issues and interests who employed different tactics to pursue their policy preferences.

The Policymaking Process

Two views of local tenurial rights dominated law reform in the New Order era, with neither challenging the state's role in terms of ultimate management control and proprietary rights. The first view was that there was nothing wrong with the 1967 BFL and its provisions for local forest access rights. The second argued that the law needed reform to its overly economic growth–oriented focus and its rigid conception of state forest control; local, particularly *adat*-based, forest management systems should be recognized, albeit within the context of state discretion. It is important to note another view that existed, though outside of formal policymaking circles. Held mostly by national and international NGOs, as well as a few academics, this view was that management and proprietary rights should be given to local, especially *adat*, communities with minimal state intervention. It was grounded in more than just the promotion of local social justice and economic benefits, being deeply rooted in the value of local self-determination. Although there was no direct political access to the policymaking process among those holding this view, the ideas it expressed may have influenced forest bureaucrats in other, more subtle forms (e.g., through their exposure to media and public campaigns).

The New Order's first forestry law reform attempt took place after nearly two decades of forest plundering and repressive measures toward local practices. The late 1980s saw slight bureaucratic changes in policy orientation, marked by the MoF's undertaking to revise (or, as the MoF called it, "to perfect") the 1967 BFL, under the probable order of President Soeharto to Hasyrul Harahap, then the who was forestry minister (Manurung 1997). Why the president would have issued such an order is unclear (Manurung 1997). Regardless, the ensuing process uncovered hidden internal divisions between reform and status quo proponents within the MoF.

The New Order lawmaking process followed several stages, each involving different players with likely different interests. The first stage, internal to the bureaucracy, typically involved formation of a working group consisting of middle- and top-level officials to prepare a draft bill. The next step, "public" consultation, involved national- and provincial-level stakeholders (e.g., NGOs, international agencies, academics, local governments, professional associations) to comment on the draft. The intensity of this process (e.g., how many times consultations were held, the numbers of provincial governments consulted) depended on the available budget and time, and the willingness of the bureaucracy to conduct such consultations. The revised draft (based on the relevant public comments)—if indeed the bill was revised at all—was further discussed in interdepartmental meetings, and if this process went smoothly, was then sent to the president and House of Representatives for approval.

In its "academic statement" (i.e., the MoF official position chapter describing scientific reasons for the BFL's "perfection"), the working group assembled to reform the BFL argued that the law, which was formulated during Soekarno's guided democracy, was no longer germane to existing sociopolitical and ecological conditions in the country and not appropriate for the second long-term (twenty-five years) national development plan that would begin in the mid-1990s (Tim Penyempurnaan Undang-Undang no. 5 Tahun 1967, Undated). Thus, the law required refinements to enable the forestry sector to continue developing and to deal with existing and future internal and external forces (e.g., changing forest conditions, technology, demography, people's perceptions, and sectoral and regional development) (Tim Penyempurnaan Undang-Undang no. 5 Tahun 1967, Undated, 56). Some bureaucrats (largely senior foresters) opposed the idea of reform, fearing that any modification would reduce MoF's power and forest control (Sukartiko cited in Manurung 1997, 24). These foresters insisted that the 1967 BFL was a "masterpiece" of Indonesian foresters and successfully superseded antiquated Dutch forestry legal frameworks (Kamdiya, cited in Manurung 1997, 24). Any revision of the BFL, they believed, should be only minor.

The reform process took a long time for both technical and political reasons. The members of the working group team changed several times, partly because some could not function properly (e.g., too busy with other tasks, posted to other areas) and partly because of differing viewpoints (anti and pro reformers) that could not be resolved. After several changes, the final team consisted entirely of reform proponents (Manurung 1997, 25). In 1996, seven years after the inception of the process, the tenth draft of the revised bill was completed. Although the *domein* doctrine and scientific forest management principles remained, the proposed changes were quite comprehensive. They focused on clarifying, expanding, elaborating, and adding definitions (e.g., forest, forestry, state forest, forest products, forest function), procedures (e.g., forestry planning, people participation), and rights and responsibilities (e.g., district and provincial government rights and responsibilities), which the 1967 BFL only broadly and at times inconsistently addressed. One major change was the inclusion of nonstate (i.e., non–central MoF) players (e.g., regional governments, private companies, cooperatives) in forest exploitation. *Adat* forests were still claimed by the state, but unlike in the 1967 BFL, *adat* forest territory and management were recognized under conditions that would be later detailed by the MoF (Manurung 1997, 31).

This tenth draft was discussed with stakeholders (e.g., representatives of NGOs, professional associations, provincial governments, and provincial forestry and other sectoral offices) in four provincial cities: Medan, Banjarmasin, Ujung Pandang, and Surabaya (Manurung 1997). Major issues raised by NGOs and other critics were primarily related to fairer forest benefit sharing and more balanced decision making between central and regional governments and local communities, more democratic access to forest exploitation (particularly for small- and medium-scale enterprises), greater public participation in forest planning, recognition of local and *adat* forestry rights, and mechanisms for more-competitive forest industries.

Instead of revising the draft based on these public inputs, however, the MoF abruptly suspended the process. The reason given for the suspension was not NGOs' or other outside criticism. The MoF was well aware that nonstate actors were not a threat to its power. Rather, it was apprehension over the Agrarian Ministry/National Land Agency's dissenting opinions. Before the draft bill went to interdepartmental scrutiny (after public consultation), some top-level MoF officials discovered that key Agrarian Ministry figures and leading agrarian experts had publicly declared that the Agrarian Ministry should be the highest public institution to wield state *domein* authority, although such authority can be delegated to other institutions under the Agrarian Ministry's consent and supervision.[27] These statements were made during the 1996 Yogyakarta Seminar on Intersectoral Coordination Policies on Land Conflict Management (Koordinasi Lintas Sektoral dalam Penanganan Konflik Pertanahan), sponsored by Gadjah Mada University and the National Land Agency. Pursuant to these statements, the MoF secretary general sent an official letter (no. 2767/II-Kum/96, dated November 28, 1996) to the forestry minister requesting suspension of the BFL reform process to prevent the interdepartmental forum from being used by the Agrarian Ministry to challenge the MoF's *domein* forest control.[28] The MoF feared that loss in the debate would result in Agrarian Ministry jurisdiction over state forest land (75 percent of the country's territory), which would emasculate the MoF. Internal divisions, coupled with outside or nongovernmental criticisms, led the MoF to believe it would not fare well in interdepartment debates on the matter. Accordingly, in March 1997, an official letter (no. 248/I/Kum-I/97, dated August 7, 1997) was sent by the head of the MoF's Law and Organization Bureau to the reform working group to suspend the reform process indefinitely and instead switch to building internal and external support for MoF's territorial authority. The reform process abruptly resumed a year later in response to mounting public demands triggered by the New Order's collapse.

The dynamics of BFL reform in the New Order period demonstrates that diverse viewpoints (both within and external to the state), again, exist over who should control the forest and how. Even among the MoF foresters, who are well known for their belief in state sovereignty over "its" forests, different viewpoints arose over how to translate this dogma into practical policies. Formal forest tenurial arrangements are a result of how these viewpoints are negotiated at every stage of the policy process. Political configurations provide a framework within which each policy player can assess the others' role and status (e.g., whether a particular group's voice needs to be considered) and act accordingly. Although the New Order's policy formulation process included several players (i.e., the People's Consultative Assembly, the House of Representatives, the Supreme Court, the bureaucracy, the public), in practice, it was the bureaucracy that held primary decision-making power. Unsurprisingly, the MoF anticipated the Agrarian Ministry's dissent with considerable anxiety.

In the New Order's political structure, the 1945 constitution mandated that the People's Consultative Assembly formulate state general policy guidelines, that the House of Representatives pass laws and legislations, that the Supreme Court conduct judicial review, and that the government formulate rules and regulations. Subsequent subordinate laws and informal political practices, however, systematically weakened the legislative and judicial bodies, instead championing the executives and bureaucracy as the primary law- and policymaking agents. At the same time, through the "floating mass" doctrine, citizens' political participation was systematically crippled. People were disconnected from their political parties, which were more accountable to the government than to the people they were supposed to represent. The system created a hierarchial power configuration in which the president, as the highest executive and military leader, was at the core, followed by the president's closest state and nonstate benefactors (e.g., families, friends), the bureaucracy and military, other political parties, and lastly ordinary citizens (Sanit 1998). While the line of command was from the core to the far end, the line of accountability was the reverse.

The fall of the New Order in 1998 was accompanied by efforts to break the elites' economic and political power base and restore popular political authority. In the forestry sector, it was seen by many as a window of opportunity to transform MoF policies from ones that benefited the privileged few to ones centered on transparent, democratic, equitable, and sustainable forest governance. Despite its weakened position, the MoF still attempted to ensure that its own policy interests would not succumb to popular demands for reform. The result was BFL no. 41/1999, which incorporates some of the elements that the public had been demanding during the review process for the last draft. The majority of the 1999 BFL's forest management principles, however, are similar to those in the previous draft BFL.

The Post–New Order Period (1998–)

Political Economic Trends

Regaining political legitimacy, badly damaged with the New Order's fall, was an additional post–New Order government policy considerations.[29] Following the devaluation of the Thai baht in July 1997, Indonesia's currency began to depreciate, and within a year had decreased by over 70 percent.[30] Whereas other neighboring countries such as Thailand, Malaysia, and the Philippines began to recover soon after, Indonesia went deeper into economic crisis, triggering bloody political upheavals that eventually forced president Soeharto to resign. It soon became apparent that the country's ongoing financial disaster was closely linked to weak political economy structures and poor governance systems. Pressures for democratic public decision making mounted. As economic growth—the basis of the New Order's strength and "legitimacy"—reversed, political doors began to open to previously excluded social groups. Many formerly powerless nongovernment interest groups now believe they are better able to influence government policies, including those involving the forestry sector.

In the meantime, attempts to overcome the central government's grip are occurring in virtually every province, either through calls for independence or through those for local government autonomy. This previously unimaginable social transformation has manifested itself in dramatic changes in the forest landscape. Villagers' demands for full forest ownership are escalating, as are local governments' determination to have decentralized resource management. Rapidly increasing societal pressures for democracy, together with long-standing dissatisfaction over destructive state-sponsored forest exploitation and highly unequal distribution of forest benefits, have resulted in expressions of alternative forestry paradigms. "Forests for people," instead of for big business, has become a common slogan and is expected to become the guiding paradigm of any new policies. The MoF itself has adjusted (at least on paper) its new development vision in favor of people-oriented forestry, democratic forest access, and more just distribution of forest generated benefits. Yet—and differing from the demands of many NGOs and peoples' organizations—the ministry's idea so far is to limit large concessionaires' forest control and to grant local forest management rights under the state's supervision and discretion. How and to what extent this fast-changing social and political economic situation will affect de jure forest tenure depends on the manner in which the contending parties pursue their forest management ideals.

The Policy Process

The post–New Order government does not possess the near absolute control of the previous administration's bureaucracy to screen out undesirable policies. The previous three views on local forest access rights now all have political access to engage and influence the formation of forestry law.[31] On June 29, 1998, a month after Soeharto stepped down, the MoF (now the Ministry of Forestry and Estate Crops, or MoFEC) established a Reformation Committee with members drawn from itself, NGOs, academics, and business. The committee's primary task was to redesign the MoFEC's development vision, provide recommendations for organizational reform, and reformulate the 1967 BFL and its primary implementing regulation (Government Regulation no. 21/1971 on management of forest production). Despite some conceptual inconsistencies (especially on property regimes), the committee's BFL draft (BFL1) greatly differed from its predecessor, particularly in terms of providing a more favorable legal political climate for local community forest access rights. *Adat* customary forest practices were formally recognized, although *adat* lands remained under state tenurial control. Public opinions and criticisms of the new proposed BFL were invited. Many welcomed this initiative as a positive gesture toward a more democratic policy process. A series of public hearings involving various stakeholders were held in Jakarta and several other provinces. One major recommendation that emerged from these hearings was that *adat* forest be recognized as a separate category distinct from state forest; the MoFEC, however refused, to include such a distinction in the draft.

As time progressed, nonstate stakeholders began to accuse the government of hypocrisy in decision making, as exemplified by the BFL policy process. On the one hand, the MoFEC had established the "independent" Reformation Committee and was organizing a "democratic forum" to gather public inputs. At the same time, and behind closed doors, the Ministry was already drafting its own BFL (BFL 2). Many suspected that the government's Reformation Committee was merely a democratic camouflage to gain badly needed political support. Muslimin Nasution, head of the MoFEC at that time, justified the parallel law-making processes as a mechanism for understanding a range of policy options (Nasution 1998). As a new minister, he needed to understand policy interests both within and external to the MoFEC, so that he could then act based on the best available option (Nasution 1998). In addition to those from NGOs and certain people's organizations, criticisms of this nontransparent policy process also came from retired top government officials. Djamaluddin Suryo Hadikusumo and Emil Salim (former ministers of forestry and environment, respectively) publicly asked the Parliament not to approve BFL 2 since it still was not clear on *adat* rights issues and focused on timber exploitation instead of forest management (Fay and Sirait 2002).

Some NGOs and individuals, unhappy with the policy development process, demanded a moratorium on any new BFL policy formulation until after the June 7, 1999, election. The Habibie government (May

1998–November 1999) that succeeded Soeharto's was seen as only a caretaker and had no legitimacy to pass laws and regulations. Rumors circulated that the Habibie government was attempting to pass as many laws and regulations as possible prior to the June election to preserve the New Order's political influence. Nasution acknowledged that he believed the new BFL should be passed before the elected government came into power, arguing that the new government would most likely be preoccupied with politics and would not have time to design a "good" forestry law (Nasution 1998). Several months after the BFL was first drafted, some NGOs and academics (with donor agencies' support), part of a coalition network called the Community Forestry Communication Forum (Forum Komunikasi Kehutanan Masyarakat, or FKKM), proposed another version of the BFL (BFL 3) that promoted local forest management and greater state forestry accountability. Local people were assigned to be the prime players in forest management, with *adat* communities granted full management and land ownership rights at the expense of the government's forest management authority and territorial control. The BFL3 upset MoFEC officials, as it involved a substantial reductions in the MoFEC's authoritative control (and thus its power). By late 1999, despite protests by NGOs and other community forestry advocates, the House of Representatives ratified the government version (i.e., BFL2, with some elements of BFL 1 incorporated) as BFL no. 41/1999.

Although it considered relinquishing forest property ownership to local people inconceivable, the government was in a weak position to ignore swelling demands for resource management decentralization. During 1998–1999 violent street demonstrations to protest government decisions had became an everyday occurrence. A network of eighty-two NGOs and student organizations, KUDETA (Koalisi untuk Demokrasi Pengelolaan Sumberdaya Alam, The Coalition for the Democratization of Natural Resources) besieged the MoFEC headquarters demanding sustainable natural-resource management, including benefit sharing for local communities (Fay and Sirait 2002). Reform pressures also came from the World Bank and International Monetary Fund in the form of conditions on a financial bailout package sponsored by these organizations. These agencies focused mainly on market-oriented reforms whose primary objective was to improve the efficiency and environmental sustainability of timber production and other forest industries (Seymour et al. 2000). Under such pressures, the new BFL tried to accommodate transparency, efficiency, and social justice values by, among other measures, limiting concession size and redistributing concession rights through an open bidding process. Decentralization and equity were perceived to be attained by allowing local people, including *adat* communities, to manage forest areas or to control timber concessions through government-controlled cooperatives. Many perceived this as symbolizing a shift in political orientation from favoring of a few wealthy conglomerates to control by government-sponsored cooperatives (Solomon cited in Seymour et al. 2000).

Some of the principles of BFL no. 41/1999 are similar to those in the tenth draft of the New Order BFL, although the latter is more detailed. *Adat* forests, for instance, are declared to be state forests in both versions. Yet unlike in the tenth draft, *adat* management rights are clearly articulated in the new law, including other forest users' duties to *adat* (and other forest dwellers') communities.[32] Although the government is not willing to revoke its territorial claims to the *adat* forests, there is currently legal space for local people to practice their *adat* forestry systems. This is a significant change from the government's position of the last three decades and would not have been possible without the recent transformation in the nation's political economy.

It is as yet too early to assess how the new law is being implemented. The law itself is being challenged by community forestry supporters who keep demanding full local forest ownership and control, especially for *adat* communities. Concerns have been voiced that the law requires *adat* communities to form cooperatives in order for their forest claims and management rights to be recognized. Cooperatives have traditionally had a bad reputation in Indonesia, as they were perceived to be used by the New Order to control rural socioeconomic and political life. The fear now is that the cooperatives required by the new law would be used by the government to tighten control over *adat* communities' forest rights and practices.

Currently, despite the government's weakening power and legitimacy, the bureaucracy remains powerful in policymaking, partly because mechanisms for public participation are not yet in place, and also because the House of Representatives often does not have sufficient technical or policy expertise for forestry issues especially community-based forest management. Critics also argue that the current government retains many characteristics of the New Order government (e.g., highly personalized political and economic relations with nontransparent decision making). The New Order ruling party, Golongan Karya, remains powerful (as the nation's second-largest party) and controls many strategic positions in state agencies. After more than thirty years of unchallenged government domination, 1998–1999 was the low point of government's authority, which was conceded with the adoption of many progressive policy decisions. As public euphoria over good governance and democracy gradually wanes, however, the ruling elites are reinstating their political influence. Thus, inspite of nonstate policy actors' increased power, the old political structure has not radically changed, and the executive and bureaucracy continue to predominate the process of policy formulation. Unlike under the New Order, the current political climate does allow the general populace to express political demands openly and freely, although whether the government will act on those demands is another story.

On the ground, where another forest contestation front is located and where central government control is much less prevalent, the government's diminishing legitimacy is clearly manifested through chaotic and at times unlawful de facto forestry practices (e.g., soaring rates of illegal logging, increasing forest ownership claims and counterclaims). In some areas, local people with long-standing grievances are overtaking concessionaire land, confiscating or sabotaging their equipment, or asking for financial compensations for perceived prior land expropriation. In the meantime, the number of cooperatives is increasing, with memberships including locals, urban rich, bureaucrats, ex-employees of concessionaires, etc., who are attempting to use the new law for their best self-interests. Some local governments, in the name of decentralization and power devolution, have issued forest exploitation licenses for small-sized forest parcels (approximately 100 hectares each) without the benefit of transparent processes or clear indications that ecological considerations have had their due. The forestry bureaucracy, always weak in policy implementation, for its law enforcement during the New Order was greatly aided by the military force, now has virtually no capacity and legitimacy to oversee use practices for rapidly depleted resources. The military itself is attempting to maintain a safe distance from the government while being kept busy by lingering bloody ethnic conflicts (e.g., Maluku, Aceh, Poso, West Kalimantan). The current forestry minister, without any prior forestry background, seems more preoccupied with internal consolidation (e.g., investigating departmental corruption) than with taking serious and concrete measures to stop deforestation. Illegal logging is widespread, with current deforestation rates estimated at 1.7 million hectares per year.

The nation's changing political structure has forced forestry players to redefine their roles and status at every instance of encounters between people and resources. The lawmaking process is one of these instances, with the outcome (i.e., BFL no. 41/1999) providing new legal grounds on which definitions, categories, classifications, and other forms of boundaries are being renegotiated. The BFL's implementation (or lack thereof) is another sphere of forest contestation that will determine Indonesia's future forest conditions. At the same time, demands for regional autonomy, including decentralized resource control, are escalating after three decades of thightly centralized political power. If resource management decentralization is put into effect, this may profoundly change the course of the country's forest governance and socioecological landscape, although in a way that is still unpredictable. Shifting resource control and management authority from central to local government does not automatically mean that the commons will be better managed. What is being decentralized, to whom, and how it is done are some of the many questions that will shape the fate of Indonesia's forests and their governance.

Discussion and Conclusion

The analysis presented in this chapter suggests that forest conditions and tenurial arrangements in Indonesia are highly interlinked with the broader macro political economy system. In each historical period considered (Colonial, Old Order, New Order, and post–New Order), political economy trends and state forestry policy orientations shaped forest management directions. New governments with new economic development vision, accompanied by changing political configurations among contending forestry interests, often resulted in forest management policy shifts that, in turn, affected overall forest usage and access rights patterns. This is not to suggest that every change in government automatically leads to new legal forest tenure: many other factors (e.g., historical events, changing political structures, long historical battles over forest control), as previously discussed, are involved.

Despite the changing regimes, the state *domein* dogma and Dutch-introduced forest management principles remain pivotal in shaping the postcolonial Indonesian government's forestry ideology, even though those principles are continuously reinterpreted and negotiated according to ever-changing conditions and interests. At every stage of government-sponsored policy deliberations there are challenges from both within and without. Policy outcomes are a function of how these competing ideas and interests, through conflicts, collaboration, and compromise, are resolved within the existing framework of political relations. Table 8.1 summarizes the relationships among policy ideas, interests, decision-making powers, and policy outcomes over different periods in Indonesia's history. The identified ideas and interests simplify more complex realities. In addition, it is important to note that political actors that embrace particular ideas do not necessarily have interests that allow those ideas to be adopted. During the colonial era, for example, some opponents of state forestry enforcement in the Outer Islands were official foresters who, despite their belief in the virtues of state forest control, held that *adat* forestry should remain. *Adat* forestry practices, they argued, were critical for local livelihoods and could save the government administrative and political costs. Policy choices, therefore, do not necessarily reflect political actors' belief systems. Ideas in and of themselves, no matter how strong or logical, do not guarantee policy adoption; equally important is their confluence with political actors' interests and institutional power.

During the colonial period, efforts to institutionalize the state forestry system grew in tandem with increasing colonial capitalist interests in expanding territorial control. Yet opponents (from both inside and outside government) asserted that local people should be allowed to maintain their traditional rights to forest land and products, since forest-based activities remained their main means of local livelihood. Policy disagreements could not be resolved, and most Outer Islands forests continued to be managed, de facto and de jure, by numerous autonomous *adat* communities. State forestry ideas did not find their way into policy adoption, for they did not conform well with the interests of those who made decisions.

The policymaking process during the Transition period is not well documented, although subsequent enactment of the 1967 BFL is reflective of the relatively strong position of state forestry proponents during that time. Nevertheless, looking at political conditions during the policy formation process (1966–1967), it is highly likely that state forestry ideas supported the interests of Soekarno's guided democracy (with its highly centralized state) and forest service orientation (i.e., industrialization).

During the New Order, the 1967 BFL became the legal doctrine behind state forest management and thus served to frame decision makers' policy preferences and prevent ideas unfavorable to the BFL from gaining political access. The government unilaterally declared huge forest tracts, including large amounts of community controlled forests, to be state land and granted exploitation rights to big businesses, superseding non-state-institutional resource arrangements, as practiced by many forest-dependent communities.[33] Despite state pressure and threats of legal sanction, many local communities continued to practice their locally specific forest management traditions. As a result, overlapping and often conflicting forest uses, access rights, and ownership claims—resulting from the gap between de jure and de facto forest management regimes—are at the heart of most of Indonesia's current forestry problems. These conflicts, coupled with three decades of unjust distribution of forest-generated wealth, continue to place serious social and political burdens on the current government.

Institutionalization and legal codification of state forestry ideas into the forest service's mandate, organizational norms, and standard operating procedures had long-lasting impacts on subsequent policy. Ideas, Sikkink (1991) rightly argued, acquire force when they are embodied in (politically powerful) institutions, for these institutions facilitate the implementation of these ideas by giving them organizational means of expression. Policy proposals not in line with these ideas (e.g., relinquishing forest control to local people) were likely to be rejected, for they carried serious legal, administrative, and political costs. Reform deliberations during the New Order (1989–1997) to increase the economic benefits that local peoples derived from forests faced opposition from those who adhered to absolute state forest control. Status quo proponents were

Table 8.1
Policy ideas, interests, decision making, and outcomes of forestry law formations

	Period			
Policy outcome variables	Colonial late 19th century–1942	Transition 1942–1967	New Order 1967–1998	Post–New Order 1998–
Policy ideas	State-controlled forestry Adat-controlled forestry	State-controlled forestry	State-controlled forestry State-controlled forestry with limited management rights for local people	State-controlled forestry State-controlled forestry with management rights for local people *Adat*-controlled forestry
Interests	Expanding colonial political, economic, and territorial control Saving state resources Maintaining and improving local peoples' livelihoods and	Unknown	Maintaining state political, economic, and territorial control over forests Improving forest-based economic benefits for local people	Maintaining state political, economic, and territorial control over forests Political legitimacy Reducing government forest control

				Democratic and transparent forest use and access Social, economic, and political justice
Actors and decision-making power	Bureaucracy (i.e., Agrarian Ministry, Agriculture Department, Forest Service) People's Consultative Assembly	Bureaucracy Mass organizations Academics House of Representatives	Bureaucracy (i.e., Forest Service and, to a limited degree, other relevant departments) Few academics	Bureaucracy (i.e., Forest Service and, to a limited degree, other relevant departments) House of Representatives Academics NGOs Business
Policy outcomes	Forestry law formation failed for the Outer Islands; *adat*-based forestry system remained	1967 BFL with rigid adoption of state forestry system and restrictive local forest tenure	Reform failed; 1967 BFL continued with restrictive local forest tenure	BFL no. 41/1999, adopting more flexible interpretation of state forestry system, and recognition of local and *adat*-based management rights

ultimately victorious because of institutional "survival" issues: it was perceived that adoption of forest management power sharing policies would open up opportunities for departments in other sectors of the government (e.g., the National Land Agency) to wrest absolete forest control away from the forest service.

For the past decade, the centralized state ideology has been contested by those who advocate a decentralized, community-oriented approach, including relinquishing forest ownership to local and *adat* communities. Even though the government is not monolithic, as some within it support a limited degree of decentralization, nonstate players and their contesting views have remained, until recently, on the periphery of the public policymaking stage. With the government's recently diminished power and legitimacy, those whose policy ideology was previously marginalized have gained more political power and have been able to boost their agenda into the public policy sphere. Yet Indonesia's experience in formulating BFL no. 41/1999 suggests that, in the absence of institutionalized public participation and transparency in decision making, the bureaucracy largely remains in control of the policymaking process. Recent adoption of progressive community forestry policies is attributable more to unusual circumstances and the government's attempts to win back some of its heavily damaged political support than to any sincere desire for power sharing. This, coupled with the MoFEC's internal divisions over resource management devolution, creates a big question mark in terms of how, or whether, implementation of BFL no. 41/1999 will occur. At a minimum, however, it is safe to say that a more democratic regime, with an open bureaucracy and policymaking process, is more likely to allow more ideas to gain political access. This, in turn, increases the likelihood of policy outcome variations.

Forests in Indonesia represent a battleground on which multiple and often conflicting interests—local, national, and international—interact mediated by a complex grid of formal and informal sociopolitical institutions. The lawmaking process represents one of these spheres, a battleground, with forest-dwelling communities historically being the weakest, politically and economically. Even with the current government, direct representation of community groups in policy debates remains uncommon, although some NGOs and academics often claim that they represent "local communities." The discussion in this chapter demonstrates that the marginalization of local communities' forest access and rights is related to the absence of local political rights and representation. By demonstrating the importance of supralocal forces in shaping local resource tenure and management, this chapter calls for community forestry advocates to prioritize local political empowerment. Forests will always be a contested domain in Indonesia; political empowerment allows local communities to defend their rights vis-à-vis other interests. Institutionalization of local participation at every level of the policy process, together with transparent and accountable decision making, should become the major reform agenda.

This chapter does not intend to suggest, nor to romanticize the idea, that relinquishing forest management control and assuring local *adat* tenure is a panacea for every forest management problem Indonesia may have. A policy shift to community- (*adat*- and non-*adat*-) based forest management is a good start for both pragmatic (e.g., local people deal with natural resources on a day-to-day basis) and social justice (e.g., forest-dependent communities are usually poor) reasons. Such a shift, however, is only the beginning of another complex resource management trajectory. Who are the communities, who should decide which communities are eligible for particular rights and access, what kinds of rights and responsibilities are conceded, and what trade-offs are incurred and how to address them are but a few of the numerous awaiting questions.

Acknowledgments

Most of the information in this chapter was collected during nine months (May 1998–January 1999) of field research in Indonesia. The author would like to thank the Ottawa-based International Development Research Centre (IDRC) for providing financial support, the Bogor-based Centre for International Forestry Research, and the Samarinda-based Environmental Research Centre of Mulawarman University for their institutional support. Special thanks to Togu Manurung for providing data on the New Order's 1967 BFL reform process and to Anil Gupta, who read the manuscript and corrected the grammar. Nives Dolšak, Elinor Ostrom, and anonymous reviewers have critiqued the manuscript. Errors and omissions are the author's sole responsibility.

Notes

1. Ideas may entail various aspects of belief systems (e.g., culture, political ideologies). In this chapter, "ideas" refers to what Goldstein and Keohane (1993, 10) label "causal beliefs," that is, beliefs about cause- and-effect relationship in a particular issue area, which usually provide guides for individuals on how to achieve their objectives.

2. Some examples of policy implementation disrupting local-based resource management practices are Lynch and Talbott 1995 and Moniaga 1993.

3. Forestry law reforms have dealt with various aspects of forest management. This article focuses on those that pertain to local forestry systems.

4. "Outer Islands" refers to Indonesia's territory outside Java (the "Inner Islands").

5. Peluso (1992) provided an excellent account on modern forestry institutionalization in Java. This section is built on her work.

6. The VOC was mainly interested in trade monopoly and did not directly rule villagers. Instead it entrenched its power over local rulers who, in turn, extracted produce from villagers (Robinson 1986).

7. In 1850, for instance, the government "discovered" teak forests in Muna, (Southeast Sulawesi) which the local prince had long tried to "hide," but because of a shortage of capital, exploitation of these forests began only in 1910. The Dutch had to compensate the local rulers with 5,000 guldens for the use of the forest, since the forest was in their territory. Laborers were imported from other places such as Java, Ambon, Timer, and Flores, and many of them were later inflicted with malaria and dysentery. Five years later, the government withdrew because of lack of profits as well as the colonial government's new regulations that declared that logging extraction in the Outer Islands had to be conducted by private companies.

8. *Panglong* is a timber-felling business with Chinese capital and laborers operated mainly in Riau, Lingga, Singkep, Bengkalis, and Indragiri Hilir (Departemen Kehutanan RI 1986a, 142). Forest exploitation in these areas began in the 1870s in response to soaring hardwood demands for a big port construction in Singapore. Timber harvested from *panglong* forests belongs to the colonial government, whereas other nontimber forest products are owned by the local rulers (*landschap*) (Soepardi 1974a, 43).

9. Most of big investors, however, suffered great financial lost (especially before steam-engine machine was invented in the early twentieth century), not only due to unfriendly environment but also harsh competition from small-scale local timber felling (Departemen Kehutanan RI 1986a, 142).

10. Self-governed territories refer to areas where local rulers still maintained a large degree of autonomy and self-government under overall Dutch sovereignty. Directly governed areas were under Dutch direct political and administrative control, usually as a result of military defeat.

11. Driven by the world's high demand for tropical products, while the American tropical plantation system shrank because of the abolition of slavery (costless labor) in the United States and the independence of Latin American nations, European capitalists searched for more profitable (e.g., with cheap labor) and hospitable investment areas such as Southeast Asia (Broek 1971).

12. Systematic surveys, mapping, and debates over forest institutional arrangements had already begun in the early twentieth century. Sporadic forest surveys, particularly in search of new areas for exploitation, had been conducted much earlier in several parts of Sumatra, Kalimantan, Sulawesi, and Irian.

13. This section is compiled from Departemen Kehutanan RI 1986a and 1986b.

14. Areas outside reserves were loosely managed for local subsistence needs (Potter 1988, 138).

15. By 1939, the colonial government claimed territorial control over 8 percent of approximately 122 million hectares of Outer Islands forest. This comprised 7,726,800 hectares of forest in the directly governed areas and 2,591,600 hectares in the self-governing regions (Departemen Kehutanan RI 1986a, 104).

16. A majority of information in this section is derived from Sejarah Kehutanan Indonesia II (Departemen Kehutanan RI 1986b).

17. The war devastated the Forest Service, not only because most Dutch foresters left the service, but also because of massive destruction of forest industry facilities. Aside from the intentionally aggressive tree-cutting of the Japanese, Dutch foresters' "scorched-earth" policy (just before the Dutch surrendered to the Japanese) caused heavy damage to Java's forests and industrial forestry. The damage was estimated at around 17 percent (500,000 hectares) of Java's total forest land (Departemen Kehutanan RI 1986b, 60).

18. Soekarno was in power from 1945 to 1967, during which three political periods can be distinguished: revolution (1945–1949), liberal democracy (1950–1958), and guided democracy (1959–1965). During guided democracy, in which democratic rights were weaker than in previous periods, the national economy deteriorated, with inflation reaching 650 percent. Communist and noncommunist distinctions made here are merely to simplify the political situation at the time; detailed accounts of that situation are provided by Mortimer (1974) and Crouch (1973).

19. The political climate was particularly bad at the time that PKI-backed peasants launched a radical land reform movement or "unilateral action" (*aksi sepihak*), through which all landlord lands would be appropriated and redistributed to poor peasants without National Land Agency approval.

20. The 1960s Basic Agrarian Law required all land to be registered. Traditional lands could be titled if their status was converted to the law's modern land tenure category (e.g., private rights, user rights). Yet the possibility of titling their land in this way had almost no effect on forest dwellers and other rural people, who had little idea of what was occurring in the central capital (Moniaga 1993). As a result, the legal status of "unregistered" land under *adat* law remained largely

unclear, and such land was vulnerable to other claims, especially those of the state.

21. HPHH was intended to permit small manual logging operations and collection of nontimber forest products. The policy resulted in numerous small logging parcels (mostly funded by the urban rich), which made it impossible for the understaffed and underfunded provincial governments to police and tax them effectively.

22. Local-people ownership claims are estimated to cover 10–65 percent of total forest lands in Indonesia (Zerner 1992). Some prominent examples of conflicts between locals and state-sanctioned logging companies are found in Yamdena Island (Maluku Province), Sugapa (North Sumatra), Bentian (East Kalimantan), and Benakat (South Sumatra). For a detailed analysis of the relationship between environmental conflicts and forest degradation, see Barber 1998.

23. Analyses of various policies that undermine local customary practices are widespread. Some examples are Moniaga 1993 and Safitri 1995.

24. For different data on deforestation, see Sunderlin and Resosudarmo 1996.

25. For a detailed account of government-NGO relations under the New Order, see Eldridge 1995 (esp. chaps. 2 and 3).

26. Since the early 1990s, several social forestry programs (e.g., Forest Village Community Development, Community Forestry) have been launched with the objective of involving local people in state forest management. Implementation of these programs has been disappointing, however, since they do not address the overriding issue of peasants' land tenure.

27. Based on Official Letter no. 2767/II-Kum/96, November 28, 1996, from the MoF secretary general to the minister of forestry.

28. Rivalry between the Agrarian Department and the MoF over territorial control is nothing new. The former believes that the 1960 Basic Agrarian Law is the principal law concerning national land (including forest land) and that the *domein* authority (as the foundation of the Basic Agrarian Law) should rest with the Agrarian Ministry. The MoF, on the other hand, always defends itself with the assertion that the 1967 BFL conferred *domein* authority over forest land on the MoF. Although this conflict was materialized clearly on the ground (e.g., the National Land Agency issues land titling in forest reserve), both agencies seemingly tried not to have open confrontation.

29. Most of the information in this section on the post–New Order is drawn from Lindayati 2002.

30. Indonesia's 1998 economic contraction is estimated to have been between 10–15 percent, with the total proportion of poor people in the country rising from 25 to nearly 40 percent. The currency rate depreciated by approximately 70 percent (Sunderlin 1998), although it gradually began to recover by 2000.

31. For variations on these views, see Campbell 2002.

32. For a detailed account of the new law's provisions for local forest management, see Wollenberg and Kartodihardjo 2002.

33. The enforcement of state policies, however, was not self-evident and varied greatly from one locale to another. In areas where the state's economic stake was high, law enforcement was usually strong, whereas in other areas with less of an economy, the relationship between government and forest villagers was usually more relaxes.

References

Barber, C. V. 1989. *The State, the Environment, and Development: The Genesis and Transformation of Social Forestry Policy in New Order Indonesia*. Ph.D. diss. University of California, Berkeley.

Barber, C. V. 1998. "Forest Resource Scarcity and Social Conflict in Indonesia." *Environment* 40:4–37.

Boomgard, P. 1989. *Children of the Colonial State: Population Growth and Economic Development in Java, 1795–1880*. Amsterdam: Free University Press.

Broek, J. O. M. 1971. *Economic Development of the Netherlands Indies*. New York: Russell.

Bromley, D. W. 1992. "The Commons, Property, and Common-Property Regimes." In *Making the Commons Work: Theory, Practice, and Policy*, ed. D. W. Bromley et al., 3–15. San Francisco: ICS Press.

Campbell, J. Y. Forthcoming. "Forest for the People, Indigenous Communities (Masyarakat Adat) or Cooperatives? Plural Perspectives in the Policy Debate for Community Forestry in Indonesia." In *Which Way Forward?*, ed. C. Colfer, and I. A. P. Resosudarmo, 110–125. Washington, DC: Resources for the Future.

Cramb, R. A., and I. R. Wills. 1990. "The Role of Traditional Institutions in Rural Development: Community-Based Land Tenure and Government Land Policy in Sarawak, Malaysia." *World Development* 18:347–360.

Crouch, H. 1973. "Another Look at the Indonesian 'Coup.'" *Indonesia* no.15 (April):1–20.

Departemen Kehutanan RI. 1986a. *Sejarah Kehutanan Indonesia I* (The History of Forestry in Indonesia I). Jakarta: Author.

Departemen Kehutanan RI. 1986b. *Sejarah Kehutanan Indonesia II* (The History of Forestry in Indonesia II). Jakarta: Author.

Departemen Kehutanan RI. 1986c. *Sejarah Kehutanan Indonesia III* (The History of Forestry in Indonesia III). Jakarta: Author.

Dick, H. W. 1990. "Interisland Trade, Economic Integration, and the Emergence of the National Economy." In *Indonesian Economic History in the Dutch Colonial Era*, Monograph Series 35, ed. A. Booth, W. J. O'Malley, and A. Weidemann, 297–321. New Haven: Yale Center for International Areas Studies.

Eldridge, P. J. 1995. *Non-Government Organizations and Democratic Participation in Indonesia*. Kuala Lumpur: Oxford University Press.

Fay, C., and M. Sirait. 2002. "Reforming the Reformists: Challenges to Government Forestry Reform in Post-Soeharto Indonesia." In *Which Way Forward?* ed. C. Colfer, and I. A. P. Resosudarmo, 126–143. Washington, DC: Resources for the Future.

Goldstein, J. 1988. "Ideas, Institutions, and American Trade Policy." *International Organization* 42:179–217.

Goldstein, J., and R. O. Keohane, eds. 1993. *Ideas and Foreign Policy: Beliefs, Institutions, and Political Change*. Ithaca and London: Cornell University Press.

Guha, R. 1990. *The Unquiet Woods: Ecological Change and Peasant Resistance in the Indian Himalaya*. Berkeley and Los Angeles: University of California Press.

Hall, P. A., ed. 1989. *The Political Power of Economic Ideas: Keynesians across Nations*. Princeton: Princeton University Press.

Hall, P. A. 1993. "Policy Paradigms, Social Learning, and the State: The Case of Economic Policymaking in Britain." *Comparative Politics* 25(3):275–296.

Hansen, R., and D. King. 2001. "Eugenic Ideas, Political Interests, and Policy Variance: Immigration and Sterilization Policy in Britain and the U.S." *World Politics* 53:237–263.

Howlett, M. 1994. "Policy Paradigms and Policy Change: Lessons from the Old and New Canadian Policies towards Aboriginal Peoples." *Policy Studies Journal* 22:631–649.

Krasner, S. D. 1993. "Westphalia and All That." In *Ideas and Foreign Policy: Beliefs, Institutions, and Political Change*, ed. J. Goldstein, and R. E. Keohane, 235–64. Ithaca and London: Cornell University Press.

Lindayati, Rita. 2002. "Ideas and Institutions in Social Forestry Policy," ed. C. Colfer, and I. A. P. Resosudarmo, 36–59. Washington, DC: Resources for the Future.

Lynch, O. J., and K. Talbott. 1995. *Balancing Acts: Community-Based Forest Management and National Law in Asia and the Pacific*. Washington, DC: World Resource Institute.

Manurung, T. 1997. *Political Economy in the Formulation of Forestry Law in Indonesia*. Bogor, Indonesia: Center for International Forestry Research.

Moniaga, S. 1993. "Toward Community-Based Forestry and Recognition of Adat Property Rights in the Outer Islands of Indonesia." In *Legal Frameworks for Forest Management in Asia*, Occasional Chapter no. 16, ed. J. Fox, 131–150. Honolulu: East-West Center, Program on Environment.

Mortimer, R. 1974. *Indonesian Communism under Soekarno*. Ithaca: Cornell University Press.

Nasution, M. 1998. Personal communication to author. Jakarta, Indonesia, November 27.

Odell, J. 1982. *U.S. International Monetary Policy: Markets, Power, and Ideas as Sources of Change*. Princeton: Princeton University Press.

Peluso, N. L. 1992. *Rich Forests, Poor People: Resource Control and Resistance in Java*. Berkeley and Los Angeles: University of California Press.

Poffenberger, M. 1990. "The Evolution of Forest Management Systems in Southeast Asia." In *Keepers of the Forest: Land Management Alternatives in Southeast Asia*, ed. M. Poffenberger, 7–26. West Hartford, CO: Kumarian Press.

Potter, L. 1988. "Indigenes and Colonisers: Dutch Forest Policy in South and East Borneo (Kalimantan) 1900 to 1950." In *Changing Tropical Forests: Historical Perspectives on Today's Challenges in Asia, Australasia and Oceania*, ed. J. Dargavel, K. Dixon, and N. Semple, 127–149. Canberra: Tropical Forest History Working Group.

Robinson, R. 1986. *Indonesia: The Rise of Capital*. North Sydney, Australia: Allen & Unwin.

Sabatier, P. A., and H. Jenkins-Smith, eds. 1993. *Policy Change and Learning*. Boulder, CO: Westview.

Safitri, M. 1995. "Hak dan Akses Masyarakat Lokal pada Sumberdaya Hutan: Kajian Peraturan Perundang-undangan Indonesia" (Local Community Access and rights to Forest Resources: Analysis of Indonesia's Regulations). *Ekonosia* 3:43–59.

Sanit, A. 1998. *Reformasi Politik*. Yogyakarta, Indonesia: Pustaka Pelajar.

Seymour, F. J., N. K. Dubash, J. Brunner, F. Ekoko, C. Filer, H. Kartodihardjo, and J. Mugabe. 2000. *The Right Conditions: The World Bank, Structural Adjustment, and Forest Policy Reform*. Washington, DC: World Resource Institute.

Sikkink, K. 1991. *Ideas and Institutions: Developmentalism in Brazil and Argentina*. Ithaca and London: Cornell University Press.

Soepardi, R. 1974a. *Hutan dan Kehutanan Dalam Tiga Jaman* (Forest and Forestry in Three Eras). Vol. 1. Jakarta: Perum Perhutani.

Soepardi, R. 1974b. *Hutan dan Kehutanan Dalam Tiga Jaman* (Forest and Forestry in Three Eras). Vol. 2. Jakarta: Perum Perhutani.

Sunderlin, W. D. 1998. "Between Danger and Opportunity: Indonesia's Forests in an Era of Economic Crisis and Political Change." Available at <http://www.cgiar.org/cifor>.

Sunderlin, W. D., and I. A. P. Resosudarmo. 1996. "Rates and Causes of Deforestation in Indonesia: Towards a Resolution of the Ambiguities." Bogor, Indonesia: Center for International Forestry Research. Occasional paper no. 19.

Tim Penyempurnaan Undang-Undang no.5 tahun 1967. Undated. Naskah Akademis Penyempurnaan Undang-Undang no.5/1967, draft 7. (Academic Manuscript for the Perfection of Law no.5/1967, draft 7).

Wertheim, W. F., J. F. Kraal, C. C. Berg, R. A. M. Bergman, C. T. Bertling, W. Brand, G. H. v. d. Kolff, P. W. Milaan, H. Offerhaus, R. Roolvink, F. R. J. Verhoeven, and A. v. Marle, eds. 1985. *Indonesian Economics: The Concept of Dualism in Theory and Practice*. Vol. 6 of *Selected Studies on Indonesia*. The Hague: W. van Hoeve Publishers.

Wollenberg, E., and H. Kartodihardjo. 2002. “Devolution and Indonesia’s New Basic Forestry Law.” In *Which Way Forward?*, ed. C. Colfer, and I. A. P. Resosudarmo, 81–95. Washington, DC: Resources for the Future.

Zerner, C. 1992. *Indigenous Forest-Dwelling Communities in Indonesia’s Outer Islands: Livelihood, Rights, and Environmental Management Institutions in the Era of Industrial Forest Exploitation*. Washington, DC: Resource Planning Corporation.

9

A Framework for Analyzing the Physical-, Social-, and Human-Capital Effects of Microcredit on Common-Pool Resources

C. Leigh Anderson, Laura A. Locker, and Rachel A. Nugent

In recent years, microfinance—extending small loans to the poor for income-generating activities and other financial services—has grown enormously in its popularity as a developmental tool. The World Bank estimates that there are now over 7,000 microfinance institutions serving sixteen million poor people with an annual cash turnover around US$2.5 billion. Further, it believes that "the potential for new growth is outstanding" (Global Development Research Center 2000).

The World Bank is not the only organization that anticipates growth in microfinance. The Microcredit Summit set a goal for microfinance organizations (MFOs) to reach "100 million of the world's poorest families, especially the women of those families, with credit for self-employment and other financial and business services, by the year 2005" (Microcredit Summit 2000). If this target is reached, it would represent an almost tenfold increase over current levels in the number of clients served and would encompass one third to one half of the world's poor. Considering the financial support and international attention microfinance programs have garnered, this goal is lofty, but not impossible.

If the leaders of the Microcredit Summit reach their goal, providing credit to such a huge proportion of the poor would represent a fundamental shift in development policy. Though many people see such a shift as positive, there is also some concern that money will be diverted away from other development programs before we adequately understand the long- and short-term effects of microfinance, and in particular, microcredit (e.g., Buntin 1997; Solomon 1998; Roth 1997). Many loan recipients, for example, make at least part of their living by exploiting local common-pool resources (CPRs), but little is known about microcredit's effects on the quality and use of these resources.

It is with this lacuna in mind that we explore in this chapter the theoretical links between microcredit and CPRs. Our focus is rural microcredit and the associated CPRs: drainage and irrigation systems, fishing pools, other water sources, grazing lands, and forests. We examine how three common characteristics of microcredit programs—loaning primarily to the very poor, focusing on women, and employing group lending techniques—can affect the sustainable use of CPRs by altering the demand and supply of physical, human, and social capital. In particular, we suggest that microfinance programs can lower the costs of collective action in managing local CPRs by creating social capital in a community. We devise a framework that identifies the channels through which microcredit affects the use and management of CPRs both directly and indirectly. We provide examples of these channels from microfinance programs around the world and conclude with an example of how this framework can be applied to deforestation in Vietnam and Thailand.

Microcredit

Microcredit is extended by microfinance organizations (MFOs), which may also offer savings, insurance, or other financial services. Despite a long informal history throughout much of the developing world, microcredit was only recently popularized in the development community through the work of the Grameen Bank in Bangladesh, which today serves over two million borrowers and is the largest MFO, with organizations replicating its model worldwide. The goal of the Grameen Bank from the beginning was to target its services toward the poorest of the poor who work in the informal sector of the economy, "the hawkers and venders seen peddling their wares on street corners, the farmers and hucksters who sell locally grown foodstuffs in remote villages and towns, and the entrepreneurs who recycle used bedsprings, make brooms, sew clothing in their homes or in little shops located in alleyways and shantytowns" (U.S. House of Representatives 1986, 2).

Historically, workers in the informal sector have been excluded from formal banking for several reasons. First, these workers were overwhelmingly female, and the banking system was overwhelmingly male. Second, they lacked the literacy and other skills necessary to get loans; the paperwork required to apply for a loan was often lengthy and required the ability to read and write. Third, the very small loans needed by this population entailed high transaction costs, which would make the loans unprofitable for lenders. To cover these costs, many bankers believed the interest rates charged on such loans would have to be prohibitively high. Fourth, information asymmetries between borrower and lender (as in most credit markets) meant the lenders who might have extended loans to these workers were unable to know the true creditworthiness of this group of borrowers. Finally, workers in the informal sector were perceived as credit risks because they lacked physical collateral (FAO and GTZ 1998; World Bank 1993).

Despite these concerns, innovative techniques introduced by Grameen Bank founder Mohammad Yunus and adopted by other MFOs have been used to make workers in the informal sector eligible for credit. Loan application and servicing procedures are conducted openly in the villages where these workers live utilizing simple procedures that do not require applicants to be literate, and repayments are frequent and small, suiting the income flows of the very poor. Finally, social capital is substituted for physical capital as collateral. Participants may be required to borrow in groups, select their own group members, and act as mutual guarantors (van Bastelaer 1999, 10). This type of group lending depends upon the existence of high levels of trust and reciprocity—the essential components of social capital—among all participants to maintain high repayment rates.[1] The idea that social collateral could substitute for physical collateral in the provision of loans was an entirely novel one in the world of formal banking before Yunus used it, although it had existed in informal banking for some time.[2]

Microcredit and Common-Pool Resources

Changes in the physical, human, and social assets that arise from microcredit activities will affect a community's production, consumption, and management opportunities and decisions around CPRs. The net effect of increased demands on CPRs, which may include increased waste and byproducts compromising the quality of the resources, may be negative or positive. Which they are depends on agro-climatic zones, the particular CPRs, the design of the microcredit program, and other local and nonlocal institutions. In this chapter we suggest that the positive human- and social-capital effects of microcredit may mitigate or outweigh any damages to CPRs from changes in assets or behavior caused by the extension of microcredit. These positive effects can occur through strengthened informal agreements about managing the commons sustainably or through greater access to nondegrading production and consumption choices made available through the granting of microcredit.

Microcredit can affect CPRs through the direct, intentional efforts of lenders or indirectly through changing the constraints faced by borrowers. There are two routes through which microcredit and environmental resources are intentionally linked: MFOs with environmental goals, and environmental organizations using microcredit to promote resource goals.

Direct Linking of Microcredit and Environmental Goals

It is relatively uncommon for MFOs to tie environmental management explicitly to lending, though environmental practices are often mentioned in the members' (borrowers') conditions of lending. In large part the inclusion of these practices in the terms of lending may be due to the precedent set by the Grameen Bank in its sixteen conditions that borrowers are encouraged to meet. Grameen Bank members pledge, among other things, that "[they] we will keep [their] children and the environment clean; [they] will build and use pit-latrines; during the plantation seasons, [they] will plant as many seedlings as possible" (Khandker 1998, 113). These conditions have been copied by hundreds of Grameen replication banks worldwide.

Although they remain in the minority, there is also a small but growing group of MFOs concerned with producing "green" products or technologies and promoting ecofriendly economic activities among its loan recipients. Grameen Shakti, for example, is dedicated to providing renewable energy sources, such as solar Photo Voltaic (PV), biogas, and wind turbines, to villages in Bangladesh that are without electricity. Likewise, the Solar-Based Rural Electrification Concept (SO-BASEC) in the Dominican Republic and Honduras uses microcredit to promote solar-based renewal energy. Other examples include the Asia Institute of Technology, which promotes biotechnology-based microenterprises, such as mushroom and bioorganic fertilizer production, that reduce harm to watersheds.

Conservation NGOs or development NGOs with a conservation agenda have also used microcredit to promote their environmental agendas. In some cases these organizations have developed, or are attempting to develop, microcredit capacity themselves, whereas in others they are partnering with more specialized credit suppliers such as local or international banks or other NGOs. In Bangladesh, the Community Based Fisheries Management (CBFM) project teams the Department of Fisheries with five NGOs whose approach is to form groups of fishers, provide them with credit and other support, and help them "develop institutions and techniques for managing the fisheries" (Thompson, Islam, and Kadir 1998, 2). Conservation International hopes to reduce deforestation and poaching in Ghana and several other countries by increasing the value of the countries' forests as tourist sites. Myrada, based in Bangalore, India, organizes credit management groups of the rural poor, manages microwatersheds, and reforests arid areas.

Indirect Effects of Microcredit on CPRs

Microcredit can *indirectly* affect a CPR via changes in the behavior of CPR users elicited by three common characteristics of microcredit programs. Figure 9.1 shows how each of microcredit's program components—extending credit to the very poor, lending primarily to women, and employing group lending methods—changes an existing constraint under which the borrower lives, subsequently leading to a change in his or her productive or consumptive behavior. It is these behavioral changes, which also depend on initial conditions such as income levels, enterprise and business structures, agro-climatic zones, culture, and other local and nonlocal institutions that have implications for CPRs.

Microcredit affects CPRs through financial and human- and social-capital changes: changes in the levels, diversity, or regularity of borrowers' income; changes in the discount rate between their present and future consumption; changes in the role of women and providing them with a forum in which to exchange information; and changes in the cost of collective action for managing environmental resources.

Credit Extension to the Poor: Physical-Capital Effects Extending credit has consequences for environmental resources, both through direct

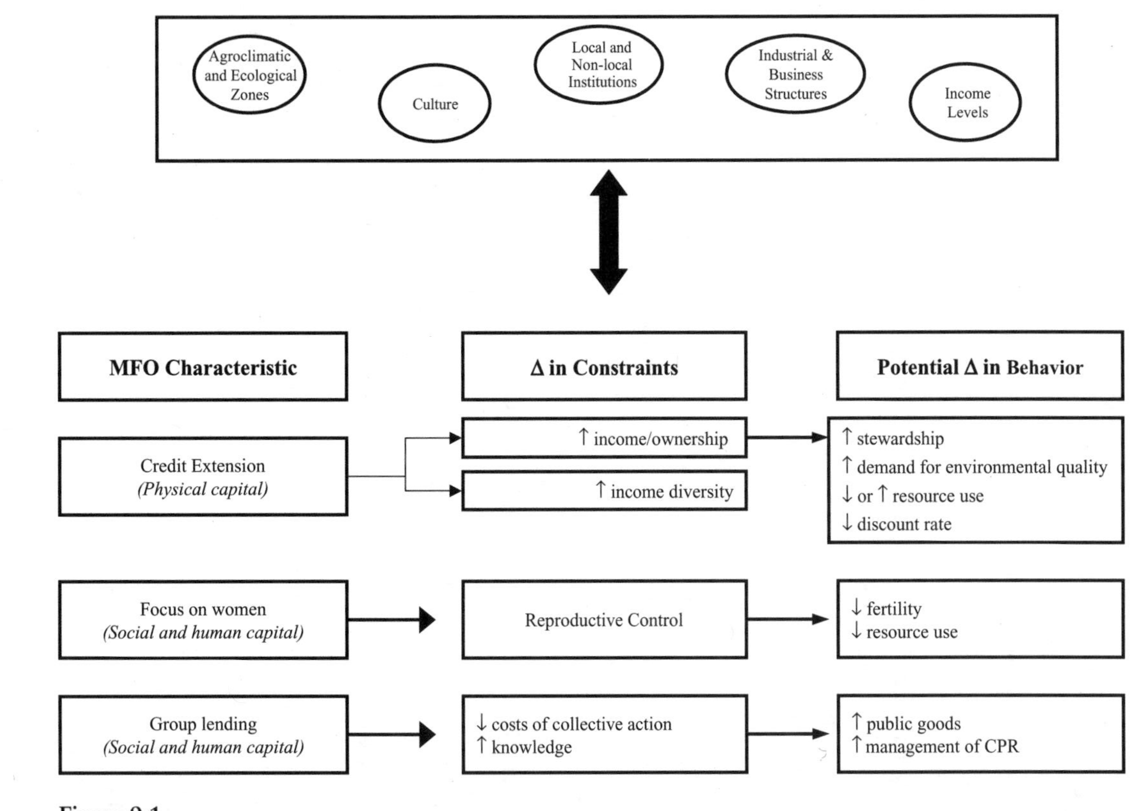

Figure 9.1
Relationship of microcredit to CPRs and the external environment.

investments in physical capital and the potential changes in borrowers' income resulting from the extension of credit. Credit allows microentrepreneurs to invest in small-scale capital such as sewing machines, looms, bicycles, rickshaws, livestock, tools, and other supplies. Without some capital to invest, the relatively large investments required to acquire these resources would be difficult for individuals living at or near subsistence levels to make. The philosophy behind microcredit is to provide access to physical capital as the key to better income opportunities. These opportunities can then free the poor from dependency on CPRs.

Income Effects We know that poverty and the environment are intimately linked (Dasgupta and Mäler 1994). The poor are less able to place resource conservation above other, more pressing survival needs, often rendering them the primary victims of their own and others' environmental degradation.

The evidence is weak, but growing, that microcredit raises the income or assets of at least some recipients (Sebstad and Chen 1996). As income increases, we expect the quantity, composition, and timing of economic activity of the poor to change.[3] Changes in consumption activities afforded by increased income have effects on the environment, which may change over time and be positive, negative, or ambiguous. The negative impacts can include increased resource consumption and generation of more, and more toxic, waste (Dasgupta and Mäler 1994). On the positive side, rising incomes tend to be correlated with improved household infrastructure, including sanitation and cooking facilities, greater access to safe drinking water, and increases in contraceptive use.

The relationship between income growth and resource use in production is more complex. For rural, largely biomass-based subsistence economies, growth involves either intensification of land use, extensification of land use, or new rural nonfarm activity, including resource extraction (Dasgupta and Mäler 1994). Some of these activities change environmental resource use directly, others through changes in property rights and ownership, and still others in the use of labor, capital, or technology that complements or substitutes for natural-resource inputs. For example, the impact of changes in fuel use depends on the fuel and fuel-burning technology used. Switching from biomass or coal to electricity, for example, may reduce the destruction of forest sinks and decrease CO_2

emissions. Intensifying agricultural efforts through the increased use of chemical fertilizers available through microcredit can contaminate local water supplies.

The extensification of land use has the most direct implications for CPRs. In some cases additional income relieves the pressure on the recipients to expand into forested areas to clear land or collect fuelwood, whereas in others it affords them the ability to do so. For example, raising livestock, a common rural microenterprise, can promote deforestation. In general, expanding the land area used to support a population, rather than using existing land more intensively, increases deforestation pressures, the use of marginal lands, and environmental impacts from migration. Marginal lands are often less productive than other lands, suffer more from soil erosion (especially if they are on hillsides), and require more fertilizer and water.

More positively, increased income may promote resource stewardship through increasing property rights, increasing access to more environmentally benign technology, and lowering discount rates. Women and the poor in particular have often been denied access to credit, and—culture, law, and regulations permitting—microcredit can improve their chances to own land and other property.

Similarly, the higher an individual's income, the less he or she must be preoccupied with satisfying current consumption needs. It becomes possible to trade off some current consumption for a higher, more sustained future return. Further, a higher income can offer an individual the flexibility to smooth consumption and expenditures to achieve better returns or to improve practices. Discounting future earnings at a lower rate can increase a borrower's willingness to spare a young tree from harvest until it reaches a more valuable size.

Income Diversification Diversifying sources of income can be an important result of microcredit, particularly for the rural poor, whose incomes depend upon the resource base and who are thus vulnerable to weather fluctuations, climate change, and pests and diseases. The consequences for CPR use and management depend on how income is diversified: through savings or other financial services offered by the MFOs, through additional resource-dependent activities such as harvesting new crops, or through new or expanded nonfarm activities.[4] Once again, the impacts on deforestation rates are ambiguous. For example, credit can provide the recipient with the opportunity to diversify by increasing livestock holdings, which can lead to overgrazing, clogging of common drainage systems, and pressures to deforest. Conversely, credit and insurance services may reduce the need to hold insurance in kind, traditionally accomplished by carrying excess livestock, and thus reduce pressure on these same CPRs (Dasgupta and Mäler 1994).

Focus on Women: Human-Capital Effects As a group, MFOs overwhelmingly focus on extending credit to women. Their reasons for targeting women may differ. Women are reputed to be better credit risks and to have higher payback rates than men, to possess more unrealized entrepreneurial capacity, and to be more inclined to use income they control for improving children's nutrition and education. Some MFOs may simply wish to increase women's economic power.[5] Regardless of the motivation, a microcredit program that provides training, increases the spread of knowledge and best practices, and requires regular group meetings for sharing business results adds to and utilizes the human capital of the borrowers and the community (Schrieder and Sharma 1999, 74). The value added may be particularly high for women otherwise unexposed to training and women in cultures that otherwise offer them few or no opportunities to leave their homes.

The environmental consequences of increased income or property rights for women may also be more pronounced than for men because women begin with more limited rights but often have a major role in natural-resource stewardship. Whether or not they have other employment, women in most developing countries are responsible for cooking and household sanitation. As such, they commonly spend many hours gathering fuelwood and nontimber forest products and collecting water (World Resources Institute 1994, 46).

Women's domestic responsibilities also mean that they suffer more than men from deforestation and desertification and have a particular incentive to maintain or improve their local environment and CPRs. This incentive has translated into some of the most creative conservation initiatives worldwide. Women's groups have organized to collectively lease and revive exhausted cropland, offer leadership in water supply and management, plant trees, construct terraces, and provide education about a variety of resource conservation opportunities (World Resources Institute 1994, 46).

Reproductive Control The evidence is growing that microcredit programs may contribute to reduced fertility (Schuler and Hashemi 1997). This is not surprising, given the higher opportunity cost of bearing children for a successful female microentrepreneur relative to a woman employed only in household or farm activities. Also, as women's incomes rise, child mortality rates usually fall, lessening their need or desire to bear as many children.

There may, however, be an even more direct avenue of influence from microcredit to reproductive control that would explain cases like Bangladesh, where fertility rates are plummeting though mortality rates are not. Some MFOs also provide, indeed encourage, family planning education as part of their program and regular meetings. Upon joining the Grameen Bank, members pledge to keep their families small. This pledge may be easier to keep with the sense of economic power experienced by taking a loan. One senior Grameen Bank official explains: "A woman who is not earning any cash cannot tell her husband that she doesn't want to get pregnant. She cannot say, 'We have to take precautions.' But suppose she is repaying a 3,000-taka loan and is hoping to get a 6,000-taka loan? Now she can tell her husband, 'If I get pregnant my group will not recommend 6,000 takas because in three or four months' time I'll be heavily pregnant.' Now, she has a bargaining position" (Bornstein 1997, 106).

Whether from new economic power, new information, or a new support system, women who have received microcredit seem to have taken more control over childbearing decisions. Borrowers with the Bangladesh Rural Advancement Committee (BRAC) and Grameen Bank, for example, are far more likely to practice contraception than the national average (Bornstein 1997, 106).

The environmental resource implications of reduced fertility are reasonably unambiguous: fewer children usually mean less resource consumption and less waste. If fertility is declining because of increasing income, however, it may be that though total consumption is falling, per capita consumption is rising. As always, the net effect also depends on how the composition of activities changes, not simply the level.

Group Lending: Social- and Human-Capital Effects

A particularly innovative feature of modern microcredit programs has been their use of group lending techniques. Group lending reduces information asymmetries common to most lending situations by aligning borrowers' incentives with lenders' and using borrowers' superior knowledge of each other to screen members, monitor repayment, and exert peer pressure. Groups utilize the networks of trust and relationships in the village, mutual guarantees, and shared knowledge about eligibility and performance to help ensure repayment of the loans that are granted to the group. Peer selection avoids adverse selection and promotes group homogeneity in gender, landholding, and income terms. Group homogeneity, in turn, is believed to promote repayment (van Bastelaer 1999, 12). Peer monitoring is encouraged when participants in a group loan are required to act as mutual guarantors (joint liability) or receive loans contingent on others in the group paying their loans back (contingent renewal). These procedures address the typical problems of moral hazard, in which borrowers may engage in riskier behavior once they have received the loan. These group incentives and dynamics are reinforced through regular, often weekly, group meetings often required under the terms of the group loan.[6]

The group lending techniques of MFOs can affect the social capital of a community in many ways: drawing upon existing social capital, creating new social capital, and providing "bridging" social capital between participants and the surrounding society. Each of these social-capital effects has direct consequences for the use and maintenance of a community's CPRs.

Like that of microfinance, the concept of social capital has only recently reached the level of common use in development efforts. Traditionally, capital has been defined as "the contribution to productive activity made by investment in physical capital (for example, factories, offices, machinery, tools) and in human capital (for example, general education, vocational training)" (Pass and Lowes 1991, 57). Social capital can likewise be considered an asset: an investment that may yield additional value in the future.

Social capital is defined most simply as "the institutions, the relationships, the attitudes and values that govern interactions among people and contribute to economic and social development" (World Bank 1999).

Whereas the creation of physical or human capital can be an individual project, social capital is, by nature, social; it is created, according to James Coleman (1990, 304), "when the relations among persons change in ways that facilitate action."

Coleman's work, *The Foundations of Social Theory*, laid out much of the technical groundwork for the current discussion of social capital. Coleman conceived of social capital as facilitating the action of two people in the following way: "if *A* does something for *B* and trusts *B* to reciprocate in the future, this establishes an expectation in *A* and an obligation on the part of *B* to keep the trust. This obligation can be conceived of as a 'credit slip' held by *A* to be redeemed by some performance by *B*" (Coleman 1990, 306). In this example, the interaction between *A* and *B* was facilitated by the existence of social capital: it provided the trust that *A* needed to believe that if *A* did something for *B* today, *B* would reciprocate tomorrow. Additional stocks of social capital can be created when *B* reciprocates *A*'s actions at a later date. At that point, *A*'s expectations are fulfilled and *A*'s level of trust grows. Similarly, existing social capital can be diminished if *B* fails to reciprocate. Coleman (1990, 306) argues that the amount of social capital in any given environment is composed of "the level of trustworthiness of the social environment, which means that obligations will be repaid, and the actual extent of obligations held." (Coleman cites rotating credit associations as an illustration of the value of trustworthiness critical to this form of social capital.)

Thus, social capital can be conceived of as an asset that arises from and enables the use of networks existing in a community in such a way that norms of trust and reciprocity are promoted. Networks are linkages between actors that can be "horizontal" or "vertical" in their orientation. Horizontal networks, according to Putnam (1993, 173), bring "together agents of equivalent status and power," whereas vertical networks link "unequal agents in asymmetric relations of hierarchy and dependence." Horizontal linkages are generally those positive social networks that contribute to the overall productivity of a community, such as volunteer associations. Coleman added the notion of vertical linkages "characterized by hierarchical relationships and an unequal power distribution among members" (Grootaert 1988, 3) and allowed for social capital to produce negative as well as positive associations. Depending upon the type of network actors use, the social capital that is produced can have negative or positive association; a given form of social capital that is valuable in facilitating certain actions may be useless or even harmful for others.

Both these views and our conceptualization of social capital fit Coleman's definition and are contrasted with the definition adopted by Birner and Wittmer in chapter 10. The latter definition is a "private perspective" of the returns from social capital, whereas we are discussing in this chapter a more "Putnamesque" public perspective (their terms) on social capital.

It is fairly well understood how microfinance programs use existing networks of horizontal associations to lower the costs of monitoring and enforcing existing rules and other transaction costs (Besley and Coate 1995; Stiglitz and Weiss 1981). van Bastelaer (1999) also notes the importance of hierarchical relationships, including relationships between borrowers and lenders that become personal through regular meetings and/or that create traditional patron-client relationships and the need to demonstrate allegiance, as well as the relationships between MFOs and local or national governments.[7] These relationships manifest the improved nature of communication that characterizes functioning networks.

Group-based microfinance can also lower the costs of crafting new rules, adding to the stock of social capital (Ostrom 1990, 1992). MFOs use existing social capital but arguably also create social capital through meetings and other services: "microfinance has the potential to enable collective action, the coming together of the community, and more sustainable community-based organizations. . . . In as far as microfinance interventions allow to invest in education and training, members of the community can acquire skills that will allow them to locally design, develop and manage community projects" (Schrieder and Sharma 1999, 74).

In villages where borrowers meet regularly to discuss lending, the costs of collective action for other village undertakings that may be CPR-based are significantly lower; communication among participants greatly increases the chances of successful collective action. Ostrom, Gardner, and Walker (1994, 167) have shown in a series of experiments that given the right institutional framework to communicate "players successfully used the opportunity (1) to calculate coordinated yield-improving strategies, (2) to devise verbal agreements to implement these strategies, and (3) to deal with nonconforming players." By the nature of its credit activities, microfinance adds further incentives for cooperation by increasing the anticipated payoffs and lowering discount rates.

As Ostrom and others have noted, it takes effort and energy to create social capital. But regular meetings, repeat interaction, and common credit goals can facilitate the communication, knowledge about fellow actors, common understanding about the incentive structure, and trust prerequisite to collective action (Ostrom 1994, 532). van Bastelaer (1999) also argues that social capital is created when MFOs such as the Grameen Bank and its replicators require all members to engage in the same behavior every week, such as repeating the list of decisions that accompany group membership. This routinization creates a corporate culture or cultural habit.

It seems less likely that microfinance meetings per se would be used to determine collective-choice rules as much as they would be used to monitor and amend the day-to-day operational rules governing CPR use (Ostrom 1994). This is because the microfinance participants may be only a subset (though arguably an important one) of the local CPR users and because in many cases they are likely to be women. If, as is often the case, the women are the primary users of the local CPR, microfinance activities may provide a conduit for changing the operational rules for using the CPR among themselves.

If, as Ostrom (1990, 21) proposes, the primary difference between those who have "broken the shackles of a commons dilemma and those who have not" is an internal difference, and, as she argues, failure may occur because "participants may simply have no capacity to communicate with one another, no way to develop trust, and no sense that they must share a common future," then finding ways to strengthen the social capital of a community may be one key to solving the complexity of the commons. Yet as Ostrom points out, even groups with sufficient trust to cooperate may be blocked in their efforts to manage the commons by more powerful individuals. In this case, she argues, "such groups may need some form of external assistance to break out of the perverse logic of their situation." It is possible that MFOs may be able to provide such external assistance ("bridging capital," as it is called in the literature). Woolcock and Narayan (2000, 233) write that "the clear challenge to social capital theory, research, and policy from the networks perspective is thus to identify the conditions under which the many positive aspects of 'bonding' social capital in poor communities can be harnessed and its integrity retained (and, if necessary, its negative aspects dissipated), while simultaneously helping the poor gain access to formal institutions and a more diverse stock of 'bridging' social capital."

Applying the Framework to Deforestation in Vietnam and Thailand

We have hypothesized that the behavioral changes resulting from microcredit programs involving women and group lending techniques can improve CPR use and management but that the effects of increased income resulting from microcredit loans can be positive or negative. Results from a 1999 nonrandom survey of 140 microcredit summit member organizations reflect a perception of this mixed effect among participants in the survey. For example, of the seventy-one organizations that reported an environmental impact as a result of microcredit, 42 percent felt that microcredit had increased local water use and 14 percent reported that it had led to a decrease in such use. Furthermore, 54 percent of respondents believed that microcredit had led to a reduction in deforestation, whereas 13 percent reported that deforestation had increased as a result of microcredit (Anderson, Locker, and Nugent 2002).

None of these organizations systematically measures and records resource use, however, nor are any of them able to attribute any changes they have observed solely to microcredit. It is possible, however, to get some sense of the linkages described in the previous section by briefly applying our theory to one CPR—forests—for several programs in Vietnam and Thailand.

Forests are critical resources for many poor communities. Deforestation leads to fuelwood shortages, adds to biomass collection times and implies a loss of bark, saps and pharmaceuticals, a loss of species habitat, soil erosion, and in watersheds, an increased runoff of rainwater leading to soil loss and clogging of common water reservoirs and irrigation systems.

In Vietnam, forests "provide subsistence products directly and indirectly to most of the rural population and generate work opportunities for more than 28 million people" (Hines 1995, 1). Since Vietnam has a total population just exceeding seventy-seven million (with almost 50 percent living below the poverty line), this means that over a third of the country's total population depends upon the forests for survival.

As early as 1992, the FAO noted that "numerous reports describe progressive reduction of forest area [in Vietnam] by rural wood harvesters who gather fuelwood for their own use and for cash sale to supplement their household income. . . . In many places, there are distinct foot tracks up steep hillsides used by the gatherers, who frequently can be seen carrying wood to roadsides from natural forests" (FAO and Ministry of Forestry, Republic of Vietnam, 1992, 68). In 1995, Deborah Hines of the United Nations Development Programme found that "an informal collection and distribution system is still very much in operation. . . . Unauthorized cutters sell to unregistered traders who then sell to private wood processors" (8). In her report on the financial viability of smallholder reforestation in Vietnam, she concluded that in Vietnam's northern midlands and central highlands, fuelwood collection is the second biggest cause—and in the rest of the country it is the single largest cause—of deforestation (Hines 1995, 8). Addressing deforestation in Vietnam therefore requires dealing with the issue of fuelwood collection.

The government of Vietnam has pursued policies of resettling minorities and privatizing parcels of land to deal with deforestation, but continued poverty and poor monitoring and enforcement of forest use have kept deforestation rates high. One suggestion has been to "draw upon the growing body of knowledge about management of common property resources to develop institutional mechanisms to facilitate acquisition and development of blocks of barren hill land by small groups of poor households. The already-demonstrated ability of small groups composed of kin and close neighbors to co-own buffalo suggests that similar small face-to-face groups might also be able to successfully manage tracts of hill land held as common property. Loans might be given at preferential rates to such groups, for example. If organizational methods similar to those pioneered by the Grameen Bank in Bangladesh were employed, the default rate on such loans might be lowered to an acceptable level" (Cuc et al. 1996, 125). That is, the group lending aspect of microcredit could work to improve the maintenance of Vietnam's forests.

In Vietnam, the use of microcredit did not become popularized until the institutionalization of *Doi Moi*, the Vietnamese government's shift from a state-planned economy to market liberalization in response to a stagnant economy and international isolation. Until 1988, the Vietnamese financial system consisted only of the state bank, which was broken down under *Doi Moi* into institutions created to deal with specific sectoral credit needs (Wolff, 1999, 53). To meet the demand for rural credit, the Bank of Agriculture (VBA) was established, and in 1995, the Bank of the Poor (VBP) was created as a nonprofit subsidiary of the VBA to target poorer clients. No collateral is required for loans granted by the VBP, which average US$96 (Fallavier 1998).

Government policymakers in Vietnam have made limited use of microcredit as a tool with which to manage Vietnam's forests more sustainably despite the widespread knowledge that deforestation occurs, in part, because of poor, landless women collecting fuelwood for sale and consumption. Several NGOs, however, have experimented with linking credit to forest conservation.

Though there are no rigorous evaluations of these programs, their application illustrates some of the linkages suggested in the previous section. We hypothesized there that extending credit to the poor would ease income constraints, thereby increasing resource stewardship, the demand for environmental quality, and the demand for resource use; would involve women in ways that decrease resource use; and would decrease the costs of collective action for managing local CPRs. These predicted outcomes can be examined with respect to microcredit programs by the Vietnamese government that do not attempt to incorporate forest management goals and with respect to those of Oxfam America and Population and Development International (PDI) that do.

Credit Extension to the Poor

Extending credit to the poor, the first characteristic of microcredit programs, may increase deforestation if the microenterprise that is financed through the microcredit requires either additional fuelwood or land cleared for enterprise activities. A 1999 survey of 220 households in Vietnam's Ha Tay province found that most of the loans from those receiving VBP credit have been used to increase and diversify recipients' income through the raising of livestock. No individuals mentioned the effect of their livestock activities on forest land, since they are still small enterprises and confined to backyards, though many mentioned the polluting of common drainage systems. If these enterprises expand, however, more land may be claimed for grazing. Several borrowers have also started making wood furniture, which increases the demand for timber.

Deforestation pressures will be eased only if the microentrepreneurs, including those raising livestock, are diverted from other activities such as collecting fuelwood for sale or burning forestland to plant crops. Providing alternative income sources has been one strategy Oxfam and PDI have pursued for accomplishing such diversion.

Oxfam has a credit and environmental restoration project in the Can Gio mangrove forests of southern Vietnam. Among other activities, Oxfam provides capital and training to poor families to start aquaculture enterprises as a means of raising income and reducing the incentive to use and sell mangrove wood. A study by the Management Board for Protected Forests and Environment reports that indicators of living conditions for the families starting the aquaculture enterprises have improved, as has management and protection of the forest (Oxfam America 1999). PDI is likewise trying to discourage deforestation in Thailand's Western Forest Complex along the Myanmar border by using microcredit to promote more sustainable means of production, such as the use of more soil-friendly and organic crops, harvesting nontimber forest products, and the use of aquaculture.

Enterprises such as furniture making or food stands that increase the demand for timber as fuel also increase incentives to manage proximate timber supplies more carefully. Furniture builders are more likely to use mature trees for furniture production, sparing immature trees from collection for fuelwood. Likewise, microentrepreneurs who start enterprises that do not use wood as an input (such as bicycle repair and sewing services) have a greater stake in managing forestland sustainably if their business activities mean that their time spent gathering fuelwood bears a higher opportunity cost.

Higher income resulting from microcredit activities can decrease deforestation if the income generated leads to a higher demand for environmental quality (such as a commitment to planting more trees or switching from biomass to electricity) or if it allows agricultural intensification through additional fertilizer or pesticide use (which has other environmental consequences for water CPRs). Evidence from Madagascar suggests that access to member-based financial institutions encouraged agricultural intensification there by increasing lowland rice yields and upland soil fertility. Increased access to capital, however, also increased upland farming opportunities, though on net, an income increase of 1 percent of households involved with MFOs decreased upland use by 0.36 percent (Zeller et al. 2000).

Finally, for individuals already raising livestock, credit extension that diversifies income can reduce deforestation pressures by reducing the need to carry extra livestock for insurance purposes.

Focus on Women

Focusing on women, the second characteristic of microcredit programs, is particularly interesting in the case of Vietnam because of the role of the Vietnam Women's Union (VWU). This mass organization was formed as part of the communist effort to mobilize and engage groups in the society not traditionally active in politics, such as women, youth, and the rural poor. Most microcredit programs, both government and NGO, work through the VWU,[8] and it is estimated that by 1996, the VWU was running 50,000 women's saving and credit groups (Fallavier 1998, 67).

VWU loan officers consistently report repayment rates of 98 to 100 percent, and the women who receive loans seem to be experiencing significant increases in their income after just a couple years of borrowing. Some questions remain, however, about the opportunities the VWU offers to poorer members of Vietnamese society.[9] Pressure to maintain high repayment rates may lead to groups' selecting only the more financially stable members of the commune for loans, thus excluding some of the most needy forest users: widows, the elderly, and the poorest of the poor.

Perhaps as a result of the egalitarian principles of communism, Vietnamese women have historically been active in the formal and informal economy. Hence the novelty of microcredit for them, vis-à-vis men, may be less than in some other cultures. They are, however, the primary gatherers of fuelwood in that country, so the opportunities provided by regular meetings to build human and social capital for managing the commons remains.

Group Lending

Like the focus on women, group decision making and regular communication, the third characteristic of microcredit programs, is expected to improve forest management and reduce deforestation. As discussed earlier, the very poor in many countries resort to collecting and burning firewood from a common source. A stronger community network among microcredit borrowers may lead to agreements about sharing arrangements for firewood sources, an increased awareness of the effects of depleting the natural resource that firewood represents, and innovations leading to substitute fuel sources: a group purchase of coal, for example, if it is available in the area.

Group decision making could also be facilitated by the Vietnamese experience with collective behavior under communism. In the villages of Ha Tay, many borrowers were aware of the effects of their microenterprise activities on the commons, but no one reported these externalities' being discussed during regular group meetings. This differs from the situation in the weekly savings meetings in the Western Forest complex of Thailand, which were used, among other things, as a means to report on, monitor, and determine sanctions for illegal forest burning. Whether this is a result of PDI's efforts, of culture, of conditions, or of history is a matter of conjecture.

Conclusion

There are important connections between microcredit programs and environmental resources and in particular CPRs. Links occur both directly and indirectly through changes in physical capital and enterprise activity brought on by the availability of credit. These impacts on income and property ownership have ambiguous effects on the use of environmental resources. Depending on borrowers' responses to the changed opportunity set resulting from the receipt of microcredit, forest use may increase or decrease.

Microcredit programs can also create and enhance human and social capital within a community, however, especially by expanding the options for women and the poorest. These impacts are postulated to improve CPR use and management and to mitigate the negative consequences of increased resource use and degradation arising from extending credit and increased economic activity.

Involving women in microcredit programs may increase their resource management and other knowledge, their economic independence, and their reproductive control. Enhanced social capital through group lending and meetings may lower the costs of collective action and hence the costs of managing CPRs. An example, based on the circumstance described for Ha Tay province, is improved management of waste from small livestock production to prevent degradation of community water bodies. Collectively, the group could organize a manure collection-to-compost scheme that would not be worthwhile on an individual basis. In this way, social-capital creation from microcredit may be the linchpin to guiding the changes in behavior that follow from credit availability. Greater stewardship of natural resources and protection of the environment are incentive-compatible behaviors under the new set of conditions existing after the granting of microcredit.

These incentives can be supported by appropriate education and training for environmental protection by sensitized MFOs. Many microcredit programs have begun to organize themselves consciously in ways that are linked to environmental resource goals. Nonetheless much remains to be understood about the connection between microfinance, the environment, and CPRs. The framework presented in this chapter describes some of these links and suggests the need for a more ambitious and long-term empirical agenda.

Notes

1. The legendary repayment rates of most MFOs tend to be exaggerated by normal accounting standards but are still impressive. Despite high repayment rates, however, most MFOs, including the Grameen Bank, still rely on subsidies. See Morduch 1998.

2. In Southeast Asia (and other parts of the world), there have long existed informal credit and savings groups based on social collateral. Two examples are Rotating Savings and Credit Associations (ROSCAs) and "chit funds." As early as the late 1800s, British administrators in India described how "people would bring their subscriptions of money, rice, or coconuts to the house of one selected to receive the prize that month; the winner would entertain them all at a party or feast. This would continue until each had his turn, then the cycle would start again" (Malooney and Ahmed 1988, 105).

3. Although the fungibility of microcredit loans has raised some serious concerns about their long-term effectiveness in promoting sustainable livelihoods, for the rural poor in particular it is difficult to separate production and consumption decisions, since labor is the main productive asset and adequate nutrition is

essential to work. The environmental implications of microcredit loans, however, do depend on whether loan funds are used for productive activities or channeled into consumption.

4. A participant's income can also be diversified through the use of savings or other financial services offered by the MFOs.

5. Women's empowerment, defined as increasing women's autonomy and control over their lives and over decision making, is believed, on net, to increase with access to credit, though some studies suggest that women do not fully control the loans they receive. There is also speculation that some forms of domestic violence may be increasing as a result of women's increasing access to credit (Goetz and Gupta 1996, 45–63).

6. The group meetings also reduce information asymmetries between lenders and borrowers, since other villagers are far more likely to understand the creditworthiness of a particular individual in their village than a nonlocal bank manager.

7. For example, the Grameen Bank's close relationship with the government of Bangladesh is maintained through government officials who sit on Grameen's board. Peter Evans (1996) writes about the synergy between local government and a Grameen replicator in Vietnam, though most microcredit programs in Vietnam also work through the powerful Vietnam Women's Union.

8. According to CGAP (1996), the only exception is ActionAid's programs. However, this is contradicted by ActionAid's program description, which states that two Vietnam Women's Union officials are responsible "for programme operations as a whole."

9. For a more in-depth examination of social capital and the Vietnam Women's Union, see Locker 2000.

References

Anderson, C. Leigh, Laura Locker, and Rachel Nugent. 2002. "Microcredit, Social Capital, and Common Pool Resources." *World Development* 30(1):95–105.

Besley, T., and S. Coate. 1995. "Group Lending, Repayment Incentives, and Social Collateral." *Journal of Development Economics* 46(1):1–18.

Bornstein, David. 1997. *The Price of a Dream: The Story of the Grameen Bank*. Chicago: University of Chicago Press.

Buntin, John. 1997. "Bad Credit: Microcredit Yields Macroproblems," *New Republic* 216(13) (March 31):10–11.

Consultative Group to Assist the Poorest (CGAP). 1996. *Microfinance in Vietnam: A Collaborative Study Based upon the Experiences of NGOs, UN Agencies and Bilateral Donors*. Hanoi: CGAP.

Coleman, James. 1990. *The Foundations of Social Theory*. Cambridge: Harvard University Press.

Cuc, Le Trong, A. Terry Rambo, Keith Fahrney, Tran Duc Vien, Jeff Romm, and Dang Thi Sy, eds. 1996. *Red Books, Green Hills: The Impact of Economic Reform on Reforestation Ecology in the Midlands of Northern Vietnam*. Berkeley, CA: East-West Center Program on the Environment.

Dasgupta, Partha, and Karl Göran Mäler. 1994. "Poverty, Institutions, and the Environmental Resource Base." World Bank paper no. 9. Washington, DC: World Bank.

Evans, Peter. 1996. "Government Action, Social Capital and Development: Reviewing the Evidence on Synergy." *World Development* 24(6):1119–1132.

Fallavier, Pierre. 1998. *Developing Micro-Finance Institutions in Vietnam: Policy Implications to Set up an Enabling Environment*. Master's thesis. Vancouver: University of British Columbia.

Food and Agriculture Organization of the United Nations (FAO) and Deutsche Gesellschaft für Technische Zusammenarbeit (GTZ). 1998. "Agricultural Finance Revisited: Why?" AGS series paper no. 1. Rome: FAO.

Food and Agriculture Organization of the United Nations (FAO) and Ministry of Forestry, Socialist Republic of Vietnam. 1992. *Tropical Forestry Action Programme: Vietnam Fuelwood and Energy Sectoral Review*. Bangkok: FAO.

Global Development Research Center. 2000. "MICROFACTS: Data Snapshots on Microfinance." Available at <http://www.gdrc.org/icm/data/d-snapshot.html>.

Goetz, Anne Marie, and Rina Sen Gupta. 1996. "Who Takes the Credit? Gender, Power, and Control over Loan Use in Rural Credit Programs in Bangladesh." *World Development* 24(1):45–63.

Grootaert, Christiaan. 1998. "Social Capital: The Missing Link?" Social Capital Initiative working paper no. 3. Washington, DC: The World Bank.

Hines, Deborah. 1995. *Financial Viability of Smallholder Reforestation in Vietnam*. Hanoi: United Nations Development Programme.

Khandker, Shahidur. 1998. *Fighting Poverty with Microcredit: Experience in Bangladesh*. New York: Oxford University Press.

Locker, Laura. 2000. "Microcredit, Social Capital and Sustainability." Honors thesis. Henry M. Jackson School of International Studies, University of Washington, Seattle.

Malooney, Clarence, and A. B. Ahmed. 1988. *Rural Savings and Credit in Bangladesh*. Dhaka, Bangladesh: University Press Limited.

Microcredit Summit. 2000. "Microcredit Summit Fulfillment Campaign." Available at <http://www.microcreditsummit.org>.

Morduch, Jonathan. 1998. "Does Microfinance Really Help the Poor? New Evidence From Flagship Programs in Bangladesh." Draft report. Cambridge: Harvard Institute for International Development (HIID), Harvard University.

Ostrom, Elinor. 1990. *Governing the Commons: The Evolution of Institutions for Collective Action*. New York: Cambridge University Press.

Ostrom, Elinor. 1992. *Crafting Institutions for Self-Governing Irrigation Systems*. San Francisco: Institute for Contemporary Studies Press.

Ostrom, Elinor. 1994. "Constituting Social Capital and Collective Action." *Journal of Theoretical Politics* 6(4):527–562.

Ostrom, Elinor, Roy Gardner, and James Walker. 1994. *Rules, Games, and Common-Pool Resources*. Ann Arbor: University of Michigan Press.

Oxfam America. 1999. Grant Application for Grant Vietnam 67/99. Boston: Oxfam America, Global Program Department.

Pass, C., and B. Lowes. 1991. *The Harper Collins Dictionary of Economics*. New York: Harper Perennial.

Putnam, Robert D. 1993. *Making Democracy Work: Civic Traditions in Modern Italy*. Princeton: Princeton University Press.

Roth, James. 1997. "The Limits of Microcredit as a Rural Development Intervention." Ph.D. diss. Manchester University, Manchester, UK.

Schrieder, Gertrud, and Manohar Sharma. 1999. "Impact of Finance on Poverty Reduction and Social Capital Formation: A Review and Synthesis of Empirical Evidence." *Savings and Development* 23(1):67–93.

Schuler, Sidney R., and Syed M. Hashemi. 1997. "The Influence of Women's Changing Roles and Status in Bangladesh's Fertility Transition." *World Development* 25(4):563–575.

Sebstad, Jennifer, and Gregory Chen. 1996. "Overview of Studies on the Impact of Microenterprise Credit." Washington, DC: Agency for International Development Assessing the Impact of Microenterprise Services.

Solomon, Lawrence. 1998. "Micro-credit's Dark Underside," *World Press Review* 45 (August):33.

Stiglitz, Joe, and A. Weiss. 1981. "Credit Rationing in Markets with Imperfect Information." *American Economic Review* 71(3):393–410.

Thompson, Paul M., and Muhammad Nurul Islam, Muhammad Monjur Kadir. 1998. "Impacts of Government-NGO initiatives in Community Based Fisheries Management in Bangladesh." Available at <www.indiana.edu/~iascp/iascp98>.

U.S. House of Representatives, Select Committee on Hunger. 1986. *Banking for the Poor: Alleviating Poverty through Credit Assistance to the Poorest Microentrepreneurs in Developing Countries*. Washington, DC: U.S. Government Printing Office.

van Bastelaer, Thierry. 1999. "Imperfect Information, Social Capital and the Poor's Access to Credit." Working paper no. 234. College Park, MD: Center for Institutional Reform and the Informal Sector (IRIS), University of Maryland.

Wolff, Peter. 1999. *Vietnam—The Incomplete Transformation*. Portland, OR: Frank Cass.

Woolcock, Michael, and Deepa Narayan. 2000. "Social Capital: Implications for Development Theory, Research and Policy." *World Bank Research Observer* 15(2):225–250.

World Bank. 1993. *A Review of Bank Lending for Agricultural Credit and Rural Finance, 1948–1992*. Washington, DC: Author.

World Bank. 1999. "The Initiative on Defining, Monitoring and Measuring Social Capital: Overview and Program Description." Social Capital Initiative working paper no. 1. Available at <http://www.worldbank.org/poverty/scapital/wkrppr/wkrppr.htm>.

World Resources Institute. 1994. *World Resources 1994–95*. New York: Oxford.

Zeller, Manfred, Cecile Lapenu, Bart Minten, Eliane Ralison, Desire Randrianaivo, and Claude Randrianarisoa. 2000. "The Critical Triangle between Environmental Sustainability, Economic Growth, and Poverty Alleviation." In *Beyond Market Liberalization: Income Generation, Welfare and Environmental Sustainability in Rural Madagascar*, ed. B. Minten and M. Zeller, 209–236. Aldershot, UK: Ashgate Publishing.

10

Using Social Capital to Create Political Capital: How Do Local Communities Gain Political Influence? A Theoretical Approach and Empirical Evidence from Thailand

Regina Birner and Heidi Wittmer

Thus, through much of western Germany by the later middle ages the peasantry had succeeded, through protracted struggle on a piece-meal village-by-village basis, in constituting for itself an impressive network of village institutions for economic regulation and political self-government. These provided a powerful line of defense against the incursions of landlords. In the first instance, peasant organization and peasant resistance to the lords appear to have been closely bound up with the very development of the quasi-communal character of the village economy. Most fundamental was the need to regulate co-operatively the village commons and to struggle against the lords to establish and protect commons rights. . . . Sooner or later, however, issues of a more general economic and political character tended to be raised. The peasants organized themselves in order to fix rents and ensure the rights of inheritance. Perhaps most significantly, in many places they fought successfully to replace the old landlord-installed village mayor (*Schultheiss*) by their own elected village magistrates.

—R. Brenner, "Agrarian Class Structure and Economic Development in Pre-industrial Europe"

Introduction

In his seminal article "Agrarian Class Structure and Economic Development in Pre-industrial Europe," Brenner (1976) describes a phenomenon that is explored in this chapter: Local communities use the social capital formed for the purpose of communal-resource management to create political capital, which they employ in their struggle against domination—with remarkable success, in the case Brenner describes. The peasants of Western Europe were able to abolish serfdom, unlike their fellow peasants in Eastern Europe, who, in the absence of communally managed natural resources, did not have such social capital, which could have been used for the creation of political capital (compare Brenner 1976, 57–58). In this chapter, we explore the analytical power of the idea of using social capital to create political capital for a better understanding of the struggles of peasants for power in contemporary societies.

The concept of social capital has received increasing attention in sociology, economics, education, and related disciplines in recent years.[1] In the development-oriented literature, the World Bank and international development research institutions have contributed substantially to popularizing this concept.[2] From the perspective of economics, social capital has two distinct advantages contributing to its increasing use: (1) As capital is essentially an economic concept, the notion of social capital enables social scientists to incorporate social factors into a coherent analytical framework based on economic, human, natural, and physical capital. (2) The concept of social capital allows scholars to analyze social issues in a quantitative way and thus to incorporate them into quantitative economic models. Social capital has also been recognized as a useful concept for the study of common-property and community-based natural-resource management (Ostrom 1994; Bebbington 1997; Grootaert 1998). Considering the different sources of social capital in the literature, one can distinguish three major approaches, which differ with respect to underlying theory and empirical application (Wall, Ferrazzi, and Schryer 1998). Referring to their most important proponents, we label them here the "Bourdieu approach," the "Coleman approach," and the "Putnam approach." Interestingly, the application of social capital in economics and natural-resource management has almost entirely neglected the approach of Bourdieu, who studied the role of social capital from the perspective of the individual and focused on exclusionary forms of social capital.

In this chapter, we propose to extend the analytical framework created by the combination of the concepts of social, economic, human, and natural capital by including the concept of political capital. This concept can be derived from political-resource theory and shares the advantages of the social capital concept mentioned earlier. We argue that this concept allows us to achieve a better understanding of the political processes that lead to or prevent the change of resource management regimes. We intend to show that the concept is particularly useful for analyzing processes of devolution, which have emerged as a major trend in natural-resource policies in recent years. The term "devolution" describes the transfer of authority, rights, and responsibilities from the state to nongovernmental bodies such as local communities or user groups, whereas the term "decentralization" refers to the transfer of decision-making authority and payment responsibility to lower levels of government (Meinzen-Dick and Knox 2001, 42).

A major objective of devolution is to improve the frame conditions for successful common-property management regimes. Devolution implies shifting power and resources from state agencies to local communities, which may evoke the resistance of the state agencies concerned. Commercial enterprises such as logging concession holders may also lose political influence and income opportunities as a consequence of devolution. Nature conservation organizations may oppose devolution, if they do not trust the capacity of local communities to manage their resources in a sustainable way. Devolution processes will therefore involve power struggles in the course of which local communities have to defend their interests. Against this background, the question is under which conditions devolution will occur at all. As Agrawal and Ostrom (2001, 76) have pointed out, the literature on devolution has focused on the normative question of why devolution should occur; the positive analysis of the political processes leading to—or preventing—devolution in natural-resource management has been relatively neglected. The present study aims to contribute to filling this gap in the literature.

Drawing on insights from a study by Booth and Richard (1998), we argue that the concept of social capital is not sufficiently refined to explain the relation between social organization and political outcomes. The introduction of the concept of political capital serves to overcome this problem. A major element in the framework proposed here is the use of social capital for the creation of political capital. We argue that to study this process, it is useful to take Bourdieu's approach to social capital into account. We stress that the objective of the proposed framework is a positive analysis of policy processes, not a normative analysis of policies. Our approach focuses on questions such as: Why and how is devolution achieved? To which extent can it be achieved? Why is it often not achieved? We assume that analytical concepts, such as the concept of political capital, that help us to find answers to these questions are relevant both for scientists and for policymakers and practitioners dealing with devolution in natural-resource management. Normative policy questions concerning the optimal level and form of devolution go beyond the scope of this chapter, but we do fully acknowledge the relevance of such normative questions (see Birner and Wittmer 2000 for a normative analysis).

The chapter is organized as follows: the next section develops the analytical framework, starting with a review of the concept of social capital, then outlining the concept of political capital, before discussing the use of social for the creation of political capital. The third section illustrates the application of the framework using the case of devolution in Thailand's forestry sector. The fourth section discusses the insights that can be gained from the case study, and the final section draws some more general conclusions.

Theoretical Framework

Concepts of Social Capital

Bourdieu distinguishes economic, cultural, and social capital. He introduces the concepts of capital and capital accumulation to be able to analyze the social world as an accumulated history that cannot be reduced to a sequence of mechanical equilibria (Bourdieu 1992, 49). He defines social capital as the totality of all actual and potential resources associated with the possession of a lasting network of more or less institutionalized relations of knowing or respecting each other (Bourdieu 1992, 63). Social capital is based on material and symbolic relations of exchange and can be institutionalized and expressed by a name or title that shows one's belonging to a family, clan, nobility, party, and so forth. According to Bourdieu (1992, 64), the amount of social capital held by an individual depends on the extent to which he or she can mobilize a social network and on the capital, including the economic and cultural capital, held by the members of that network. Bourdieu (1992, 76) makes it clear that he introduces the concept of social capital neither for pure theoretical considerations nor as a mere parallelism to economic capital. He uses the concept to explain why persons holding similar economic and cultural capital differ considerably in their achievements, depending on the extent to which they are able to mobilize the capital of a more or less institutionalized group (e.g., of the family, the nobility, alumni of an elite school, an exclusive club) for their purposes. Consequently, Bourdieu's concept of social capital has widely been used in the study of social inequality and hierarchical social structures. As Wall, Ferrazzi, and Schryer (1998, 305) note, social capital as a means of excluding others from access to resources has been a major focus of those influenced by Bourdieu.

Coleman (1988) introduced the concept of social capital as a tool in his undertaking to combine the rational-choice paradigm of mainstream economics with a sociologist focus on norms, rules, and obligations in explaining human behavior. Like Bourdieu, Coleman, in his empirical work on education, considers social capital resources for individuals, but he notes that social capital can also be defined for "corporate actors" such as purposive organizations (Coleman 1988, S98a). Likewise, he suggests that the social-capital concept can be used to explain different outcomes not only at the level of individual actors, but also at the system level (Coleman 1988, S101). Unlike Bourdieu, Coleman (1990) is concerned with a quantitative measurement of social capital and shows that relative quantities of social capital can be measured by considering the position of a particular actor within a social network.[3] Whereas Bourdieu focuses on membership in exclusive organizations as a source of social capital, Coleman differentiates various forms of social capital ranging from obligations and expectations to norms and sanctions. Coleman (1990) asserts that most forms of social capital can be considered a public good, which is characterized by the difficulty of excluding potential beneficiaries and therefore leads to free riding regarding its provision, as mentioned in chapter 1.

Putnam (1993) introduced the concept of social capital in his *Making Democracy Work* to explain how responsive and effective democratic institutions can be created. He uses the term "social capital" to refer to features of social organizations, such as trust, norms, and networks (Putnam 1993, 167). He argues that social capital, defined in this sense, can considerably improve a society's capacity to overcome social dilemmas that have been described in the literature as prisoner's dilemmas, tragedies of the commons (Hardin 1968),[4] or problems of collective action (Olson 1965). Following Coleman, Putnam (1993, 170) stresses that social capital "is ordinarily a public good, unlike conventional capital, which is ordinarily a private good." Putnam argues that "networks of civic engagement" like neighborhood associations, choral societies, and sports clubs represent horizontal interactions that essentially promote trust, reciprocity, and cooperation within a society. Unlike Bourdieu and Coleman, Putnam does not study the role of social capital as a resource for individuals. He focuses on explaining differences in institutional performance at the regional level and analyzes the influence of indicators of social capital on this performance, using various indicators of social capital such as density of sports and recreational associations. This approach of including indicators of social capital in statistical analysis to explain different outcomes at the regional level has also been used in studies on natural-resource management (Zeller et al. 1999).

In conclusion, one can see a major difference between the approaches developed by Bourdieu, Coleman, and Putnam in the perspective on social capital they apply. Bourdieu's approach can be considered the private perspective and Putnam's approach the public perspective on social capital. Coleman, as explained above, has explored both perspectives. Whereas the private perspective deals with the advantages that an individual derives from the possession of social capital, the public perspective is concerned with the advantages of the existence of social capital for the society as a whole. In this chapter, we focus on the private perspective on social capital, because we intend to study how social capital can be used to serve the interests of various actors in political processes. As our unit of analysis is not individuals but local communities and other groups—"corporate actors" in Coleman's sense—we use the term "actor's perspective" in the following instead of "private perspective." We take into account both the exclusive forms of social capital that were at the center of Bourdieu's approach and the more open types of organizations Putnam as well as Coleman dealt with, but we look from the actor's perspective at both types of organizations to identify how they can serve the interests of different actors.

Concepts of Political Capital

As an analytical concept, political capital has apparently not gained wide currency in political science, political economy, or any related discipline. The term is hardly found in any handbook or dictionary of political science,[5] but it is frequently used by journalists in the expression "to make political capital" of some event, with a connotation of taking an unfair advantage of the event.[6] Booth and Richard (1998) apply the concept of political capital in a study that reconsiders Putnam's (1993) major argument that civil society, expressed in citizen organizational activity, contributes to successful governance and democracy. The authors criticize Putnam for failing to specify the mechanisms by which civil society impinges upon government. Putnam, they assert, does not elucidate how group involvement affects citizen behavior so as to influence government performance or enhance prospects for democracy (Booth and Richard 1998, 782). They argue that associational activism, to have political significance, must foster attitudes and behaviors that actually influence political regimes. The authors introduce the concept of political capital to label such "state-impinging attitudes and activities" (782) and use four measures of it: democratic norms, voting, campaign activism, and contacting public officials. The authors conclude from their multiple-regression analysis that political capital, rather than social capital, explains how formal group activism influences democracy in Central America.[7]

The concept of political capital is closely related to the concept of political resources.[8] As Hicks and Misra (1993, 671) note, a wide range of resource theories in political science "share a focus on the empowering role of resources for the realization of outcomes that advance the actors [*sic*] perceived interests."[9] One of them is resource mobilization theory, which deals with the emergence, dynamics, and tactics of social-movement organizations. This approach examines the critical role that the mobilization of resources, for example, discretionary time and money of potential supporters among both the mass and elites, play in the emergence and success of social movements (McCarthy and Zald 1977). This concept of resource mobilization is taken up in this chapter, although the focus here is placed on the mobilization of social capital rather than economic resources, as will be discussed later in the chapter in more detail.

The concept of political capital developed here also draws on the political-resource framework developed by Hicks and Misra (1993) and Leicht and Jenkins (1998), who study the role of resources for the adoption of certain policies. Hicks and Misra (1993, 672) use political-resource theory to explain welfare spending and define "instrumental political resources" as "specific resources used by specific authors to realize their perceived interests." Examples of such instrumental political resources include interest organizations, electoral leverage, and disruptive leverage. The authors also take into account the political frame conditions that empower the actions of the interest groups or condition the effectiveness of their instrumental political resources.[10] The fiscal capacity and internal organization of the state are examples.

From the results of their multiple-regression analysis, Hicks and Misra conclude that an integrated framework based on political-resource theory has more explanatory power than pluralist, state-centered, or mass political-conflict theories, which focus on a particular set of influencing factors in explaining policy formation. Leicht and Jenkins (1998) apply a similar approach to explain the adoption of venture capital programs by state governments in the United States. They use event history methods in their statistical analysis and consider additional instrumental political resources, for example, pacts between leaders of encompassing peak associations representing different interest groups.

Based on these studies, "political capital" is defined here as the resources used by an actor to influence policy formation processes and realize outcomes that serve the actor's perceived interests. Important forms of political capital include lobbying, electoral leverage, disruptive leverage (e.g., organizing public rallies), and international support. Ideology can be considered an important source of political capital, too, because it can be used to influence political decision makers both directly and indirectly, through influencing public opinion. The studies quoted above show that the macropolitical frame conditions have to be included in an analysis of political capital, as they influence the possibilities for the diverse actors to accumulate political capital and condition the effectiveness of different types of political capital with regard to determining policy outcomes.[11]

In conclusion, the concept of political capital is introduced here because it offers the following analytical advantages:

1. The concept of capital allows scholars to study the spending of resources as an investment that improves the flow of future benefits. The issues with which we are concerned, such as lobbying, organizing demonstrations, and mobilizing voters, have to be considered investments in this sense. Related to the investment aspect is the notion that capital can be accumulated in the course of time. This makes it possible to take historical developments into account, a point that has been emphasized by Bourdieu (1992, 49).

2. As mentioned in the introduction to the chapter, one advantage of the political-capital concept is its compatibility with an already well-established framework of economic, natural, human, and social capital. Such a framework allows scholars to study how one form of capital can be used for the creation of another form of capital, a process we explore in this chapter.
3. A further advantage of the political-capital concept is the possibility it affords us of applying analytical concepts used in economics, such as "portfolio diversification" to mitigate risks. (For reasons of scope, these aspects are not further explored in this chapter.)

One might ask whether the introduction of the concept of political capital will lead to an unnecessary "inflation" of the capital concept. Would it be more useful to consider what has been identified as political capital here to be just another category of social capital? Unlike physical capital, neither social nor political capital is directly observable. Therefore, the question arises: What can be gained by postulating that one theoretical construct (social capital) is transformed into another theoretical construct (political capital)? Social and political capital are, however, defined here in such a way that they refer to distinct variables in the real world that can be empirically observed. We contend that it depends on the purpose of analysis whether the introduction of the concept of political capital has analytical advantages. We argue that this concept is particularly useful for positive policy analysis. As Booth and Richard (1998) suggest, the concept of social capital is not sufficiently refined to explain political outcomes. Conceptualizing political capital as a separate category of capital will certainly provide more analytical clarity than further broadening the concept of social capital.

Using Social Capital for the Creation of Political Capital

To explore the analytical potential of introducing the political-capital concept, we consider the use of social capital for the creation of political capital as one factor explaining policy formation. For reasons of scope, other capital transformations are not discussed in this chapter, even though they may offer analytical clues as well, especially in the field of environmental politics. For example, political decision makers who grant logging concessions to influential army figures transform (their country's) natural capital into (their personal) political capital.

Social capital can be used for the creation of political capital in different ways. Most notably, social capital helps to overcome Olson's (1965) famous collective-action problem of political engagement.[12] Social networks and organizations provide platforms for the exchange of political ideas and mutual encouragement for political participation. Social organizations can foster the development of ideological positions, an important form of political capital. In formal organizations, members have the opportunity to develop skills such as speaking in public, assuming official functions, and mobilizing others that can be used for political action (e.g., the organization of public rallies). Social capital held in the form of membership in elite groups may promote access to political decision makers, which facilitates lobbying. Not all forms of social capital will necessarily support the formation of political capital. The results of Booth and Richard (1998) suggest that informal communal organizations may lead to concentration on self-help, which does not necessarily encourage and perhaps even discourages political participation.[13]

Social capital is not expended when it is used for the creation of political capital. This does not imply, however, that the amount of political capital that can be created from a certain "stock" of social capital is unlimited. Complementary resources such as time, effort, and economic capital have to be invested (e.g., for staging rallies or engaging in lobbying). Moreover, the number of issues that a politician or administrative officer is willing to pursue in the interest of a particular group will be limited, and the number of decision makers a group can approach is limited as well. Therefore, interest groups have to set priorities when using their social capital for the creation of political capital. Investment in political capital also involves uncertainty.

A political decision maker in whom a particular social group has "invested" may be captured by other interest groups or may trade off the group's interests in political deals. Investments in political capital may have to be written off as well if a political decision maker is not successful in reelections. To mitigate against such risks, interest groups may try to diversify their investment in political capital. Investments made in political capital may depreciate over time (e.g., if the group does not continue to invest in maintaining its relations with political decision makers). To analyze the political processes leading to devolution, it is important to consider which types of social capital and other resources the proponents and the opponents of such processes hold and how effective they are in using their resources for political-capital formation.

Figure 10.1 illustrates the theoretical framework that has been developed in this section. The figure shows that a particular political outcome may have a "feedback effect" on the political capital held by the proponents and opponents of a particular policy. The ability of an interest group to prove to potential supporters that it is successful in pursuing political goals may improve its possibilities of mobilizing supporters in subsequent policy processes, whereas a negative political outcome may have the opposite effect. Actors may also strategically adjust the policies

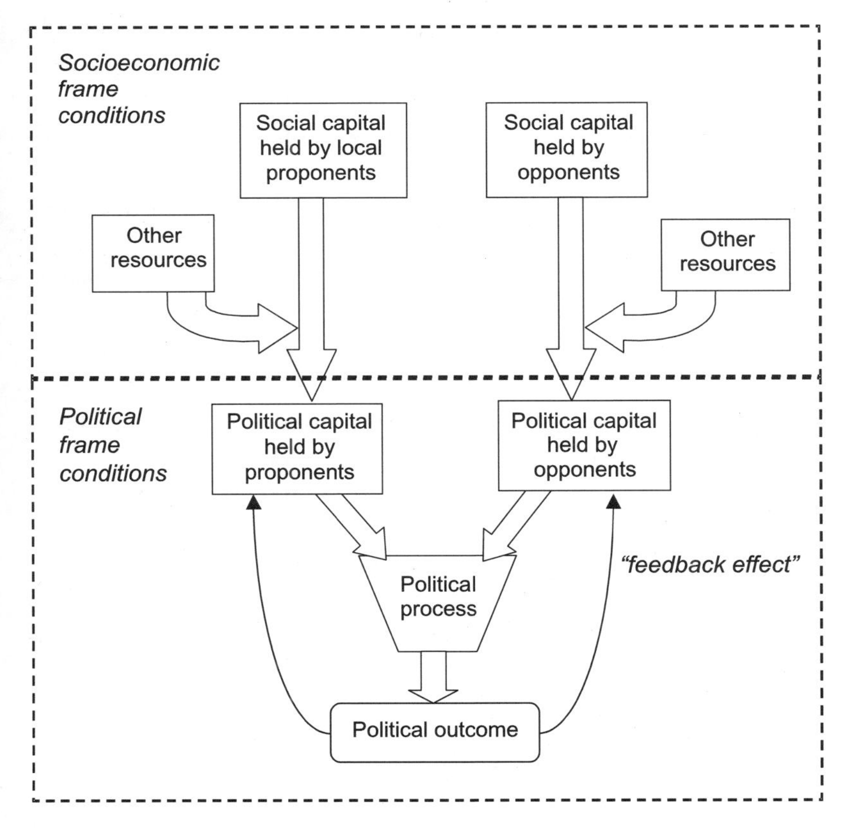

Figure 10.1
Analytical framework.

they pursue to increase the chances of a positive outcome to increase their political capital. Figure 10.1 also indicates that the socioeconomic frame conditions that influence the formation of social capital can be incorporated in the analysis. For reasons of scope, however, this chapter takes social capital as the starting point of the analysis without studying the conditions explaining its creation. The question of the creation of social capital is addressed by Anderson, Locker, and Nugent (chapter 9).

The framework presented in figure 10.1 has been developed for qualitative analyses that take as units of analysis either corporate actors or individuals. Taking the struggle for Thailand's community forestry law as an example, this chapter deals with corporate actors such as NGOs that try to achieve policy outcomes at the national level. The framework also allows us to study the interaction between corporate actors and individuals within a political process, for example, an NGO lobbying a minister. One can also use the proposed framework to study policy processes at the regional and local levels, where interactions between individuals and organized groups may be even more relevant. An example might be the interaction between a village headman who favors a logging concession and an organized group of villagers resisting it.

The framework has been designed for analyzing particular political processes in a case study approach, but not for quantitative applications of the political-capital concept. Its value can be seen in its usefulness for identifying the sources of power and the strategies that diverse actors can use to pursue their goals in a political process. It is, however, feasible to use the concept of political capital in quantitative studies, as Booth and Richard (1998) show. Even though it is not possible to measure the quantities of political capital held by different actors in a way similar to measuring quantities of economic capital, one can, as Booth and Richard show, use measurable indicators of political capital in statistical analyses that aim to explain different policy outcomes across regions or countries.

The Case of the Thailand's Community Forestry Bill

The case of Thailand's Community Forestry Bill has been selected for examination in this chapter because the struggle to pass this bill illustrates very well how the proponents and opponents of community-based forest management used their specific social capital for the creation of political capital to influence the legislative process, which has lasted for more than a decade. The empirical information presented here is derived from secondary sources, especially from reports that appeared in the press between 1997 and 2001 and from interviews with experts and actors held in July–August 1999 and March–April 2000.[14]

Overview of the Policy Process

Thailand is a constitutional monarchy with a bicameral National Assembly, consisting of the House of Representatives and the Senate. Following a national election for the House of Representatives, the leader of the party that can organize a majority coalition usually becomes prime minister. Together with the other ministers, he forms the Council of Ministers (Cabinet). The Cabinet can submit bills to the National Assembly. The new constitution, which was promulgated in 1997, allows citizens to propose a law to the National Assembly by way of collecting 50,000 signatures for a law petition (Constitution of the Kingdom of Thailand 1998, section 170).

Efforts to enact a Community Forestry Bill in Thailand can be traced back to the resistance of local communities and NGOs to government-supported commercial forest plantations in the 1980s. A major triggering event for the emergence of the Community Forestry Bill on the political agenda was the famous Huay Kaew case: the wife of a member of Parliament (MP) leased supposedly degraded forest land from the Royal Forestry Department (for brevity, hereafter referred to as the Forest Department) for reforestation. The land, however, was located in a forest area that had been well managed and maintained by a local community. In 1989, after public protests, the director-general of the Forest Department eventually withdrew the lease contract and—for the first time—publicly granted the village the right to manage its forest (Brenner et al. 1998, 16; Sukin 1997). In the same year, a national NGO meeting formulated the demand for a Community Forestry Bill for the first time (Brenner et al. 1998, 16).

During the 1980s, the need for community participation in forest management was also increasingly recognized within the Forest Department because it became obvious that the department's manpower and budget were insufficient to rehabilitate and protect the country's rapidly declining forest resources (Pratong and Thomas 1990, 177). Thailand's National Forest Policy of 1985, widely considered a milestone document of government forest policy, clearly directed the Forest Department to encourage community participation and collaborate with civil society, the private sector, and other government agencies in forest management. Against this background of increasing public pressure for community forestry rights and changing perceptions within the bureaucracy, the Forest Department drafted the first version of a Community Forestry Bill in 1991–1992. One year later, local groups prepared their own draft, which became known as the People's Draft (Support Group of the Community Forestry Bill [by the People], 1999). In 1995, a committee consisting of government officials, experts, NGOs, and local representatives was appointed to produce a joint draft, which was approved by the Cabinet of Prime Minister Banharn Silapa-Archa in 1996. Because of the dissolution of the Parliament, however, the bill could not be passed.

Under the People's Draft of the Community Forestry Bill, villagers could request the establishment of a community forest from a Provincial Community Forest Committee, comprising members of the provincial government and representatives of the Forest Department, NGOs, and local communities, as well as academics. Upon approval, the members of a community forest would elect a Community Forestry Management Board, which would be responsible for the management of the community forest according to a plan submitted to the provincial committee. The draft envisages that the rights to use, manage, and protect the forest would be issued to the community, but it does not foresee a transfer of full ownership rights. The institutional arrangements of community forest management suggested by the People's Draft were not controversial. The debate, rather, concentrated on the possibility of establishing community forests in protected areas (national parks, wildlife reserves, and critical watershed areas) and on the activities to be allowed in such areas. This question is of particular importance for the mountain areas of northern Thailand, where ethnic-minority groups live in areas that have been classified as protected areas by the state. NGOs and community-based organizations representing these groups campaigned for the possibility of establishing community forests in protected areas, whereas conservation-oriented NGOs and an organization representing lowland farmers opposed this possibility.

In view of the growing controversy about these issues, the newly elected government of Prime Minister Chavalit Yongchaiyuth decided to propose the Community Forestry Bill for consideration by the House of Representatives only after a public hearing (Inchukul 1997). The hearing was held in 1997 and was attended by more than 250 people, including academics, Forest Department officials, and representatives of NGOs and community-based organizations. The hearing did not lead to a consensus (Hongthong 1997). Nevertheless, a revised version of the draft was approved by the Cabinet in the same year. In response to protests by NGOs, the Prime Minister of the following government Chuan Leekpai appointed a committee to revise the draft. After including the comments by the Forest Department, this version was approved by the Cabinet in October 1999. According to the Cabinet spokesperson (Srivalo 1999) and additional information from the Forest Department, this draft allows communities to establish community forests in conservation areas if they can prove that they have conserved the forest area for at least five years before the bill was enacted. This was an agreement reached by the revision committee and agreed upon by the Forest Department. The draft approved by the Cabinet also stipulates that the director-general of the Forest Department will have to approve each provincial committee's decision to declare a community forest. According to this draft, he will also have the authority to rescind designation as a community forest area. The original draft prepared by the Forest Department at the beginning of the 1990s and the draft of the revision committee had envisaged a more decentralized arrangement that did not require the Forest Department's director-general to approve the decisions of the provincial-level committees to declare community forests. The draft of the revision committee also included a provision for commercial plantations, which was heavily opposed by the supporters of the People's Draft (see Sukin 1999a).

Making use of the above-mentioned law petition regulation that was introduced with Thailand's new constitution in 1997, NGOs and community-based organizations that opposed the government's draft submitted their own draft, an updated version of the earlier People's Draft, together with 50,000 signatures to the Parliament on March 1, 2000. This act was described in the press with headlines such as "Citizens Draft Historic New Forest Law" (Tangwisutijit 2000) and

"Landmark Public Bill Submitted" (Atthakor 2000). As three of Thailand's political parties also prepared their own drafts, five different drafts were eventually submitted to the House of Representatives in 2000. The House agreed to use the Cabinet's draft as the basis of the bill and appointed a committee to scrutinize the bill. The legislative process was, however, interrupted by the dissolution of the Parliament prior to the general elections in January 2001. Only one week after those elections, the supporters of the People's Draft announced that they would lobby the New Aspiration Party MPs to table all the Community Forestry Bill drafts as soon as the House of Representatives session started (Kongrut 2001). The New Aspiration Party, which had supported the People's Draft, became a coalition partner of the Thai Rak Thai Party, which won the elections. The supporters of the People's Draft envisioned that they could also win the support of Thai Rak Thai, a comparatively new party with an outspoken populist agenda[15] that had not previously been involved in the community forestry debate. Representatives of the organizations supporting the People's Draft became members of an ad hoc panel of the House of Representatives to scrutinize the bill, whereas major opponents of the People's Draft, especially the "deep green" NGOs and the director-general of the Forest Department, were excluded. In November 2001, the House of Representatives finally passed a version of the bill that met major demands of the supporters of the People's Draft, such as the prohibition of the commercial use of community forests and the inclusion of a provision to establish community forests in protected areas (Susanpoolthong 2001). Environmental groups criticized this provision, and lobbied the Senate to reject the bill (Khuenkaew 2001). The Senate deliberated the bill in March 2001 and decided to amend it by a clause that prohibits community forests in protected areas before accepting it (Sattha 2002). In case of such an amendment, the bill has to be returned to the House of Representatives. If the House of Representatives does not approve the amendment, the bill will be reconsidered by a joint committee of the House of Representatives and the Senate. If either House disapproves the bill after the reconsideration by the joint committee, the House of Representatives has after a time lapse of six months the possibility to reaffirm the original bill or the bill considered by the joint committee by the vote of more than one-half of its members (Constitution of the Kingdom of Thailand, sections 175 and

176). When we submitted the final draft of this chapter in May 2002, the outcome of this process was still open.

The following subsection applies the framework developed in the chapter's second section to identify the factors that influenced the political feasibility of a more community-oriented version of the Community Forestry Bill, represented by the People's Draft.

Social Capital

Community-Based Organizations A remarkable feature of local communities[16] in Northern Thailand is their high degree of organization related to natural-resource management. In addition to traditional or customary institutions of natural-resource management, which are especially prevalent among the different ethnic minorities living in the mountain areas of Northern Thailand, forest and watershed management groups or committees at the village level have increasingly been formed during the last few decades. Watershed network organizations have also been created that allow for cooperation among villages in the same microwatershed.

Interviews with representatives of watershed management groups and networks in 1999 provided evidence that the formation of organizations and organizational networks for watershed and forest management was substantially promoted by (1) an increasing shortage of irrigation water and (2) efforts to protect villages' forest resources against claims by private investors or conservation purposes of the state. The development of formalized village regulations concerning forest and watershed management, including enforcement mechanisms such as payments to village funds, and the use of three-dimensional watershed models play an important role in these organizations (see, e.g., Upper Nan Watershed Management Project 1997). The formation of such organizations took place both without intervention from outside and with the support of governmental and nongovernmental organizations and development projects. Various foreign-funded "highland development projects" in northern Thailand that were set up primarily to promote opium replacement have increasingly supported institution building for natural-resource management (Roongruangsee 1994; Poffenberger and McGean 1993).

Ethnic minorities also created more politically oriented organizations to defend their interests with regard to civil rights. Prominent examples include the Northern Farmers' Network and the Northern Tribal People's Network. Hill tribe people in over one hundred villages located in protected areas in the Country's north have joined the Assembly of the Poor, a nationwide network that includes both rural and urban grassroots organizations. The Northern Community Forest Assembly, comprising more than 730 communities, has been formed with a specific focus on the community forestry issue.

Lowland farmers in northern Thailand also voice their interests with respect to the community forestry issue. They claim that hill tribe settlements in critical watershed areas are responsible for water shortages and that their inhabitants should be resettled. Even though they engage in public actions such as road blockades ("Battle for Inthanond Affects All Parties" 1998), their degree of organization is comparatively low. The Chom Thong Watershed and Environment Conservation Organization is apparently the only local organization opposing the People's Draft of the Community Forestry Bill.[17]

Nongovernmental Organizations As indicated above, two groups of NGOs in Thailand can be distinguished that differ in their position concerning the Community Forestry Bill.[18] The NGOs promoting the People's Draft are to a large extent engaged in community-based rural-development activities in the northern provinces and have played a supportive role in the emergence or strengthening of the community-based organizations described above. They typically exercise advocacy for ethnic minorities living in protected areas. Prominent examples of this group of NGOs include the Northern Development Foundation, the Inter-Mountain People Education and Culture in Thailand, and the NorthNet Foundation. The position of these NGOs as promoters of the People's Draft has to be seen in the broader perspective of the NGO movement in Thailand to which they belong. Predominant orientations of this movement, as described by political observers (Connors 1999), include a close relation to rural grassroots organizations, advocacy of civil rights and minority rights, an explicitly critical position toward the state and the bureaucracy, promotion of decentralization and political reform, and a pronounced critique of commercialization and Western-style capitalist development.

The second group of NGOs, which opposes the People's Draft of the Community Forestry Bill, is much smaller. These NGOs are often referred to as "deep green" in the public discourse. The group comprises around twenty-five NGOs, which have formed the Center for Watershed Forest Conservation. Prominent members include the Seub Nakhasathien Foundation and the Dhammanaat Foundation. A leading member of the Dhammanaat Foundation stressed in an interview with one of the authors that the opposition started with only three NGOs. Social capital in the form of family relations and personal friendships was initially used to extend this network.

Elites Following Bourdieu (see "Theoretical Framework" section), membership in elites and alliances with elites represent another form of social capital. The supporters of the People's Draft were able to build alliances with academics who act as free-of-charge consultants and advisors for them (Tangwisutijit 1998). Academics also play an important role in the conservation-oriented NGOs. A leading member of the Dhammanaat Foundation possesses social capital in the form of belonging to the Thai nobility. One can also consider the relationships with religious groups that are found among both supporters and opponents of the People's Draft a form of social capital.

According to the framework proposed in chapter 9, the social capital held by other actors, such as commercial interest groups and the Forest Department, would have to be identified as well. The information on this question available to the authors of this chapter is, however, limited. It appears justified to assume that entrepreneurs interested in commercial forest plantations hold social capital in the form of belonging to national business circles and having close linkages with political circles.

Use of Social Capital for the Creation of Political Capital

In this subsection, we first discuss the types of political capital that the proponents of the People's Draft Preferred to use (electoral leverage, direct participation in the legislative process and public protest) in their attempts to influence the form of the final Community Forestry Bill and continue with a discussion of the types of political capital that were preferred by the opponents (lobbying of administrative and political decision makers). Finally, we deal with types of political capital that were used by both groups (use of ideological resources and scientific knowledge in the public discourse and international influence). The effectiveness of the different types of political capital depends largely on the macro-political frame conditions that are discussed later in the chapter.[19]

Electoral Leverage Actors such as NGOs and community-based organizations that possess a high level of social capital in the form of members and networks can try to convert it into political capital in the form of electoral leverage. The Northern Community Forestry Assembly has stated that it will pursue such a strategy through pressuring political parties to announce their stand with regard to the Community Forestry Bill before the next election so that it can mobilize its members to vote only for parties that support the People's Draft (Sukin 1999b).

As noted earlier, the opponents of the People's Draft do not hold comparable social capital in form of large organizations and networks that can be used for mobilizing voters. They emphasized in discussions with political parties and in the public discourse, however, that the People's Draft would not be in the interest of the majority of the Thai population, indicating that the majority of the voters would not support the draft. They argued that community forestry, as foreseen in the People's Draft, would lead to the deterioration of the watersheds in northern Thailand, which are of crucial importance for irrigation and water supply throughout the country ("Keep the Watershed Free of Inhabitants" 1998).

Direct Participation in the Legislative Process As mentioned earlier, the new constitution has created the possibility of citizens' participation in the legislative process by proposing a law. The social capital of the supporters of the People's Draft allowed them to make use of this possibility for creating political capital. They were able to collect more than 50,000 signatures in support of the People's Draft, which definitely requires a high degree of organization and logistical support.[20]

Public Protest The organization of public rallies has been an important strategy of the supporters of the People's Draft of the Community Forestry Bill. Such rallies serve both the function of raising public attention and that of urging politicians to enter into direct negotiations with NGOs and community-based organizations. In 1997, the Northern Farmers' Network, as part of the Assembly of the Poor, representing more than one hundred villages located in protected areas, staged a protest in front of the government house that went on for several months.[21] The primary objective of the protest was to avoid the eviction of the villagers from the protected areas, but the Community Forestry Bill was also raised as a closely related issue.

The protest rallies led to direct negotiations with Prime Minister Chavalit Yongchaiyuth and resulted in three Cabinet resolutions that allowed the villagers to stay in protected areas under certain conditions. One of the three resolutions not only dealt with the protesting villagers but intended to benefit villagers living in protected areas nationwide. The issuance of these Cabinet resolutions was used to mobilize opponents of the community forestry concept. In 1998, the Chom Thong Watershed and Environment Conservation Group, claiming that ethnic minorities settling in upland areas were responsible for their water shortages, organized a rally involving thousands of lowland villagers against the Cabinet resolutions. For several months, the group also organized road blockades that affected ethnic minorities living in upland areas ("Battle for Inthanond Affects All Parties" 1998).

Lobbying Political and Administrative Decision Makers According to an interview with the vice president of the Dhammanaat Foundation, this NGO saw itself "forced" to enter the political arena by its concern about the ecological implications of the three Cabinet resolutions of 1997 and by provisions in the Community Forestry Bill that made it possible to establish community forests in protected areas.[22] The foundation tried to build a network of conservation-oriented NGOs (see earlier discussion) and, as a major strategy, lobbied political and administrative decision makers. It appears justified to assume that the social capital held by this group in the form of belonging to elites was particularly useful for creating political capital through lobbying.[23]

Interviews with members of NGOs that support the People's Draft left the impression that during the 1990s, lobbying of political decision makers and party members was a less important element in their strategy.[24] They concentrated instead on the Forest Department. Whereas lobbying members of the Forest Department had played a role in their strategy in the first half of the 1990s, the second half of the 1990s saw an increasing public confrontation between the supporters of the People's Draft and the Forest Department.[25] In 2001, after an outspoken populist party had won the parliamentary election in a landslide victory, lobbying of MPs became more important for the supporters of the People's Draft (compare Kongrut 2001).

Ideological Resources Used in the Public Discourse All actors involved in the process could obviously make political capital of the high "politicizability" of the forestry issue in Thailand. Forestry-related issues generally receive a high level of coverage in the press.[26] Both the supporters and the opponents of the People's Draft were able to link their standpoint in the community forestry debate to broader value and belief systems or ideological positions. The supporters of a more decentralized forest policy associate the community forestry issue with minority and civil rights and with the related social and economic problems faced by ethnic-minority groups. A considerable proportion of the ethnic-minority population in northern Thailand does not have Thai citizenship and is therefore marginalized. The opponents of the People's Draft link their position to the welfare of "the nation" by stressing the national concern for water. In view of frequent water shortages in Thailand, the water question can easily be politicized because of its importance both for the rural lowland Thai population depending on irrigated agriculture and for the urban population, whose demand for water is increasing.

Associating a political opponent with a specific political position in the public discourse can be seen as another way of making political capital. The proponents of the People's Draft were often accused of being "leftist" and described as "melons": outside green (environmentalist), inside red ("communist"). The draft's opponents were accused of being dominated by neocolonial Western influence and described as "bananas": outside yellow (Asian), inside white (Western).

Strategic Use of Scientific Knowledge in the Public Discourse In environmental politics, scientific knowledge plays a particularly important role (Keeley and Scoones 1999). The strategic use of scientific knowledge in political discourse can therefore be considered an important form of political capital. Both the proponents and the opponents of the People's Draft used scientific knowledge to gain political capital. The alliance between academics and the grassroots-oriented NGO movement, a distinctive feature of Thai politics, represents an important source of social capital that was used in the debate over the Community Forestry Bill to create political capital, as the following quotation shows: "Dr Anan Kanchanaphan of Chiang Mai University explains that academics have a role to play because Thais trust their status as aajaan (teachers), the kind of respect which NGO workers are not accorded. 'Sometimes we haven't said anything new or anything different from the activists and villagers. The problems tackled by the Assembly of the Poor, for example, did not receive much attention from the public until we came out to stress the very same points,' he said" (Tangwisutijit 1998). The involvement of academics in the debate provoked heavy criticism by the Chom Thong Watershed and Environment Conservation Organization. In 1998, they even burned in effigy several professors from Chiang Mai University who had supported the position of the minority groups living in the upstream areas.[27]

Actors opposing decentralization in forest management can also rely on academics who support their position. In 1999, the dean of the Faculty of Forestry of Kasetsart University advised the Forest Department not to propose a Community Forestry Bill at all. He expressed the opinion that decentralizing forest management to the community level would cause many practical problems and make monitoring impossible.[28] The conservation-oriented NGOs in Thailand base their position mostly on arguments drawn from natural sciences, especially hydrology and ecology. In particular the Dhammanaat Foundation frequently issues public statements drawing on such arguments ("Keep the Watershed Free of Inhabitants" 1998). Empirical evidence on the role of agriculture in watershed degradation based on long-term studies, however, is rather limited for northern Thailand. Even in the area of Doi Inthanon, where the most serious upstream downstream conflicts in northern Thailand have occurred, no appropriate data exist that can show to what extent water shortages are caused by deforestation in upland areas, or increased water use for irrigation in lowland areas, or natural processes ("Battle for Inthanond Affects All Parties" 1998). Forsyth (1999) summarizes the results of three recent hydrological, pedological, and ecological research projects in northern Thailand that suggest that much of the so-called watershed degradation is actually the result of long-term naturally occurring biophysical processes.

International Influence International organizations play an indirect role in the policy formation process by providing funds, both to the government and to the nongovernment sector, for community forest–related activities. In view of their opposition to international financial institutions, many NGOs in Thailand do not accept funds from the World Bank or the Asian Development Bank. They rely mostly on European donors, especially the Danish Environment and Development Co-operation, which prefer to fund activities related to community-based natural-resource management. The conservation-oriented NGOs receive international funds as well. The international influence on policy formation in Thailand appears to be limited to this indirect way through financing. With regard to the Community Forestry Bill, international organizations and foreign-funded development projects appeared not to be prominent actors in the political arena.[29]

Discussion

Table 10.1 presents the types of social and political capital identified from the case study as relevant for natural-resource policy.

Comparative Advantages in the Use of Social Capital for the Creation of Political Capital

The case presented in this chapter shows that three different types of social capital are useful for the creation of political capital: community-based organizations (e.g., forestry user groups), public interest groups (NGOs), and exclusive groups or networks (e.g., the nobility, academia). As outlined above, the supporters of a decentralized community forestry policy in Thailand had by far more of the first two types of social capital than their opponents.[30]

The coalition formed between the community-based organizations and NGOs supporting the People's Draft was certainly a key factor in explaining the creation of political capital in support of a decentralized community forestry policy in Thailand. Both types of organization

Table 10.1
Types of social and political capital relevant for natural-resource policy

Types of social capital	Types of political capital
Community-based organizations, user groups (including umbrella organizations)	Electoral leverage (e.g., mobilizing voters)
NGOs/public interest groups (including umbrella organizations)	Direct participation in the legislative process (e.g., law petition)
Coalitions of different types of organizations	Public protest (e.g., public rallies)
Elites (e.g., academia, upper-class circles, nobility)	Lobbying of political and administrative decision makers
	Ideological resources used in the public discourse
	Strategic use of scientific knowledge in the public discourse
	International influence (e.g., donor funds)

concentrated on creating the types of political capital for which their dense organizational network and the existence of umbrella organizations provided them with a comparative advantage: public protests, popular participation in political decision making, and electoral leverage. The fact that the supporters of the People's Draft were the first to make use of the new law petition regulation of the new constitution demonstrates the "political entrepreneurship" of their leaders. The community-based organizations certainly benefited from the political experience and skills that the NGOs could gain more easily, as they were involved in diverse political fields. Because of their involvement in national networks, they could facilitate the appearance of the regional community-based organizations in the national political arena.

Lacking a dense organizational network with a broad popular base, the opponents of a decentralized community forestry policy in Thailand primarily used the social capital they held in the form of membership in elites, which gave them a comparative advantage for creating political capital in the form of lobbying. They could use this advantage effectively, for example, to convince leaders of political parties to support their positions and to build a coalition with the Forest Department. Unlike the supporters of the People's Draft, the opponents of a decentralized community forestry policy did not belong to an NGO movement that had opposition of bureaucracy and state institutions as an important goal on its political agenda. This probably increased its comparative advantage in creating political capital in the form of lobbying.

Although there were thus important differences between supporters and opponents of a decentralized community forestry policy in the use of social capital for the creation of political capital, there were also remarkable similarities. Both actors had close connections with academics, who possess social capital in the form of deep social respect. This facilitated the strategic use of scientific knowledge in the public debate as one form of political capital. The proponents of the People's Draft could draw on sociocultural scientific knowledge concerning ethnic minorities, whereas the opponents referred to hydrological knowledge concerning land use and water supply. In both cases, stylized general models were more useful as political capital than differentiated, site-specific empirical evidence. The proponents and the opponents were equally able to link their position to more general issues of national concern, which also constitutes a source of political capital.

On balance, it is remarkable that the opponents of the People's Draft, as a comparatively small group, managed in a relatively short time to create enough political capital to prevent the implementation of a decentralized version of the Community Forestry Bill for many years. Their influence can largely be attributed to their ability to use the exclusionary type of social capital highlighted by Bourdieu: they belonged to the upper-class elite and effectively used this social capital for the creation of political capital.

Macropolitical Frame Conditions of Political Capital Formation

Macropolitical frame conditions have an important influence on the opportunities of different actors to create and use political capital. The existence of the following macropolitical variables (see table 10.2) can be derived from the case study presented in this chapter.

Political Regime In spite of the problems discussed in what follows, Thailand is a democracy, in which the government can be changed by elections, demonstrations are possible, and the media can serve as a forum for controversial political discussions. This made it possible to

Table 10.2
Macropolitical frame conditions influencing the creation of political capital

Type of political regime
Political party system/electoral system
Participatory elements in political decision making
Political relevance of regional/rural issues on the national level and national issues on the regional/rural level
Prevalence of "money politics" and vote buying
Possibilities of achieving goals through lobbying
Role of the bureaucracy
Freedom of the press
Political culture
Opportunities to politicize certain issues
Scope of international influence

create political capital in the form of electoral leverage and public protest and in the form of strategically using scientific knowledge and ideological arguments in the public discourse. The new constitution of 1997 opened up new possibilities for the creation of political capital in the form of popular participation in the legislative process, which can be used by actors who have the appropriate resources to mobilize enough people to sign a law petition. Obviously, in more authoritarian regimes, different strategies would be required to create political capital.

Political Party System Opportunities to create political capital in the form of mobilizing voters are conditioned by the political party system. Thailand's political parties emerged after World War II, with numerous interruptions by military governments that have hindered the emergence of a well-functioning party system (Limmanee 1998; Connors 1999). The system shows a considerable fluctuation: political parties frequently emerge and decline. The parties are not clearly differentiated from one another in terms of programs and ideological orientations. Prior to elections, leading politicians and large numbers of their supporters change their parties. This creates a considerable uncertainty concerning the question of into which party or politician actors should "invest" their resources to create political capital. Another limitation to creating political capital in this way is the fact that business circles that finance the parties have a high influence on party politics in Thailand. The term "money politics" is often used to characterize such a system.[31] Vote buying has been a common practice in Thailand (Limmanee 1998) and remained "rampant" in the parliamentary elections of 2001 in spite of all measures to reduce it (Sukpanich 2001). Efforts to mobilize people to vote for parties that promise to achieve certain political goals such as a decentralized community forestry policy have to compete with efforts to simply buy votes. Another macropolitical frame condition limited the opportunities for the proponents of the People's Draft to create political capital in the form of electoral leverage: a considerable proportion of the ethnic minorities in Thailand do not have Thai citizenship and are therefore not entitled to vote.

A frame condition that has been conducive to the formation of political capital by mobilizing voters is a comparatively strong regional representation in Thai party politics (Limmanee 1998, 418). Consequently, it is a major strategy of MPs, especially in rural areas, to address in their election campaigns what they perceive as the current most popular demands in their constituencies rather than referring to general party principles (Limmanee 1998, 416, 419). The regional element in Thai politics probably promoted the involvement of some political parties in the debate over community forestry policy, which is an important regional issue in the northern and northeastern regions of the country. As noted earlier, three major political parties that have a power base in the north and the northeast submitted their own draft of the Community Forestry Bill to the Parliament.

Role of the Bureaucracy The role of the bureaucracy, in this case the Forest Department, constitutes an important macropolitical frame condition as well. In the 1960s, Riggs (1966) characterized Thailand as a "bureaucratic polity" in which political decisions are made by the civil administration, the police, and the armed forces, rather than by political parties operating under democratic rules. At the end of the 1980s, Egedy (1988) found that the basic features of Riggs's characterization were still evident in Thailand, but that the power of the bureaucracy had increasingly been challenged by extrabureaucratic forces, especially by political parties representing commercial and industrial groups and, to a certain extent, the urban intelligentsia as well as workers' and peasants' unions. In the 1990s, this trend continued, especially because of the political reform movement, which was provoked by the military coup of 1991 and its defeat in 1992 (Connors 1999).

In the case of the Forest Department, the influence of politics on the bureaucracy is clearly indicated by the fact that the director-general is politically appointed. This increases the possibility that the political party in charge of the Ministry of Agriculture can influence decision making in the Forestry Department.[32] Nevertheless, it appears reasonable to assume that in spite of the increasing political control of the bureaucracy, the influence of the Forest Department on the political process remains considerable. The Forest Department prepares the draft of the Community Forestry Bill that is submitted by the Cabinet to the Parliament, and this draft is usually the core draft on which the Parliament deliberates. Moreover, the Forest Department will be in charge of implementing the Community Forestry Bill once it is passed.

Against this background, a promising strategy of political-capital creation is the combination of lobbying both political party leaders and officials of the Forest Department. The interviews held for this chapter provided evidence that the conservationist NGOs, especially the Dhammanaat Foundation, effectively pursued this strategy. The effectiveness of this strategy for creating political capital was further increased by the fact that the interest of the conservationist NGOs in having a more centralized version of the Community Forestry Bill coincided with the interests of officials in the Forest Department, who would have lost power in the process of devolution and decentralization. As newcomers in the political arena, the conservationist NGOs could thus combine their social capital with that of members of the bureaucracy, whom the "bureaucratic polity" system traditionally allowed to use exclusionary social capital (e.g., in the form of family relations) to create political capital (compare Riggs 1966). Providing environmentalist arguments for a more centralized Community Forestry Bill, the conservationist NGOs could at the same time win the support of political parties and make use of their above-mentioned increased influence on the bureaucratic system.

In contrast to the conservationist NGOs, the actors supporting the People's Draft increasingly engaged in confrontation with the Forest Department toward the end of the 1990s. This can be considered a strategy for creating political capital by mobilizing their members. Such confrontation was also stimulated by problematic actions of the Forest Department affecting ethnic minorities.[33]

Political Culture The creation of political capital also depends on the political culture, another aspect of the macropolitical frame conditions. An example is the use of lobbyism: in Thailand, it has been a widely accepted practice to use personal relations to political or administrative decision makers to pursue business interests.[34] This frame condition probably facilitates the use of lobbying to pursue noneconomic goals, too. However, in the case of the Community Forestry Bill, this opportunity could be fully exploited only by the conservationist NGOs, which do not challenge the current political culture that accepts the use of personal relations to political decision makers to achieve personal goals. The political culture conditions, of course, the creation of other types of political capital, too. For example, the political culture influences the issues that are politicizable and the arguments that carry weight in the public discourse.

Political Stability The degree of political stability in a country is another macropolitical frame condition that influences the creation of political capital. Frequent changes of the government interrupted the process of passing a Community Forestry Bill several times. After each change, the actors had to write off the political capital they had invested in decision makers who lost office. Several times the legislative period did not last long enough to pass the bill after a revised draft reflecting the new power relations had been submitted to the Parliament.

Outlook

An assessment of the political process in Thailand suggests that a "window" for a more decentralized system of forestry management was opened in the early 1990s. The former coalition between business interest groups favoring commercial forest plantations and the military lost influence when democracy was reestablished after the 1991 military coup. At that time, the need to integrate the local population into forest management gained acceptance in the Forest Department. The legislative process needed to accomplish this integration was, however, retarded, among other reasons because of political instability. By the end of the 1990s, it had became increasingly difficult to make the decentralized option a legislative reality. The NGO movement had become divided over this issue, and actors who stressed the value of natural resources as a public good to be protected by the nation-state had accumulated political capital. In 2001, however, the victory of a new populist party in the election of the House of Representatives increased the value of the political capital held by the proponents of the people-centered version of the Community Forestry Bill, and the House of Representatives passed the bill in a form that met their major demands. Nevertheless, after intensive lobbying by conservationist groups, the Senate disapproved the crucial provision that made the establishment of community forests in protected areas possible. Since the final decision on the bill was still pending when we submitted the draft of this chapter, however, it remained to be seen whether the political capital of the opponents would finally be outweighed by the political capital of those promoting a version of the bill that places more trust in civil society and local communities than in the nation-state and the public administration.

Conclusions

The case study presented in this chapter makes it possible to draw more general conclusions about political participation of local communities and about the potential of political capital as an analytical concept.

Positive Effects of Political-Capital Creation

As has been pointed out earlier in the chapter, Thailand's political system in recent years has been characterized by bureaucratic dominance and weak political institutions. Military coups, the last of which occurred in 1991, have frequently interrupted the nation's democratization process. "Money politics" and vote buying have dominated the political scene. Against this background, it is a challenging task to promote democratization and political participation, as envisioned in the new constitution of 1997. This chapter's case study suggests that natural-resource management issues, such as community forestry, can support democratization processes, if these issues are easy to politicize. The high level of political mobilization of the rural population of Thailand in connection with the community forestry law has certainly helped to strengthen the nation's democratic institutions. It has created awareness about the needs and opportunities for political participation among the rural population. The proponents of a decentralized forest management policy were the first to make use of the new constitution's provisions for popular participation in political decision making, such as the law petition provision. One can conclude that the creation of political capital for specific purposes, such as community forestry, can at the same time help promote political participation in a more general sense.

The fact that community-based organizations and NGOs in Thailand have had to fight for their version of the Community Forestry Bill in a political struggle for more than a decade has created a certain amount of pressure to prove to their critics that community forestry will indeed lead to sustainable management of Thailand's forests. This pressure may have been conducive to the creation of functioning institutions of community-based forest management. Therefore, one can expect that, after its eventual approval, the implementation of a Community Forestry Bill in Thailand will be more successful than in countries where a community forestry law has been passed without much public participation.[35] This implies that the need to create political capital for achieving political goals can have positive feedback effects on the formation of sustainable resource management institutions.

Ambiguous Effects of Political-Capital Creation

The case study presented in this chapter also highlights the problems involved in the creation of political capital. The fear of reducing one's stock of political capital by making concessions to opponents has certainly contributed to the impasse in developing a community forestry management policy in Thailand. Efforts to create political capital by mobilizing the local population bring with them the danger that local communities will become "instrumentalized" by one or the other side of the debate. Actions such as burning political opponents in effigy, which occurred during demonstrations related to the community forestry law, may create political capital, but they are not necessarily conducive to the development of a democratic political culture. The possibility of making political capital by opposing state institutions has ambiguous effects in a situation in which a transformation from authoritarian to democratic institutions is already taking place, as in the Thai case, where the new constitution promoted this process. On the one hand, creating political capital by criticizing state institutions can help to develop the healthy skepticism essential to making political and administrative decision makers accountable for their actions (Finkel, Sabatini, and Bevis 2000). On the other hand, an undifferentiated opposition to an entire government department does not leave space for supporting reformist forces within that department's bureaucracy and may destroy any basis for future cooperation between local communities and that department.

The problem of becoming instrumentalized in the course of creating political capital also applies to the conservationist NGOs in Thailand. By struggling for environmental reasons against a decentralized version of the Community Forestry Bill, they also served the interests of Forest Department officials and business circles, for whom decentralization and devolution implied a loss of opportunities for power and income.[36]

The strategic use of scientific knowledge in the public discourse, as one form of political capital, has ambiguous effects as well. On the one hand, it creates a demand for policy-relevant scientific knowledge. On the other hand, it may promote "orthodoxies" in both the natural and the social sciences. This problem is not confined to the case study presented here. Leach and Mearns (1996), Hajer (1995), and others have shown that environmental discourses are often dominated by "environmental orthodoxies" concerning deforestation, land degradation, shifting cultivation, and so on that, on closer examination, are contradicted by site-specific empirical evidence. Agrawal and Gibson (1999) observed that social studies of local communities often paint a romanticized picture of the community. For analytical purposes, they argue, it is more useful to take into account the socioeconomic differentiation and multiple interests of local communities as well as the role of local elites and micropolitics in those communities. Therefore, it appears necessary that natural and social scientists be sensitized about the problems involved in the role of scientific knowledge as political capital.

On the Potential of Political Capital as an Analytical Concept

To conclude, we hope to have shown in this chapter that the concept of political capital has potential as an analytical tool both for practitioners and for scientific purposes. As the term "political capital" is used in everyday language in a sense very similar to that proposed here, it may be easier to understand and apply this concept intuitively than other more abstract or unfamiliar concepts. The case study presented in this chapter has shown that the concept of political capital can be used to interpret a wide array of strategies used by community-based organizations and NGOs to promote their goals in political processes.

Analyzing the strategies of different actors using the political-capital concept may promote a better understanding of particular policy outcomes, such as an impasse, as in the Thai case. The concept may also help to identify the problems involved in various strategies. With regard to the discussion in the social sciences on how to explain policy outcomes, introducing the concept of political capital has analytical advantages in that it allows scholars to integrate the arguments of different pluralist, state-centered, and mass political-conflict theories in political science. It also makes it possible to accommodate the role of knowledge, ideology, and discourse, which are particularly relevant for environmental policy formation.

In our case study, we used community rights to forest resources as an example to apply the political-capital concept. The approach can, however, also be used to study the struggles of local communities to gain access to other resource systems, such as irrigation water or fisheries. We expect the same types of political capital and political frame conditions discussed here for the forestry case to be relevant for other resource systems as well, even though their relative importance may differ. For example, for resource systems that are less politicizable at the national level than forestry, the possibility of mobilizing voters or using ideological resources in the public discourse may be less relevant compared to the lobbying of political and administrative decision makers.

The case study presented in this chapter has also aimed to identify the key variables to be used in future qualitative and quantitative analyses, which are necessary to explore further the analytical potential of the political-capital concept. We hope that the use of this concept can contribute to a better understanding of how policies can be pursued that contribute to the goal addressed in this volume: crafting sustainable commons in the new millennium.

Acknowledgments

We thank Daniel Bromley, Clark Gibson, Matthias Nott, Prasnee Tipraqsa, Manfred Zeller, two anonymous referees, and the editors for helpful comments. We also thank all persons who provided information on the empirical case. Financial support by GTZ (German Agency for Technical Co-operation) for the fieldwork is gratefully acknowledged.

Notes

1. Wall, Ferrazzi, and Schryer (1998, 301) found that the number of journal articles listing social capital as an identifier increased from 14 in 1981–1985 to 109 in 1991–1995.

2. See, for example, the Social Capital homepage of the World Bank at <http://www.worldbank.org/poverty/scapital/index.htm>. See Fox 1997 for a critical assessment of the impact of World Bank projects on social capital.

3. Coleman doubts whether social capital will become as useful a quantitative concept as are the concepts of financial or physical capital and sees the current value of the concept in "its usefulness for qualitative analyses of social systems and quantitative analyses that employ qualitative indicators" (Coleman 1990, 307).

4. It is meanwhile widely acknowledged that Hardin described a "tragedy of open access."

5. "Political capital" as an entry is not found in any of the following handbooks or dictionaries of political science: *Oxford Concise Dictionary of Politics* (McLean 1996), *A New Handbook of Political Science* (Goodin and Klingemann 1996), *International Encyclopedia of Government and Politics* (Magill 1996), *A Dictionary of Modern Politics* (Robertson 1993), *Dictionary of Politics* (Raymond 1992), *Encyclopedia of Government and Politics* (Hawkesworth and Leogan 1992), *The Blackwell Encyclopedia of Political Science* (Bogdanov 1991), *The Blackwell Encyclopedia of Political Thought* (Miller 1991), *The Public Policy Dictionary* (Kruschke and Jackson 1987), *The Dictionary of Political Analysis* (Plumo, Riggs, and Robin 1982), and *A Dictionary of Political Thought* (Swinton 1982).

6. This expression is found in *Safire's Political Dictionary* (1978, 547–548). The dictionary traces the phrase back as far as 1842 and points out that "its frequent current use makes this phrase an important political Americanism."

7. Booth and Richard (1998) do not elaborate on the concept of political capital in a theoretical perspective, nor do they relate it to other literature sources. The term "political capital" is also used by other authors. Kessler (1998) uses the term in the title of his paper "Political Capital: Mexican Financial Policy under Salinas." The author explains certain policy contradictions in Mexico's financial

policy as a response to the electoral challenges confronting the ruling party, but he does not explain or apply "political capital" as a theoretical concept. Grossman (1994) uses the concept of political capital in a study of intergovernmental grants. In his model, federal politicians use grants to buy the political capital of state politicians and state interest groups, which can be used to increase the support of state voters for federal politicians. Coats and Dalton (1992) investigate the question whether the "brand name capital" of politicians results in barriers to entry in political markets. Unlike Booth and Richards (1998), these authors do not relate the concept of political capital to the concept of social capital.

8. This is a parallelism to the concept of social capital, which is also related to the concept of social resources (Wall, Ferrazzi, and Schryer 1998, 301).

9. The volume *Macropolitical Theory* of the *Handbook of Political Science* (Greenstain and Polsby 1975) refers to political resources in this sense in the entry by Dahl (1975) on "Governments and Political Oppositions."

10. These frame conditions can be considered as infrastructural political resources. The authors use the term "infraresources." The distinction between instrumental resources and infraresources was introduced by Rogers (1974).

11. Like social capital, political capital is considered in this chapter from the actor's perspective. The focus is placed on local communities and NGOs that are able to create such political capital. The concept of political capital can also be applied from the perspective of political and administrative decision makers. Politicians hold types of political capital that differ from the types that NGOs or community-based organizations hold. For example, campaign funds and voters' support are important types of political capital for politicians. They may also be able to create political "brand-name" capital that can be inherited (Coats and Dalton 1992). For reasons of scope, however, the political capital held by political and administrative decision makers is not explored in this chapter.

12. In essence, Olson (1965) argues that there is no incentive to participate in a political-pressure group if those who do not participate cannot be excluded from the benefits that the group wants to achieve. He uses this argument for cases in which the number of group members is not very small, coercion is absent, and the group is striving for the provision of a public good.

13. La Due Lake and Huckfeldt (1998) have coined the term "politically relevant social capital" to describe those types of social capital that foster political engagement. They measure politically relevant social capital in terms of communication about politics within an individual's recurrent network of social relations.

14. The interviews were held in connection with a consultancy to the development project Sustainable Management of Resources in the Lower Mekong Basin and in connection with the research project Development of a Watershed Information System for the Assessment of Land Use Systems and Conflicts in the Mountain Regions of the Lower Mekong Basin: Case Studies in Vietnam

and Thailand. Both projects are financed by the German Agency for Technical Co-operation (Deutsche Gesellschaft für Technische Zusammenarbeit GTZ).

15. Examples of Thai Rak Thai's populist policies include a debt suspension for farmers and a village fund scheme that provides a locally managed development fund of one million baht to each of the more than 70,000 villages in the country.

16. The term "local communities" is used here in a general sense to refer to the village population. On the problems with this concept, see Agrawal and Gibson 1999.

17. According to an interview in the journal *Watershed* with the first chairman Thong-In Namthep, the organization emerged from a local water users' organization. Its members include farmers cultivating paddy and other crops, but most are owners of longan orchards who live in the big cities ("The Hilltribes Need to Move Down to the Lowlands" 1998). Therefore the organization can not necessarily be considered community-based, even though it is a local organization.

18. The community-based organizations described in the previous section can also be considered NGOs. The term "NGO" is used here, however, to refer to organizations that pursue public interests and not to organizations that pursue their interests as resource users.

19. As the final outcome of the case is still open, the potential feedback effects of the political outcome indicated in figure 10.1 are not included in the analysis.

20. In July 2000, the supporters of the People's Draft also demanded that five activists, including leaders of community-based organizations and NGOs, should participate in the committee of the Parliament that was in charge of scrutinizing the Community Forestry Bill. The discussion of the draft in Parliament was then, however, interrupted by the general elections. (See "Panel Urged to Include Five Activists" 2000, 18.)

21. This paragraph is based on interviews and various articles in *The Nation* and the *Bangkok Post*, which reported on these events.

22. The vice president of the Dhammanaat Foundation also expressed her views in an interview with *Watershed*: "Q: When did Dhammanaat make the decision to begin to act at a national political level? A: We didn't make any kind of decision, it was made for us, by the Community Forestry Bill proposing that community forests might be set up in these fragile areas. We have not been and never wanted to be involved in politics. The only reason we're doing it is to ensure the survival of everyone in the nation. That's our only reason. I really must emphasize that" ("No One Should Live and Farm in the Upper Watershed Forests" 1998, 13).

23. According to the interviewed Dhammanaat leader, the group was, for example, able to convince leading members of the Democrat Party to support its position. The fact that it used its social capital for lobbying the Forest Department is illustrated by the following statement: "It's easy for Dhammanaat to work closely with people like the RFD, because they see things in the same way. The former director of the Royal Forest Department (RFD), Phairot

Suvannakorn, was very close to Dhammanaat, and people in the same social circle brought them together" ("Participation Has to Come" 1998, 27).

24. One can assume that the entrepreneurs interested in commercial forest plantation also possess social capital in the form of membership in elites that is well suited for lobbying. In the 1980s, forest plantations were at the center of the public struggles for forest land in Thailand. It appears that in the 1990s, commercial interest groups primarily acted behind the scenes. They were not noted as having issued a public statement in any of the numerous newspaper reports on the Community Forestry Bill. Nevertheless, they may have used lobbying to defend their interests. An indication for this can be seen in the fact that the government draft of the Community Forestry Bill issued in 1999 included a provision for commercial plantations, as noted in "Overview of the Policy Process."

25. Important issues that promoted these confrontations were the resettling of villagers living in protected areas and whether plantations should be allowed in community forests. See, e.g., Hongthong 1999. The appointment of an explicitly conservation-oriented director-general of the Forest Department provoked the confrontations. See, e.g., Hongthong 1998.

26. As one article stated: "Improving forestry management, along with the related issue of water management, is probably the biggest environmental issue facing Thailand today" (Fahn 1999).

27. See "Participation Has to Come" 1998.

28. According to a newspaper report, "Dr Uthid Kut-in, Kasetsart University faculty of forestry dean, . . . suggested the community forest be included in the Forest Reserve Act or the National Park Act rather than creating a Community Forest Bill, because the new bill would give more authority than the two current Acts. Decentralising forest management to community level will cause many practical problems, and monitoring will be impossible, because local communities lack academic support, he said. Dr Uthid said he did not recommend improving the bill and it would be unacceptable to resubmit it to the Cabinet" (Sukin 1999a).

29. This situation is different from that in other countries in the region, for example, Vietnam, where international organizations and bilateral development projects in the forestry sector played an important role in establishing a policy-oriented National Working Group on Community Forestry. See the Website of this group at <http://www.mekoninfo.org>.

30. The Northern Community Forest Assembly comprises more than 730 communities. The NGOs supporting the People's Draft are part of the national NGOs Coordinating Committee on Development, formed by more than 200 NGOs. The opponents of the People's Draft relied on only one community-based organization, and their NGO network (Center for Watershed Forest Conservation) comprised twenty-five NGOs. Social capital held in the form of membership in exclusive networks is obviously more difficult to assess in empirical investigations. The case study presented in this chapter used interview information and evidence in the form of titles and positions indicating elite membership. For more detailed studies, network analysis would be a useful tool for assessing this type of social capital.

31. As Connors (1999, 205) describes: "The buying of influence extended from the lowest level—the buying of a citizen's vote—to purchasing a support base in a political party to secure a Cabinet seat. Once there, the great expenditure would be recouped by any (and most illegitimate) means."

32. The following newspaper report on the appointment of Plodprasop Suraswadi as director-general of the Forest Department in 1998 illustrates this point: "The Cabinet Tuesday approved the proposal of the Agriculture Ministry to appoint Plodprasop as RFD chief, replacing Sathit Sawinthara who was moved to the post of inspector general of the ministry after the Salween scandal. . . . The new director general admitted that his transfer was backed by Chat Thai Party leader Banharn Silapa-archa. 'I accept that he is the one who put me in this position. He gave me a lot of good advice which I agreed to follow,' he said. Plodprasop said he will change the image of RFD to one of protector of the forests" (Hongthong 1998).

33. For example, in March 1998, Forestry Department officials arrested fifty-six people in an ethnic-minority village on charges of encroachment and clearing forests. The way in which the action was taken (arrest without court order and not in the act of committing a crime) generated criticism by advocacy groups (see "Forestry Officials" 1998). Another example was reported in *Watershed* ("This is Like Dying" 2001). In August 2000, Forestry Department officials and police passively watched as lowland people destroyed the fruit orchards of the Hmong people in Pa Klan village. The houses on the farmland were also destroyed and burned.

34. An illustrative example is the case of Thaksin Shinawatra, who was elected prime minister in 2001, even though he had obviously used personal relations (e.g., to receive licenses from state agencies) in a way that allowed him to become one of the richest persons in Thailand ("Thaksin Shinawatra" 2001).

35. Cameroon is an extreme example of a case in which a forestry law including a community forestry regulation has largely been imposed by international donors, especially the World Bank. As a consequence, implementation has proved to be difficult (Ekoko 2000).

36. After the House of Representatives passed the people-oriented version of the Community Forestry Bill in November 2001, senators reported that there was heavy lobbying of the Senate to reject the bill (Khuenkaew 2001). In this context, one journalist argued that "by delaying the Community Forestry Bill, the well-meaning environmentalists become the tools of money barons and forest officials. Guess who's smiling now?" (Ekachai 2001).

References

Agrawal, A., and C. Gibson. 1999. "Enchantment and Disenchantment: The Role of Community in Natural Resource Conservation." *World Development* 27(4):629–649.

Agrawal, A., and E. Ostrom. 2001. "Collective Action, Property Rights and Devolution of Forest and Protected Area Management." In *Collective Action,*

Property Rights and Devolution of Natural Resource Management—Exchange of Knowledge and Implications for Policy, ed. R. Meinzen-Dick, A. Knox, and M. Di Gregorio, 75–109. Eurasburg, Germany: Deutsche Stiftung für internationale Entwicklung (DSE).

Atthakor, P. 2000. "Landmark Public Bill Submitted. *Bangkok Post*, March 2, 2000. Available at <http://www.bangkokpost.net>.

"Battle for Inthanond Affects All Parties." 1998. *The Nation*, June 9. Available at <http://www.nationmultimedia.com>.

Bebbington, A. 1997. "Social Capital and Rural Intensification: Local Organizations and Islands of Sustainability in the Rural Andes." *Geographic Journal* 163(2):189–197.

Birner, R., and H. Wittmer. 2000. "Co-management of Natural Resources: A Transaction Costs Economics Approach to Determine the 'Efficient Boundaries of the State.'" Paper presented at the annual international conference of the International Society for New Institutional Economics, Tübingen, September.

Bogdanov, V. 1991. *The Blackwell Encyclopedia of Political Science*. Oxford: Blackwell Publishing.

Booth, J. A., and P. B. Richard. 1998. "Civil Society, Political Capital and Democratization in Central America." *Journal of Politics* 60(3):780–800.

Bourdieu, Pierre. 1992. "Ökonomisches Kapital—Kulturelles Kapital—Soziales Kapital" (Economic Capital—Cultural Capital—Social Capital). In *Die verborgenen Mechanismen der Macht* (*The Hidden Mechanisms of Power*). Schriften zu Politik und Kultur 1, 49–79. Hamburg: VSA-Verlag.

Brenner, R. 1976. "Agrarian Class Structure and Economic Development in Pre-industrial Europe." *Past and Present* 70:30–75.

Brenner, V., R. Buergen, C. Kessler, O. Pye, R. Schwarzmeier, and R. D. Sprung. 1998. "Thailand's Community Forestry Bill: U-Turn or Roundabout in Forest Policy." Working paper no. 3. Freiburg, Germany: DFG Graduate College, Socio-Economics of Forest Use in the Tropics and Subtropics.

Coats, R. M., and T. R. Dalton. 1992. "Entry Barriers in Politics and Uncontested Elections." *Journal of Public Economics* 49:75–90.

Coleman, J. 1988. "Social Capital in the Creation of Human Capital." *American Journal of Sociology* 94:S95–S120.

Coleman, J. 1990. *Foundations of Social Theory*. Cambridge: Harvard University Press.

Connors, M. K. 1999. "Political Reform and the State in Thailand." *Journal of Contemporary Asia* 29(2):202–226.

Constitution of the Kingdom of Thailand. 1998. Tentative translation by Foreign Law Division, Office of the Council of State. Bangkok, Thailand.

Dahl, R. 1975. "Governments and Political Oppositions." In *Macropolitical Theory.* Vol. 3 of *Handbook of Political Science*, ed. F. Greenstain and N. Polsby, 145–147. Reading, MA: Addison-Wesley.

Egedy, G. 1988. "Thailand: Stability and Change in a Bureaucratic Polity." Discussion paper no. 248. Brighton, UK: Institute of Development Studies.

Ekachai, S. 2001. "Even Rustics Can Save Forests." *Bangkok Post*, December 20. Available at <http://www.bangkokpost.net>.

Fahn, J. 1999. "Creating a Buffer for Forests." *The Nation*, October 4. Available at <http://www.nationmultimedia.com>.

Ekoko, F. 2000. "Balancing Politics, Economics and Conservation: The Case of the Cameroon Forestry Law Reform." *Development and Change* 32:131–54.

Finkel, S. E., C. A. Sabatini, and G. G. Bevis. 2000. "Civic Education, Civil Society, and Political Mistrust in a Developing Democracy: The Case of the Dominican Republic." *World Development* 28(8):1851–74.

"Forestry Officials Arrest 56 People in Pang Daeng Village." 1998. Watershed 4(1):7.

Forsyth, T. 1999. "Historical Evidence for Watershed Degradation: How Important Is Agriculture?" Paper presented at workshop "Environmental Services and Land Use Change," Chiang Mai, Thailand, May 21–June 2.

Fox, J. 1997. "The World Bank and Social Capital: Contesting the Concept in Practice." *Journal of International Development*, 9(7):936–971.

Goodin, R., and H.-D. Klingemann. 1996. *A New Handbook of Political Science*. Oxford and New York: Oxford University Press.

Greenstain, F., and N. Polsby, eds. 1975. *Macropolitical Theory*. Vol. 3 of *Handbook of Political Science*. Reading, MA: Addison-Wesley.

Grootaert, C. 1998. "Social Capital: The Missing Link?" Social Capital Initiative working paper no. 3. Washington, DC: World Bank.

Grossman, P. J. 1994. "A Political Theory of Intergovernmental Grants." *Public Choice* 78:295–303.

Hajer, M. 1995. *The Politics of the Environmental Discourse*. Oxford: Clarendon.

Hardin, G. 1968. "The Tragedy of the Commons." *Science* 162:1243–1248.

Hawkesworth, M., and M. Leogan. 1992. *Encyclopedia of Government and Politics*. London and New York: Routledge.

Hicks, A., and J. Misra. 1993. "Two Perspectives on the Welfare State: Political Resources and the Growth of Welfare in Affluent Capitalist Democracies, 1960–1982." *American Journal of Sociology* 99(3):668–710.

Hongthong, P. 1997. "No Consensus on Forest Bill at Public Hearing." *The Nation*, May 16. Available at <http://www.nationmultimedia.com>.

Hongthong, P. 1998. "RFD Chief May Stir a Hornet's Nest with Move." *The Nation*, April 8. Available at <http://www.nationmultimedia.com>.

Hongthong, P. 1999. "Plantations Spark RFD Conflict with Villagers." *The Nation*, July 6. Available at <http://www.nationmultimedia.com>.

Inchukul, K. 1997. "Forestry Bill Ready for Consideration." *Bangkok Post*, May 10. Available at <http://www.bangkokpost.net>.

Keeley, J., and I. Scoones. 1999. "Understanding Environmental Policy Processes: A Review." Working paper no. 89. Sussex, UK: Institute of Development Studies, University of Sussex.

"Keep the Watershed Free of Inhabitants." 1998. *The Nation*, June 3. Available at <http://www.nationmultimedia.com>.

Kessler, T. P. 1998. "Political Capital: Mexican Financial Policy under Salinas." *World Politics* 51:36–66.

Khuenkaew, S. 2001. "Farmers Rally to Demand Bill Be Passed—Senators Tell of Heavy Lobbying to Reject Law." *Bangkok Post*, December 21. Available at <http:www.bangkokpost.net>.

Kongrut, A. 2001. "Bid to Promote Forest Living—Activists Aim to Seek NAP Support for Bill." *Bangkok Post*, January 14. Available at <http://www.bangkokpost.net>.

Kruschke, E., and B. Jackson. 1987. *The Public Policy Dictionary*. Clio Dictionaries in Political Science no. 15. Oxford: Clio Press.

La Due Lake, R., and R. Huckfeldt. 1998. "Social Capital, Social Networks, and Political Participation." *Political Psychology* 19(3):567–584.

Leach, M., and R. Mearns, eds. 1996. *The Lie of the Land: Challenging Received Wisdom on the African Environment*. Oxford: James Currey.

Leicht, K., and J. C. Jenkins. 1998. "Political Resources and Direct State Intervention: The Adoption of Public Venture Capital Programs in the American States, 1974–1990." *Social Forces* 76(4):1323–1345.

Limmanee, A. 1998. "Thailand." In *Political Party Systems and Democratic Development in East and Southeast Asia*, ed. W. Sachsenröder, 403–445. Ashgate, U.K.: Aldershot.

Magill, F. 1996. *International Encyclopedia of Government and Politics*. London and Chicago: Sale Press.

McCarthy, J. D., and M. N. Zald. 1977. "Resource Mobilization and Social Movements: A Partial Theory." *American Journal of Sociology* 82(6):1212–1241.

McLean, I. 1996. *Oxford Concise Dictionary of Politics*. New York: Oxford University Press.

Meinzen-Dick, R., and A. Knox. 2001. "Collective Action, Property Rights, and Devolution of Natural Resource Management: A Conceptual Framework." In *Collective Action, Property Rights and Devolution of Natural Resource Management—Exchange of Knowledge and Implications for Policy*, ed. R. Meinzen-Dick, A. Knox, and M. Di Gregorio, 41–73. Eurasburg, Germany: Deutsche Stiftung für internationale Entwicklung.

Miller, D. 1991. *The Blackwell Encyclopedia of Political Thought*. Oxford: Blackwell Publishing.

"No One Should Live and Farm in the Upper Watershed Forests." 1998. *Watershed* 4(1):10–13.

Olson, M. 1965. *The Logic of Collective Action—Public Goods and the Theory of Groups*. Cambridge: Harvard University Press.

Ostrom, E. 1994. "Constituting Social Capital and Collective Action." In *Journal of Theoretical Politics* 6(4):527–562.

"Panel Urged to Include Five Activists—People's Draft Bill Left in 'Dim Light'." 2000. *Thai Development Newsletter* 38/39:18.

"Participation Has to Come from People Both in the Lowlands and the Highlands in the Same Watershed." 1998. *Watershed* 4(1):26–28.

Plumo, J., R. Riggs, and H. Robin. 1982. *The Dictionary of Political Analysis*. Clio Dictionaries in Political Science no. 3. Oxford: Clio Press.

Poffenberger, M., and B. McGean. 1993. *Community Allies: Forest Co-management in Thailand*. Southeast Asia Sustainable Forest Management Network Research Network Report, Vol. 2. Berkeley, CA: Center for Southeast Asia Studies, University of California.

Pratong, K., and D. E. Thomas. 1990. "Evolving Management Systems in Thailand." In *Keepers of the Forest—Land Management Alternatives in Southeast Asia*, ed. M. Poffenberger, 167–186. Manila, Philippines: Ateneo de Manila University Press.

Putnam, R. D. 1993. *Making Democracy Work: Civic Traditions in Modern Italy*. Princeton: Princeton University Press.

Raymond, W. 1992. *Dictionary of Politics*. 7th ed. Lawrenceville, VA: Brunswick Publishing Corp.

Riggs, F. W. 1966. *Thailand: The Modernization of a Bureaucratic Polity*. Honolulu: East-West Center Press.

Robertson, D. 1993. *A Dictionary of Modern Politics*. London: Europa Publications.

Rogers, M. F. 1974. "Instrumental and Infra-Resources: The Bases of Power." *American Journal of Sociology* 79:1418–1433.

Roongruangsee, C. 1994. "Development of Community Institutions and Networks: Village Organization in the Highland Development Process." Paper presented at United Nations International Drug Control Programme / Office of the Narcotics Control Board (UNDCP/ONCB) seminar "Two Decades of Thai-UN Co-operation in Highland Development Areas under Drug-Control," Chiang Mai, Thailand June 20–22.

Sattha, C. 2002. "Farmers Protest during Talks—Eight Demands to be Discussed at Meetings." *Bangkok Post*, March 17. Available at <http://www.bangkokpost.net>.

Safire, W. 1978. *Safire's Political Dictionary*. New York: Random House.

Srivalo, P. 1999. "Controversial Forest Bill Gets Cabinet Okay." *The Nation*, June 19. Available at <http://www.nationmultimedia.com>.

Sukin, K. 1997. "Christians Join in Ordination of Conservation Site." *The Nation*, July 23. Available at <http:www.nationmultimedia.com>.

Sukin, K. 1999a. "RFD Chief Proposes Economic Forest Policy." *The Nation*, January 13. Available at <http:www.nationmultimedia.com>.

Sukin, K. 1999b. "Northern Villagers Seek 50,000 Signatures for Forest Bill." *The Nation*, January 18. Available at <http:www.nationmultimedia.com>.

Sukpanich, T. 2001. "Vote Buying Remains Rampant." *Bangkok Post*, January 14. Available at <http:www.bangkokpost.net>.

Support Group of the Community Forestry Bill by the People. 1999. "Sign the Petition for the Community Forestry Bill by the People." Information brochure for the collection of 50,000 signatures (Original in Thai. Unofficial translation). Chiang Mai, Thailand: Community Forestry Bill by the People Support Group.

Susanpoolthong, S. 2001. "House Okays Several Contentious Issues." *Bangkok Post*, November 1. Available at <http:www.bangkokpost.net>.

Svasti, Smansnid M. R. 1998. "No One Should Live and Farm in the Upper Watershed Forests." *Watershed* 4(1):10–13.

Swinton, R. 1982. *A Dictionary of Political Thought*. London: Macmillan Press.

Tangwisutijit, N. 1998. "Academic-NGO Alliance: The Third Force." *The Nation*, August 31. Available at <http:www.nationmultimedia.com>.

Tangwisutijit, N. 2000. "Citizens Draft Historic New Forest Law." *The Nation*, March 1. Available at <http://www.nationmultimedia.com>.

"Thaksin Shinawatra—Ein umstrittener 'Erneuerer' Thailands." 2001. *Neue Züricher Zeitung*, no. 34 (February 10/11):4.

"The Hilltribes Need to Move Down to the Lowlands." 1998. *Watershed* 4(1):19–22.

"This is Like Dying While Still Alive." 2001. *Watershed* 6(2):43–47.

Upper Nan Watershed Management Project. 1997. *Village Rules and Regulations*. Project document, without author. Nan, Thailand.

Wall, E., G. Ferrazzi, and F. Schryer. 1998. "Getting the Goods on Social Capital." *Rural Sociology* 63(2):300–322.

Zeller, M., C. Lapenu, B. Minten, E. Ralison, D. Randrianaivo, and C. Randrianarisoa. 1999. "Pathways of Rural Development in Madagascar: An Empirical Investigation of the Critical Triangle between Environmental Sustainability, Economic Growth and Poverty Alleviation." *Quarterly Journal of International Agriculture* 28(2):105–128.

V

Conclusions

11

Adaptation to Challenges

Nives Dolšak, Eduardo S. Brondizio, Lars Carlsson, David W. Cash, Clark C. Gibson, Matthew J. Hoffmann, Anna Knox, Ruth S. Meinzen-Dick, and Elinor Ostrom

Research regarding CPRs appears to give two different answers to the pertinent question of survival of such resources into the new millennium. On the one hand, multiple studies tell the same old stories. Central governments initiate the dismantling of local, well-functioning and self-governing systems, leading to governance failure at the coarser scale. The introduction of private property and market economy leads to the deterioration of common-pool resources and communities. These repetitive stories suggest that the prospects for local and sustainable management of common-pool resources are more or less dictated by macro-level interventions. Presumably, "the state" decides the degree of freedom for all its subordinates.

On the other hand, a most interesting line of research paints an alternative picture, that of sustainable management of natural resources, over years and centuries, despite such restructuring on the macro level. For example, Ostrom (1990) has drawn our attention to the fact that some institutional arrangements, for example, those of irrigation systems on the eastern coast of Spain, survive and produce benefits for their users despite massive changes at the macro level. Further, we see local resource users gaining access to institutions at the macro level and enforcing their local customary rights to resources. We see international governmental and nongovernmental organizations assisting local communities in these efforts. Clearly, the challenges to the commons in the new millennium are multifaceted. The ways in which resource users adapt to these challenges are as unique as the circumstances in which they occur. Some general lessons, however, need to be considered in devising new institutions to adapt to these challenges. Understanding these lessons is the major task of this volume.

The contributors to this volume have tried to answer the following questions: (1) What contemporary developments challenge traditional common-property institutions, and how are these institutions adapting? (2) How is the ever-increasing scale of human interactions affecting the governance of larger-scale common-pool resources? (3) What progress is being made in the design of institutions that "privatize" some rights to individuals for their use of a common-pool resource? This chapter synthesizes our learning as presented in this volume. It then discusses what remains missing in our theoretical understanding of the new challenges to the commons, explores the need for development and use of new analytical concepts and methods, and suggests some ideas for expanding future empirical research.

The Lessons

Several themes recur across the chapters of this volume analyzing challenges of governing the commons in the new millennium and various strategies of adaptation:

1. The increased interconnectedness of the biophysical world across scales and institutions across levels requires that adaptation to challenges occur at multiple levels.
2. The interests of resource users at these multiple levels are often in conflict.
3. Allocation of rights to resources (individual rights for privatization of a resource or community rights in the process of devolution) is a political process.
4. Access to this political process is limited by the structure of the macro institutions and also by the human, political, and social capital available to each group of actors.
5. More open political systems and more interconnected economies provide a larger set of adaptation strategies.
6. Adopted policy solutions are incremental and not linear.

We now turn to a more detailed discussion of challenges and various adaptation strategies, organized around three major topics of the volume: multiple scales in resource use and governance, privatization of common-pool resources, and the forms of capital that local resource users build and mobilize to govern their common-pool resources sustainably.

Multiple Scales: Conflict and Cooperation

The chapters in part II address multiple challenges that resource users face when different aspects of resource use shift from one scale to another. Technological development and changes in the natural environment may shift the "allocative" aspects of resource use (using terminology employed by Geores in chapter 4) to coarser scales. Alternatively, technological development enables resource users at coarser scales to notice and measure negative externalities that are caused by resource use at smaller scales. The challenge, then, is how to devise appropriate institutions to take into account resource use at different scales and resolve potential conflicts that occur among users at different scales with respect to their perceptions of the actual resource stocks and the most appropriate resource use.

"Authoritative" aspects of resource use (again, using terminology employed by Geores in chapter 4) also shift between larger and smaller scales, for example, in the devolution of authority over common-pool resources from national-government to local communities. Like the shift from traditional chiefs to national-government control, this reverse shift transfers some benefits of resource use to new users. Chapter 10 by Birner and Wittmer explicitly addresses how resource users generate the political capital required for such a shift. Once the authoritative aspects of resource use shift to more local levels, new users may face different challenges than they experienced before the shift, such as providing sufficient capital, producing or getting access to useful information, and developing new institutions for local management and enforcement.

The chapters in part II explicitly discuss the implications of scale for the management of common-pool resources. Cross-scale interactions among social and ecological systems help us understand the feedback effects of resource management and thresholds of change and resilience. Consequently, these chapters recognize the need for multiscale research that is spatially and temporally explicit. In these chapters, we see contestation of the different concepts of scale (e.g., differentiating levels versus scales of analysis), sampling procedures, methods (e.g., interviews, fieldwork, searches of archives), applications (e.g., forest conservation plans, regulation systems), and theories (e.g., multilevel institutions and institutional connections, social-ecological feedback interactions).

All of the chapters in this part recognize the importance of scale and the challenges posed by the management of common-pool resources, which have implications at different levels. In discussing contrasting definitions of "forest," Geores (chapter 4) explicitly notes the difficulty inherent in managing a resource (forest) that has diverse meanings and policy implications at global through local levels. How can the global resource of rain forests be reconciled with the local resource of a fruit-bearing tree, which provides subsistence resources for an individual family? The analysis in chapter 4 points out the institutional conflicts that occur when resource users at different scales do not have the same understanding of the geographic area of the resource. For example, conflicts arise when global users of forests view the forests as a sink for global pollution and attempt to exert their authority over vast forested areas.

Acheson and Brewer (chapter 2) address a case in which initial conflict over resource boundaries at a local level and the inability of local resource users to defend these boundaries results in a shift of authority over a resource from a local to a coarser scale. Here, the conflict is over areas along the boundaries of adjacent users of the resource. In this case, the challenge is to defend the boundaries and to reduce incentives for neighbors to infringe upon the territory held by a particular community of resource users. In some cases, the costs of defending the boundaries are too high, and resource users willingly shift the authority that they had over the resource to a coarser scale.

Acheson and Brewer provide an account of the Maine lobster fishery that illustrates an evolving symbiotic relationship between state and local actors, neither of which could manage the resource effectively without the other. The state resource agency would have little political legitimacy to impose state-mandated regulations on local fishers without their explicit agreement to give up some of their authority, and local fishers would have little ability to control access to local fisheries in the face of changing technology and economy without the enforcement and organizing ability of the state. The result is a hybrid system of state and local control that takes advantage of complementarities at different levels.

In chapter 3, Hanna describes a transition of the fishing commons characterized by implications at the national through local levels. Large-scale depletion of stocks has significant economic and political implications at the community and individual levels. Conventional regulation provides classic incentive problems that can lead to exploitation of open-access resources. More recent legislation explicitly recognizes the local impacts on, and attendant federal responsibility toward, communities made vulnerable by federal restrictions.

The environmental social sciences community consistently agrees on the need to pay special attention to scale dependence factors in studies of human-environmental interactions. It is relatively well understood, for example, that the characteristics of an observation are a function of the scale of data collection. Similarly, the scale of an observation alters our ability to explain and interpret processes and patterns about local social and environmental conditions. Appropriate scale depends on the study case in question, data availability, and researchers' familiarity with different methodologies (Cash 2000; Gibson, Ostrom, and Ahn 1998; Wilbanks and Kates 1999). The chapters in this volume and a host of other literature, however, point to the need for further refining the terminology we use in discussing scale. For instance, the use of the terms "local," "regional," and "landscape" levels often erroneously implies that these are nested entities.

As noted in all the chapters in part II, scholars frequently combine two categories of scale. We tend to define scale as a function of social and political construction, for instance, as it relates to jurisdiction lines and local history of space occupation, as well as of scientific construction, such as that defined by resource distribution and temporal dynamics. In most cases, our selection of scale for research purposes is based on sociopolitical construction or scientific construction, but we still lack conceptual tools with which to integrate the biophysical and the sociopolitical across multiple scales. For instance, how do we accommodate fuzzy boundaries in resource distribution (space, time) and regulation (articulation of users and political actors defining allocation rules)? In this sense, the chapters in part II propose provocative concepts and analytical tools and offer insights for further work.

Privatizing Commons: When and How?

The three chapters in part III examine challenges facing privatization of common-pool resources. In their analysis of very recent experiences in privatization of common-pool resources, such as fisheries and the atmosphere in its use as a pollution sink, they identify some of the most pressing problems involved in using individual rights to manage a resource.

Their analysis begins with the challenges of creating institutions for privatizing common-pool resources and extends to the challenges of mitigating the negative effects that these institutions may have on the resource itself as well as its users.

Though tradable permits and common-property regimes may seem to be diametrically opposed to one another, they are not mutually exclusive (McKean 2001). Tradable-permit systems are usually designed and implemented by a local, national, or state government, as discussed in chapters 5, 6, and 7. This is, however, not the only option. Blomquist (1992) illustrates that resource users themselves can initiate an institutional design that allows for temporary or permanent transfer of rights to withdraw groundwater. Similarly, fishing communities in Nova Scotia have designed systems of transferable rights to withdraw from commonly owned fisheries (McCay 2000). International negotiations for reducing overuse of the atmosphere—a global common-pool resource—also indicate that a community without a supranational government may consider designing a system of tradable permits. Furthermore, the Hague stage of negotiations in connection with the Kyoto Protocol on the use of the atmosphere as a global pollution sink illustrates that the two aspects—negotiation of rules for reducing the overuse of the atmosphere and the design of rules that allow for a transfer of allocated rights—may be intrinsically connected. Resource users may face different costs of adopting a given policy depending on whether trading of unused rights is allowed.

New Zealand was the first country to institute ITQs for its fishing industry, but the law supporting ITQs has been in place only since 1983. Results of a panel survey on fishers' opinions of ITQs are analyzed in chapter 5. Iceland's experience dates back to 1991 and is the focus of chapter 6. A highly politicized debate has emerged in that country as a result of a shift from a multistakeholder comanagement regime to implementation of tradable quotas. Chapter 7 examines efforts to establish emission trading programs in the United States and in turn discusses the potentials and challenges of creating a market mechanism for global trading of greenhouse gases. Whereas various agreements have been reached among different clusters of U.S. states since 1994 in an effort to control nitrous oxides emissions from coal-fired power plants, interstate emission trading has not been adopted.

The main challenges of designing market mechanisms to regulate rights to natural resources are related to the presence of the following characteristics: (1) large-scale and mobile common-pool resources, (2) difficulties of allocating rights efficiently and equitably, (3) substantial heterogeneity in interests among resource users, and (4) potential negative social consequences of trading in these rights.

Challenges of Devising Market Mechanisms for Large-Scale and Mobile Common-Pool Resources Both fish and the atmosphere can be described as "fugitive" resources in that they are highly mobile and, to different degrees, unpredictable in their movement. Efforts to regulate such resources (i.e., to assure that rights are not exceeded and that the rules governing rights are not violated) are complicated by this characteristic of fugitiveness.

Air pollution externalities have far-reaching effects that may be impossible to localize because of their fluidity. Regulation is challenged by the difficulties involved in ascribing the source of any particular pollution. By contrast, the impacts of fisheries exploitation, though extensive, tend to be more localized, such that fisheries are not as plagued as emissions of pollutants by an attribution problem. The observability of fish extracted and the source of extraction, as well as the spatial proximity of the two, makes monitoring easier. Limited scientific knowledge about the resource itself, however, and the results of different types of anthropogenic activity have hampered management of both types of resources. Critical thresholds might be reached beyond which fisheries may collapse or global climate may change irreparably. These potential thresholds lead scholars to recommend cautionary principles, but without clear and shared understanding of the relationships involved, it is difficult to reach a consensus on the need to restrain use.

With emissions, a disjoint between the source and the impact confounds attribution and the ability to penalize or assign payments to the correct actor. Furthermore, administrative regulation of an externality that is not confined within administrative borders weakens the potential for enforcement. Even if one could make an accurate link between the source of emissions and the impact, the rules set by one administrative unit cannot be imposed on industries located in another. Farrell and Morgan (chapter 7) illustrate this conundrum with their analysis of attempts of multiple states in the United States to enforce air quality standards: "The states are thus put into a very odd position in which they are required individually to meet an externally imposed environmental standard for a pollutant over which they (in many cases) have only partial control" (186). Yet decentralization of regulation is likely to be more cost-effective and in accordance with the individual needs of the state.

Not only are fish and pollutants mobile, but they also cross international boundaries. Global negotiations, whether on trade or the atmosphere, affect the management of these resources. Many types of institutions and actors are involved. Definitions of "stakeholders" are contested, for it is difficult to place clear boundaries on who is affected by the use of the resource. Despite (or maybe because) of this complexity, polycentric governance structures are called for, with state as well as users' participation (McGinnis 1999a, 1999b, 2000).

Challenges of Allocating Individual Rights Both the fisheries case and the case of the atmosphere illustrate how the increasing scarcity of a resource has prompted greater attention to property rights over that resource. First, national and state governments have asserted their rights over common-pool resources. This is illustrated most clearly in the Icelandic case, in which the government declared an economic exclusion zone and closed the area to foreign trawlers. The way in which air quality has been placed under management by state, national, and even international bodies, although less tangible, has similarly converted an open-access resource into something that can be regulated and treated as property. But rather than an attempt to manage the resources directly by the state or under common-property regimes, there has been a move toward private property rights.

In chapters 5, 6, and 7, the initial allocation of private rights appears as an important factor in the acceptability of the tradable-rights regime as well as the equity of outcomes. Negotiation with resource users has been critical to the establishment of ITQs as well as emission permits. This generally means that rights are allocated only to current users, usually based on existing levels of use, which favors big producers and polluters and generates problems for new entrants. If historic uses form the basis for allocating private and tradable rights in a particular resource, those individuals or companies that have historically used the resource capture the "rent" on that resource. Opposition may be generated to allowing some individuals or groups to benefit in this way. Alternatively, if resource use rights are given out on a shorter term (as opposed to permanently) and require payment of a rent to the government, there are fears that it will make the users of the resource less secure in their rights and less competitive internationally. Such a dilemma applies not only to the cases of fisheries or atmospheric pollution presented in this volume, but also to allocation of rights to water, forests, and many other natural resources that may be considered public or common.

Challenges of Designing Processes for Equitable Allocation of Transferable Rights Politics occupies center stage in establishing and enforcing regulations related to common-pool resources. The visibility and strength of different interest groups shape the structure and participation in stakeholder forums, the content of the debate, and eventual outcomes. In a democratic society, the legitimate tool for establishing political strength and gaining the support of government is acquiring public support, which is typically expressed through equal-representation voting. Other laws, values, socially established practices, and economic systems, however, can create other means of garnering political support. Although the topic is not addressed in chapter 7, U.S. laws enabling the financing of political campaigns may play a role in the considerable bargaining power of large industries when it comes to regulating pollution. Likewise, the country's federal structure provides incentives for states to place the priorities of their residents over those of the rest of the nation. Virginia's capacity to meet clean-air standards while its industries contribute considerably to the pollution problems of other states makes it rational for its elected state officials to opt out of pollution agreements to protect the employment and economic growth interests of their constituency. Whereas decentralized governance certainly has a host of virtues, it can also serve as an impediment to meeting the needs of a broader society.

The case of Icelandic fisheries regulation is remarkable for the high level of public trust in the government shown during the establishment of ITQs. Structural factors (e.g., the perceived shared interests of all Icelanders in the fishing industry), individual leadership (e.g., from the Fisheries Minister), and historic circumstances (e.g., the expulsion of the British fleet from Icelandic waters) contributed to this high degree of trust in the government. Such levels of trust are vital not only for the establishment and legitimacy of new property rights regimes but also for ongoing monitoring and dispute resolution. This is especially important for resources like ocean fisheries and the atmosphere, where users cannot directly observe others' behavior. Even under the private-property regimes that have been created, state regulation remains important. But for this regulation to be effective, the state agencies must be trusted to measure, monitor, and enforce objectively.

Challenges of Designing Institutions in the Presence of Substantial Heterogeneity of Interests Both the Icelandic fisheries and the emission trading studies offer insight about factors shaping the capacity for collective action among different interest groups when a common resource is being regulated. In Iceland, relatively successful collective action breaks down because of opportunities created by ITQs for large vessel quota owners to prosper economically relative to other industry participants. Prior to the introduction of ITQs, the government was trusted as a neutral arbiter and regulator. Public outcry against the quota system has been met by the reluctance of political parties to abandon ITQs in light of the support they had previously given to them. Government stagnation on the issue as well as its reduced role in negotiation and regulation have eroded the trust that less-powerful industry stakeholders and the public once had in the government.

Trust is also highlighted as being essential for cooperation among U.S. states in emissions compliance. The prior history of interaction and cooperation among the OTC states contributed to the relative success of the NO_x Budget compared to the OTAG process, where upwind states lacking this history of cooperation feared that they would be taken advantage of by the downwind (OTC) states. Equitable burden sharing was also seen to be a key element in states' willingness to participate. Perceptions of fairness reinforce a climate of trust. Therefore, the success of democratic mechanisms for regulating common resources, whether market-based or not, will depend on whether the various parties can establish sufficient trust among themselves to enable cooperation and political consensus.

Yet heterogeneity of interests among actors may be so extensive that it is impossible to achieve cooperation or collective action even where it was once successful. Emission trading among all of the OTC states has not become a reality because of the disparity in (dis)incentives between upwind and downwind states. In Iceland, a culture of consensus dissolved when the introduction of ITQs fractured stakeholder interests. Such collective-action failures that compromise the public good may call for intervention at the central level in setting and enforcing desired outcomes, though methodologies for achieving those outcomes may be conducive to more localized decision making.

Another approach may rest with efforts to (re)build trust and to reduce interest heterogeneity and thereby pave the way for successful collective action. The Icelandic fisheries study suggests that great caution should be used in moving to market mechanisms for regulating a common resource because of their potential to undermine cooperation and sharpen heterogeneity of interests. Such lessons are important for policymakers who have to make decisions about applying market mechanisms to resources where they are mostly untried, such as the atmospheric commons.

Challenges of Mitigating Negative Social Aspects of Privatization
Through the allocation of individual rights and market exchanges of these rights, the rights can get concentrated in the hands of a few rights holders, causing equity concerns. Chapters 5 and 6 address these issues in the case of individual transferable rights in fisheries. Survey results for small-scale New Zealand fishers show a dramatic increase in those believing resource allocation in New Zealand fisheries to be problematic under the country's quota management system. By 1999, 70 percent of the sample of small-scale fishers highlighted their dissatisfaction with resource allocation, whereas only 18 percent of surveyed fishing company managers thought it a problem. High levels of attrition in the panel survey respondents are likewise indicative of industry consolidation, as growing numbers of small-scale fishers exit the trade.

The situation appears to be even starker in Iceland. Over the years, the marketability of quota shares has led to their concentration in the hands of large vessel owners. Eythórsson (chapter 6) reports that crewmen who work on fishing vessels are increasingly vulnerable to exploitation and downward wage pressure and that fishing communities suffered when local vessel owners sold their quotas to nonlocal buyers. Thus, not only are big players dominating the small owners through their greater capacity to assume the transaction costs of tradable rights, but non–rights holders may be hurt too as a result of industry restructuring.

Chapter 6 also highlights how the adoption of market mechanisms can impersonalize an industry through the creation of anonymous forums for trading rights. The danger here is that participants do not come face to face with the consequences of their actions and therefore do not feel responsible for them. Moreover, markets have the potential to polarize stakeholder interests by favoring more powerful actors over less powerful interests. Both of these outcomes can have the subsequent effect of reducing the capacity for coordination among different stakeholders and interest groups. This effect is discussed in greater depth in the last section of this chapter.

External Economic, Legal, and Political Environments: Affecting and Being Affected by Resource Users and Their Various Forms of Capital

Society does consist of layers of institutions. Constitutional rules establish a foundation for lower-level institutions, for collective choice, and for the establishment of operational rules (Kiser and Ostrom 1982). As Lindayati (chapter 8) illustrates, however, changes at the community level are not simple reflections of macro-level adjustments. In fact, for a long period, many of the *adat* systems in Indonesia seemed to function as before despite shifts of political regimes. This might be the case when the “long arm of the state” is too short, or, as Lindayati correctly concludes, because the long arm of the State is many-handed. All centrally initiated policies have problems in penetrating all the way through the institutional layers of society. This is true not only in developing countries, such as Indonesia, but also in mature Western democracies. Ethnic issues, history, culture, all manifested in rules in use (those rules that Lindayati calls de facto rules) structure society and cannot be replaced overnight, independent of changes in official policy doctrines (North 1991).

Many developing countries are still fighting the legacies of previous colonial powers that focused on effective extraction of local resources, many of which were initially governed by local users. Following independence, the national governments of these countries developed new institutions whose goal was not necessarily restoring the precolonial governance of common-pool resources. In some instances, these new national governments embraced ideas of returning the resources to "the people." Ideas, however, as Lindayati clearly illustrates, influence policy outcomes only if the actors holding these ideas also have political power. The realities of the need for industrial development and of the centralization of power in the hands of national bureaucrats, who object to any devolution or decentralization of power, outweigh the ideas of assigning rights over local resources to local communities.

The processes for changing external political and legal environments enabling local communities to gain rights to local resources can take a very long time; they are incremental and often not linear. During one period, it may seem that the external environments are changing in the right direction, but then a shift in priorities of the actors at the national levels may slow down or even stop these processes. For example, as Lindayati illustrates, the process of devolution of power from the Indonesian Ministry of Forestry to local *adat* communities was stopped by an unrelated attempt of a different ministry to gain control over national forested land. The forestry bureaucracy at the national level was willing to give some limited authority to local communities but unwilling to engage in debate about such a transfer of authority when another player stepped in that could gain control over the entire resource.

Human beings seem to have an intrinsic drive to organize, to build institutions, and to invent new systems of self-governance. Thus, even if institutions at the level of national government can indeed be nasty creatures, there are still hopes for the future. Resource users invest their capital in two directions. First, they use financial capital made available by external sources to build physical capital in their communities and even to build their social capital (as defined by Coleman [1986]). This may protect their local common-pool resources. Second, they use their social capital (as defined by Bourdieu [1992]) to affect the external legal and political environments granting them the right to devise their own institutions for governing their common-pool resources.

The concept of social capital has opened new avenues for exploration in the social sciences (Coleman 1986; Putnam 2000). Some of the foundational work in the body of literature studying common-pool resources either refers to social capital directly or explores many of its characteristics and functions without employing the actual concept. Given the role of social relationships in the management and use of natural resources at the local level, social capital will remain of central concern to scholars of the environment.

Research presented in chapters 8, 9, and 10 extends conventional approaches to social capital by examining how social capital is linked to other forms of capital and how changes in multiple types of capital may interact and affect the use of common-pool resources. These chapters address the challenges of developing social capital and mobilizing it to protect local common-pool resources. In particular, they examine the use of social capital under two different circumstances: (1) how social capital may be created in the interaction between local communities and external financial aid and (2) how, through various channels, social capital can be used with external financial aid and how it can be used to create political capital. The chapters postulate that different forms of capital (physical, economic, political, and social) are intrinsically linked and that one form can be used to create others or to mitigate negative effects of other forms on governance of common-pool resources.

In chapter 9, Anderson, Locker, and Nugent base their analysis on the "Putnamian" approach to social capital. They explore how the social capital associated with microcredit schemes generates other types of capital or more social capital, as well as how social capital might also affect the use of common-pool resources. First, they build a feedback framework whereby microcredit schemes capitalize and depend on existing social capital to make the credit schemes viable. Individual members of a community, who do not have the ability to prove their creditworthiness to formal lending institutions and lack financial collateral, can use the community's financial ability as collateral to obtain loans. Being a member of a community of regular loan payers enables them to gain access to loans from the formal financial sector. Social capital is thereby used to increase individual financial capital.

Second, Anderson, Locker, and Nugent examine whether the credit schemes themselves may then serve to build further social capital. For example, if a group that obtains a loan is required to meet regularly to discuss loan repayment, those meetings may be increasing social capital in two ways. First, belonging to a group requires that a member meet certain expectations of the group. When these expectations are met, this increases trust among the members of the group. Second, the meetings provide a regular venue where other issues of common concern are discussed, thereby reducing transaction costs.

How do these changes lead to new institutions' being devised to protect local common-pool resources? Anderson, Locker, and Nugent postulate that these effects operate through two channels. First, some microcredit schemes explicitly require natural-resource stewardship and prescribe ways in which the loan funds are to be used. There is, however, also an indirect way through which these effects can come about: members of the group use the loan repayment meetings to devise institutions for better governance of common-pool resources. Some communities in Thailand use these meetings to devise rules for sustainable use of common-pool resources and to discuss monitoring and sanctioning of illegal behavior. This does not, however, occur in all communities. Communities in Vietnam did not utilize these meetings to discuss the management of common-pool resources. Anderson et al. build on the work of Ostrom (2001) to examine under which conditions group members are more likely to devise such rules.

In chapter 10, Birner and Wittmer extend the analysis into a political sphere. Building on the work of Bourdieu, which emphasizes resources "associated with the possession of a lasting network of more or less institutionalized relations of knowing or respecting each other" (Bourdieu 1992, 63), Birner and Wittmer examine how social capital can be used to create political capital. They begin by demonstrating that there are multiple understandings of social capital found in extant analyses. Their approach is similar to the ones used in sociology, especially by those working in the area of networks (see, for example, Lin 2001). Social capital in this light is a private, individual resource resulting from social relations. This view is different from the concept most used in discussions about natural-resource management (often associated with

Putnam) that tends to examine the more public goods aspect of social capital (e.g., trust, norms, reciprocity).

Armed with this view of social capital, Birner and Wittmer bring formal politics into the commons literature by exploring the roles of politicians, NGOs, and businesses in governance of common-pool resources. Birner and Wittmer's framework links the possession of social capital at the individual and local levels to the possession of political capital in a formal political process, ultimately resulting in policy outcomes. This is a new approach in the field, which has previously focused on dynamics at the local level. It points toward the larger political context within which local communities are embedded. It forces us to consider the relationships between local communities and regional and national political systems and the role that social capital plays in mediating and shaping them, thus tying in with the work presented in the section of the volume addressing multiple scales of analysis and management.

Future Research

Now we need to address the interesting questions raised by the chapters in this volume that may lead to productive research in the future. The broad question of many chapters in this volume is: Under what conditions will social capital lead to sustainable use of local common-pool resources? The research presented in this volume has not provided a unified answer. We do have evidence that natural resources are often destroyed more by those with larger amounts of financial capital than by users of the same common-pool resource who have less financial capital. It is not clear under what circumstances increased social capital can lead to higher levels of preservation of common-pool resources. Increased financial capital may lead to poor environmental outcomes (you can use your extra financial capital to buy chain saws as well as more efficient stoves). Increasing agricultural productivity may lead to increased pressure on forests, not less. Likewise, an increase in social capital may allow a community to come together more easily to plan to cut its forests and extend its fields. Social capital can lead to collective action, but collective action can also produce outcomes that do not conserve the natural environment.

Nor can we assume the easy transfer of social capital from a women's microcredit group (or other form of social capital) to the resources of the greater community, since the characteristics of the goods in the two cases are different, as are the actors involved. The incentives for devising rules for managing microcredit are likely to differ from those for governing common-pool resources. There may also be a negative relationship between the two, rather than the positive ones postulated by Anderson, Locker, and Nugent in chapter 9. As microcredit brings new influxes of capital into a community, this may exert pressure on its common-pool resources and also increase income heterogeneity among resource users. This increased heterogeneity may actually reduce social capital in the community. Changes in resource management systems and the privatization of the commons can also lead to destruction of social capital and trust, as exemplified in the Icelandic fisheries case.

These relationships are complex, as Anderson, Locker, and Nugent demonstrate. They are also crucial to scholars interested in common-pool resources. Microcredit is a popular policy and will no doubt affect the direction of many communities' development and thereby influence the management of common-pool resources. Tracing the threads of social capital through these processes is thus a challenging task.

Another challenge of capital transformation is presented in chapter 10, in which Birner and Wittmer conceptualize social capital as the private capital of individuals who can use it to create beneficial links to others in society. Key questions for future research are: How can the social capital of individuals translate into political capital of the group? If individuals reap political benefits from their connections to other individuals, what incentives do they have to use their "private social capital" to foster group goals, such as devolution of authority over natural resources from the government to communities?

The chapters in part IV clarify the obstacles and complexities that lie ahead for future research. Further explanatory leverage now depends on increased efforts to clarify analytic frameworks along the paths begun by the authors in this volume—operationalizing social capital and its links to policy and management outcomes is crucial—together with systematic comparative empirical work.

Other cases examined in this volume present challenges of privatizing common-pool resources through allocation of individual rights in the presence of important social variables affecting the design as well as resulting from this institutional design. The work presented in this volume discusses challenges in devising individual rights for resources with high levels of variability over time or high levels of uncertainty about their stocks or flows. Institutional design in these contexts requires both technical and institutional research. Sound management may require that resource use allocations be changed as new information becomes available. Frequent changes in the rules may, however, reduce familiarity with them, resulting in lower compliance.

Further, some institutional designs require a certain level of stability of rules to be effective. In particular, institutional designs that permit rights to a common-pool resource to be exchanged over time or over space may be less effective when the rules are frequently changed. In this case, it may be more effective to establish an interval during which the rules will not be changed. The length of this interval may have to correspond to the life span of technologies used in appropriation of the resource. If this interval is too long to ensure protection of the resource, then tradable-permit institutions should not be used to manage it. In this case, common property rather than individual property may ensure sustainable use of the resource (Rose 2002). In either case, however, a dynamic view of property rights is likely to be more appropriate to ensure sustainable and fair use of the resource than one that is static. Creating forums for negotiation and reallocation of such rights may be more important than laying down rigid rules and resource allocations.

The work presented in this volume addresses institutional design for resources that are used at multiple scales and studies how changes in national economic and political environments affect local institutional levels. The work presented here does not, however, examine how other environments at higher scales penetrate to the local level. In particular, in the recent decades, we note an increased linkage of economic and cultural environments across levels. Globalization results in changes in management of local common-pool resources.

To study the effects of globalization on local common-pool resources, we first need to define what type of globalization we are considering. The term "globalization" has multiple meanings (Prakash and Hart 2000). Depending on what type of globalization we examine, we can see positive or negative effects of globalization processes on local common-pool resources. Globalization is often considered to refer to an increased flow of production factors and final goods across national borders (*economic globalization*). Globalization can also refer to an increasingly unified *mindset*. Further, globalization can refer to the increasingly connected natural environments and our understanding that local actions often have global consequences.

Future research needs to address the effect of more widely integrated markets upon local resources' value and control. As new economic actors and transformation industries add complexity to resource management and commercialization, it is important to understand how these changes affect the authority and negotiation power of local stakeholders vis-à-vis national and multinational corporations. On the other hand, are there opportunities to use globalization to craft new resource management institutions to deal with the increasingly connected global environment? We have seen how limited territorial authorities have been unable to deal with environmental externalities. At the same time, new communication technologies provide mechanisms through which stakeholders with common concerns can build coalitions across state or national boundaries. For example, new technologies for measuring effects of (and causes of) global climate change also provide an important impetus to international negotiations on reducing emissions of greenhouse gases and increasing their sinks. The question is how to ensure that the interests of poor and marginalized groups are included in such institutional adaptations.

Globalization as a process of creating a global mindset seems to have only negative effects on local commons. Globally uniform ways of viewing a common-pool resource, identifying problems of its use, and devising solutions to those problems are unlikely to reflect either local environmental characteristics or local valuation of the resource. Developing a global mindset therefore results in the loss of local specificities. Further, cultural globalization with its preponderance of Western values of consumption may increase pressure on common-pool resources everywhere. On the other hand, while evolving this global mindset, cultural globalization may actually trigger reactions to homogenization; understanding how changes in cultural boundaries of identity and ideology relate to social practices, particularly collective action, becomes paramount to understanding diversity in global forms of resource governance (Foster 1991; Wilk 1996).

Developing New Methods

Research presented in part II on multiple scales of use of common-pool resources clearly indicates that the level at which a particular challenge of governing a common-pool resource is analyzed depends on the particular research question and requires particular methods. A key challenge is to obtain data and develop methods that enable us to analyze conflicts in resource use across scales. This is likely to call for both technical and institutional analysis, as well as interdisciplinary methods. One possible approach to devising such methods is to develop a method at a coarse scale and impose this method on all smaller scales—itself a case of imposing a global mindset or global rules on local resource management. This approach creates two problems. First, it potentially imposes high costs of method development on resource users at lower scales who may have developed equally reliable methods at their own scales. Second, imposing one standardized method creates the potential for losing necessary detail that is obtained through the methods developed by users at lower scales. Difficulties in devising comparable methods of measuring global carbon dioxide emissions and storage for the purpose of implementing the Kyoto Protocol illustrate the complexity of these efforts. Data compiled for the purpose of governing global climate as a sink for carbon dioxide further illustrates that these methods will produce data that will have to be revised and recalculated. Therefore, we need to develop rules for revising these methods and collecting data.

Furthermore, a need exists to extend research to a larger number of cases to be able to generalize the findings of the research. In particular, work on the use of human and social capital in preventing overuse of common-pool resources has suggested that the outcome of using such capital in this area depends on many factors, such as agro-climatic zones, the existing economic situation of the resource users, and the availability of alternatives. Work on the ability of communities to exert influence in the political environment at a higher level suggests that the outcome of efforts to exert such influence depends on the degree to which political institutions grant people access to decision-making forums at a national level as well as the extent of vertical linkages or "bridging" social capital.

Work in this volume and elsewhere (Bauer 1998; Boelens and Davila 1998) on individual transferable rights to common-pool resources (water, air, forest, fisheries) suggests that implementation of these rights may bring negative social consequences. These include increased income diversification, loss of access to the common-pool resource for some users, and reduced community cohesion around the resource. It would be useful to look across the resources for which transferable-rights programs have been implemented and the countries that have implemented them to identify conditions under which negative social consequences are most pronounced. For example, are these likely to be of greatest concern where there is a high degree of heterogeneity among resource users, or where resource users have no alternative sources of livelihoods?

References

Bauer, Carl J. 1998. "Slippery Property Rights: Multiple Water Uses and the Neoliberal Model in Chile, 1981–1995." *Natural Resources Journal* 38 (Winter):109–155.

Blomquist, William. 1992. *Dividing the Waters: Governing Groundwater in Southern California.* San Francisco: ICS Press.

Boelens, Rutgerd, and Gloria Davila, eds. 1998. *Searching for Equity: Conceptions of Justice and Equity in Peasant Irrigation.* Essen, the Netherlands: Van Gorcum.

Bourdieu, Pierre. 1992. "Ökonomisches Kapital–Kulturelles Kapital–Soziales Kapital" (Economic Capital, Cultural Capital, Social Capital). In *Die verborgenen Mechanismen der Macht* (The Hidden Mechanisms of Power), 49–79. Schriften zu Politik und Kultur 1. Hamburg: VSA-Verlag.

Cash, D. W. 2000. "Distributed Assessment Systems: An Emerging Paradigm of Research, Assessment and Decision-Making for Environmental Change." *Global Environmental Change* 10(4):241–244.

Coleman, James. 1986. "Social Theory, Social Research and a Theory of Action." *American Journal of Sociology* 91(6):1309–1335.

Foster, Robert. 1991. "Making National Cultures in the Global Ecumene." *Annual Review of Anthropology* 20:235–260.

Gibson, Clark, Elinor Ostrom, and T. K. Ahn. 1998. "Scaling Issues in the Social Sciences." Working paper no. 1. Bonn, Germany: International Human Dimensions Programme on Global Environmental Change.

Kiser, Larry, and Elinor Ostrom. 1982. "The Three Worlds of Action: A Metatheoretical Synthesis of Institutional Approaches." In *Strategies of Political Inquiry,* ed. Elinor Ostrom, 179–222. Beverly Hills, CA: Sage.

Lin, Nan. 2001. *Social Capital: A Theory of Social Structure and Action.* New York: Cambridge University Press.

McCay, Bonnie. 2000. Personal communication with Nives Dolšak. June 5.

McGinnis, Michael, ed. 1999a. *Polycentric Governance and Development: Readings from the Workshop in Political Theory and Policy Analysis.* Ann Arbor: University of Michigan Press.

McGinnis, Michael, ed. 1999b. *Polycentricity and Local Public Economies: Readings from the Workshop in Political Theory and Policy Analysis.* Ann Arbor: University of Michigan Press.

McGinnis, Michael, ed. 2000. *Polycentric Games and Institutions: Readings from the Workshop in Political Theory and Policy Analysis.* Ann Arbor: University of Michigan Press.

McKean, Margaret A. 2001. "Common Property: What Is It, What Is It Good For, and What Makes It Work?" In *People and Forests: Communities, Institutions, and Governance,* ed. Clark Gibson, Margaret McKean, and Elinor Ostrom, 27–56. Cambridge: MIT Press.

North, D. 1991. *Institutions, Institutional Change and Economic Performance.* Cambridge: Cambridge University Press.

Ostrom, Elinor. 1990. *Governing the Commons: The Evolution of Institutions for Collective Action.* New York: Cambridge University Press.

Ostrom, Elinor. 2001. "Reformulating the Commons." In *Protecting the Commons: A Framework for Resource Management in the Americas,* ed. Joanna Burger, Elinor Ostrom, Richard B. Norgaard, David Policansky, and Bernard D. Goldstein, 17–41. Washington, DC: Island Press.

Prakash, A., and J. A. Hart, eds. 2000. *Globalization and Governance.* London and New York: Routledge.

Putnam, Robert D. 2000. *Bowling Alone: The Collapse and Revival of American Community.* New York: Simon and Schuster.

Rose, Carol M. 2002. "Common Property, Regulatory Property, and Environmental Protection: Comparing Community-Based Management to Tradable Environmental Allowances." In *The Drama of the Commons,* ed. Elinor Ostrom, Thomas Dietz, Nives Dolšak, Paul C. Stern, Susan Stonich, and Elke U. Weber, 233–257. Washington, DC: National Research Council, National Academy Press.

Wilbanks, T. J., and R. W. Kates. 1999. "Global Change in Local Places: How Scale Matters." *Climatic Change* 43(3):601–628.

Wilk, R. 1996. "Emulation and Global Consumerism." In *Environmentally Significant Consumption*, ed. Paul Stern, Thomas Dietz, Vernon Ruttan, Robert Socolow, and James Sweeney, 110–115. Washington, DC: Committee on Human Dimensions of Global Change, National Research Council, National Academy Press.

Index

4198